ELECTROANALYTICAL CHEMISTRY

CHEMICAL ANALYSIS

A Series of Monographs on Analytical Chemistry
and Its Applications

Series Editor
MARK F. VITHA

Editorial Board
STEPHEN C. JACOBSON
STEPHEN G. WEBER

VOLUME 187

A complete list of the titles in this series appears at the end of this volume

ELECTROANALYTICAL CHEMISTRY

Principles, Best Practices, and Case Studies

Gary A. Mabbott

Registered Office
John Wiley & Sons, Inc., 111 River Street, Hoboken, NJ 07030, USA

Editorial Office
111 River Street, Hoboken, NJ 07030, USA

For details of our global editorial offices, customer services, and more information about Wiley products visit us at www.wiley.com.

Library of Congress Cataloging-in-Publication Data

Names: Mabbott, Gary A., 1950- author.
Title: Electroanalytical chemistry : principles, best practices, and case
   studies / Gary A. Mabbott.
Description: First edition. | Hoboken, NJ : Wiley, 2020. | Series: Chemical
   analysis : a series of monographs on analytical chemistry and its
   applications | Includes bibliographical references and index.
Identifiers: LCCN 2019048343 (print) | LCCN 2019048344 (ebook) | ISBN
   9781119538592 (hardback) | ISBN 9781119538608 (adobe pdf) | ISBN
   9781119538585 (epub)
Subjects: LCSH: Electrochemical analysis.
Classification: LCC QD115 .M325 2020 (print) | LCC QD115 (ebook) | DDC
   543/.4–dc23
LC record available at https://lccn.loc.gov/2019048343
LC ebook record available at https://lccn.loc.gov/2019048344

Cover Design: Wiley
Cover Image: © agsandrew/Shutterstock

Set in 10/13pt PalatinoLTStd by SPi Global, Chennai, India

Printed in the United States of America

V10017023_011720

CONTENTS

PREFACE

Although electroanalytical techniques are among the oldest instrumental methods used in chemistry, they continue to evolve. In the past two decades, there have been several exciting developments in the field that will ensure their relevance to chemical measurements for decades to come.

One of the growing areas in which electrochemical methods will continue to play an important role is in sensor technology. Electrochemical devices are relatively simple in terms of instrumentation and can be miniaturized. Both of these attributes help keep their costs down and make them candidates for applications such as remote sensors, personal health care monitors, and implantable devices. Some of the newer developments in both ion selective electrodes and voltammetric devices are making these sensors more selective, more robust, applicable to a wider range of analytes, and capable of lower detection limits. As simple as they may be in terms of associated hardware, these devices take advantage of a range of physical and chemical phenomena and are notable intellectual achievements. Advances in this area will require a firm grasp of the underlying science, imagination and hard work, but the possibilities are plentiful.

This book is primarily a textbook for instrumental analysis courses. As an academic subject, instrumental analysis encompasses an enormous field. It is not surprising, then, that university textbooks for instrumental analysis courses are also enormous. No one can expect to cover half of the material contained in them in a single semester. Mark Vitha has initiated a series of monographs as an alternative approach in which instructors can choose only those volumes covering topics that they intend to use in their own classes. The purpose of this book is to provide an option for teaching electroanalytical methods as a part of that series.

Space and instruction time allow for the inclusion of only a fraction of the interesting material in the electroanalytical field here. I have made some compromises in order to make covering the content manageable within a few weeks as well as provide a glimpse at some of the intriguing applications of electrochemical measurements. I have tried to emphasize the conceptual models of physical phenomena and make clear connections to mathematical descriptions that are useful. It is important to see how practicing scientists have used the math to extract useful information about chemical systems. I have tried to guide the reader with an overview of the most important ideas at the beginning of each chapter. Some basic concepts in electrical phenomena are introduced in Chapter 1, and the fundamentals of electrochemical cells are introduced in Chapter 2. Chapter 3 describes potentiometry, and Chapter 5 lays out the principles of voltammetry. Both Chapters 3 and 5 include example applications, but Chapters 4 and 6 provide case studies that demonstrate a lot of the best practices that good chemical analysis depends upon. Chapter 7 describes basic electrical circuitry and the use of operational amplifiers that are essential parts of electrochemical instrumentation.

Although this appears to be more material than is easy to cover in a few weeks of a single course, the overview explains which sections to concentrate on, if time is limited. I wanted to let instructors decide on what supporting material and case studies to cover to meet their needs. I hope that students will find the application material engaging. The material should also be relevant to scientists from other fields who need an introduction to the area of electrochemical analysis.

I want to thank Mark Vitha for including me in this project. His energy, insight, and tenacity for getting things done have been an inspiration for me for many years. Several people have read early drafts of various material for this book. I am particularly grateful to Mark, Larry Potts, Maggie Malone-Povolny, Wayne Boettner, and Joe Brom for their feedback. Each of them has had a different perspective and has made helpful comments. I am also grateful to Phil Bühlmann for his encouragement and insightful comments. I want to thank all the people at Wiley who have helped me in many ways, but Gayathree Sekar deserves special thanks for answering my questions and managing a myriad of things to see this book through production.

Finally, I want to thank my wife, Ann, for all her love and patience. This project would not have been possible without her support and understanding.

Gary A. Mabbott
St. Paul, Minnesota, USA
June 9, 2019

BASIC ELECTRICAL PRINCIPLES

Electrochemical methods of analysis measure electrical quantities in order to yield chemical information. In some cases, the measurement is an electric current (the movement of charge). In other cases, the measurement is a voltage (the amount of energy available to move a charge). Both of these techniques are useful for quantitative analysis of a chemical species, but they can also be used to determine characteristic properties that are useful for qualitative analysis. Some types of qualitative information can be useful for evaluating new materials, such as catalysts.

In describing the fundamentals of electroanalytical methods, this book emphasizes conceptual models. An effort has been made to tie conceptual models of phenomena to basic mathematical relationships in order to provide a foundation to use in reasoning through new situations. Greater insight into electroanalytical phenomena is the intended result. As with other branches of science, new developments displace older techniques. A fundamental understanding of the phenomena upon which electroanalytical tools operate enables one to appreciate the basis for new techniques and related progress in the field. A conceptual understanding also provides a good starting point for learning about other areas of science and technology that involve electrochemical processes. Electrochemical principles play important roles in many natural phenomena and in modern technology [1]. Among these fields are the subjects of energy storage and conversion; biological processes such as cellular action potentials, tissue repair, and growth [2]; electrochemical synthesis; separation technology, nanoparticles, and materials processing in the electronics industry.

Electroanalytical techniques are among the oldest instrumental methods of chemical analysis. They are still widely used for important analyses and are likely to continue to be important for many more decades. Although electroanalytical chemistry is a mature field in many ways, new developments in the realm of selective sensors and the application of electrochemical methods to demanding tasks, such as *in vivo* monitoring of neurotransmitters and remote environmental analysis continue to make instrumental analysis based on electrochemistry relevant. Some attributes of electrochemical analysis that lead to special advantages are summarized in Table 1.1.

Improved detection limits and greater selectivity have led to a greater range of applications. Some methods are capable of quantifying specific analytes down to the picomolar level. Another appealing attribute of electrochemical sensors is that they are relatively easy to miniaturize making them adaptable to a variety of new situations such as *in vivo*

Electroanalytical Chemistry: Principles, Best Practices, and Case Studies, First Edition. Gary A. Mabbott.
© 2020 John Wiley & Sons, Inc. Published 2020 by John Wiley & Sons, Inc.

TABLE 1.1   Attributes of electroanalytical techniques

Attribute	Makes possible
Sensitive	Low detection limits
Small	*In vivo* monitoring
	Measurement in tiny volumes
Simple construction	Implantable devices
Inexpensive	Mass production
	Use in poor communities
Simple operation	On-site operation
	Health care monitoring
	Remote sensing

monitoring [3]. The sensing element can be very small making it possible to measure quantities of chemical species in tiny volumes or in precise locations, such as at the terminus of a single neuron. Electrochemical methods usually require only very simple accessories. That makes them portable and, in some cases, it makes medical implantation of the sensor possible. Sensors can often be made of inexpensive materials that can be mass produced making them attractive for personal healthcare monitors, such as the handheld glucose monitor used by millions of people to manage their diabetes [4]. Other electroanalytical instruments are capable of a wide range of experiments making them well-suited to studying organic reaction mechanisms associated with electron transfer.

Before launching into the principles of electrochemistry, it is appropriate to say a word about the structure of this book. Chapter summaries appear at the beginning of each chapter in the form of an overview. Unlike reading a novel, here it is helpful for you to know the plot in advance. It helps you to know what to take away from the story. It is worthwhile to read the overview both before and after reading the other sections of the chapter. This book is aimed at students of instrumental analysis, but it is also intended to be a solid introduction to electroanalytical principles for any professional scientist. A lot of care has gone into explaining physical mechanisms and underlying concepts. Recent developments leading to new and interesting methods with better performance characteristics and a wide range of applications are described in most chapters. However, there is much more material than can be reasonably absorbed during a typical two-to-three-week unit of a college instrumental analysis course. Therefore, in addition to summarizing the major ideas, these chapter briefings tell you what sections to read, if time is short.

1.1. OVERVIEW

This first chapter is a bit different. It serves as an introduction to basic electrical phenomena and should be read in its entirety. Among the important ideas discussed in this chapter are a few definitions. The term "voltage" refers to the electrical potential energy of a charged particle. It is a measure of the force of attraction or repulsion on a single charged particle by the local density of charges in the neighborhood. The density of charges in the earth itself is thought to be reasonably constant and, as such, provides a local reference point for electrical energy. Instruments are often attached to a conductor in contact with the earth. This reference point is often referred to as "ground," and it is considered to be a point representing $0\,V$.

One volt equals one joule per coulomb of charge. The charge on a mole of electrons is $9.6485 \times 10^4\,C/mol$. This number shows up in a lot of electrochemical relationships and is called the Faraday, F, after the nineteenth-century scientist, Michael Faraday. Faraday established the relationship between charge, Q, transferred in an electrochemical reaction, such as the reduction of silver ions to silver metal, and the number of moles of reactant, N. This is Faraday's law: $Q = nFN$, where n is the number of moles of electrons transferred per mole of reactant. Another important concept is the free energy, ΔG, that drives an electrochemical reaction. The free energy of an electrochemical system is proportional to the voltage, E, and is a measurable quantity, $\Delta G = -nFE$.

Electrical current is the movement of charge and is analogous to current in a river. While a river's current is measured in the volume flow rate of the water, electrical current is measured in amperes. One ampere is equivalent to a coulomb of charge moving past a given point per second. Electrons carry charge in electrical circuits. Ions carry charge in solution. Although electrons are negatively charged, current is defined as though positive charges are moving in a circuit. The direction of the current, then, is defined as movement of charge from a higher potential to a lower potential.

Electrochemical experiments are performed in containers called cells in which two or more electrodes connect the cell to an outside electrical circuit that allows one to measure the voltage and/or the current during the experiment. Potentiometric methods measure the voltage (that is, potential) between electrodes without the passage of a significant amount of current. No significant chemical changes occur in a properly performed potentiometric experiment. In Chapter 2, the Nernst equation that relates potential in a potentiometric experiment to the activity of an analyte is discussed. An activity is the effective reactant concentration of a species. An activity of a species is proportional to its concentration, C_i. $a_i = \gamma_i \cdot C_i$, where the proportionality constant, γ_i, is known as an activity coefficient and is dependent upon the ionic strength of the solution. In contrast, voltammetric methods deliberately apply energy in the form of a voltage from an outside source to a cell in order to drive a chemical reaction at a working electrode. In these experiments, the current is related to the number of moles of reactant that is converted in the process. This current can be used to quantify the concentration of the original reactant.

In addition to the working electrode (or an indicator electrode in a potentiometric experiment), a second electrode is needed to transfer electrons into or out of the cell in order to counterbalance the charge going into or out of the solution at the working or indicator electrode. This second electrode is a reference electrode. It exploits a simple, reliable electron transfer process that occurs at a well-established voltage. The reference electrode is designed to maintain its potential (voltage) in the process. Consequently, all of the energy applied to the cell from the outside is focused onto the working electrode. Whenever the current level or the cell's electrical resistance, R, is high, some energy is lost as heat in overcoming the electrical resistance of the solution. This causes an error in voltammetric experiments because some voltage is lost from the voltage that was intended to be applied to the working electrode. This error can be calculated from Ohm's law, $V_{lost} = iR_{cell}$.

Interesting things happen at the boundary of any two phases. Charges, either electrons or ions, can cross these boundaries leading to an excess of electrical charge accumulating on one side and a layer of charge of opposite sign accumulating on the other side. This double layer of charge leads to a difference in electrical potential energy across the interface. This is the potential energy measured in potentiometric experiments that is related to the activity of the analyte ion. In voltammetric experiments, the boundary potential between an electrode and the solution controls the rate of the electron transfer between the analyte in solution and the working electrode.

An electrical capacitor serves as a good model for many aspects of the electrical double layer. The charge, Q, on either side of the double layer can be calculated from $Q = CV$, where V is the voltage or potential difference across the double layer and the coefficient, C, is the capacitance. There are subtleties to the structure of the double layer that have significance to electron transfer studies, but most of the charge on the solution side accumulates in a layer called the outer Helmholtz plane (OHP), where ions are separated from the electrode by a layer of one or two water molecules.

The conductance of a solution is the reciprocal of the solution's electrical resistance. Its magnitude depends on the type and concentration of the ions. The measurement of the conductance of a water sample is a semiquantitative measure of ionic concentration. Conductance is also used as a special detector for ionic solutes in ion chromatography.

Mass transport is a term for the movement of a chemical species in solution. Two mechanisms for material movement are very important to electroanalytical chemistry. The net movement in a given direction that is due to a concentration gradient and is characterized by a random walk of the molecule or the ion in an unstirred solution is known as diffusion. The flux, J_i, of a species is a measure of the net movement of material across a plane perpendicular to the direction of movement. It has units of mol/cm^2/s. Fick's first law of diffusion associates the flux to the concentration gradient for the species. $J_i = D_i(\partial C_i / \partial x)$. This is a key concept in electron-transfer experiments. The other mechanism for mass transport is convection or stirring of the bulk solution.

In both voltammetry and potentiometry experiments, a difference in rates of diffusion associated with salt bridges used with reference electrodes leads to a higher flux for either positive or negative ions over those of the opposite charge. The excess of charge "pushes back" against continuing build-up of charge leading to a steady state situation. The result is a net separation of charge and a junction potential or diffusion potential. Junction potentials are generally small, but they can be serious errors in potentiometric experiments. Later chapters discuss this issue in depth.

$$E_{\text{measured}} = E_{\text{indicator electrode}} - E_{\text{reference electrode}} + E_{\text{junction}}$$

1.2. BASIC CONCEPTS

Electrical phenomena are associated with charged particles. Electrons are the most common charge carriers that one encounters, but ions in a solution are also important charge carriers. The purpose of this chapter is to define some electrochemical terms and introduce some fundamental concepts associated with electrical charge and phase boundaries.

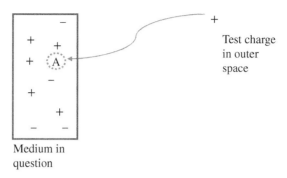

FIGURE 1.1 The electric (or electrostatic) potential energy at a point, A, in a given medium is a measure of the net energy required or released in moving a test charge from outer space (where it is assumed to be free of forces to interact with) to point A where other charges attract or repel it.

All electrochemical techniques involve measuring (and sometimes manipulating) the voltage at an electrode. What is voltage? Voltage is a measure of the electrical energy available to do work on a charged particle. A charged particle has an electric field associated with it that interacts with its environment. An electric field is the force that two charged particles experience as a function of distance between them. Charges with the same sign repel each other and charges of opposite sign attract. Consequently, the arrangement of charged particles surrounding a given location will determine whether a charged particle coming into that place from the outside will be stabilized by net attractive forces or will be destabilized by net repulsive forces. The electric potential energy for a charged particle is defined as the energy spent or released in the process of inserting a positive test charge into a specific environment. For example, consider an arbitrary location in some material, such as point A shown in Figure 1.1.

There exists some collection of charges surrounding the point in question (point A in Figure 1.1). If one were to bring a positively charged particle from outer space, where it is assumed the test charge is free from the influence of any outside electromagnetic fields to point A, one would have to do work (energy would be spent) to overcome other positive charges in the neighborhood. However, if negative charges dominate the neighborhood at point A, there would be a net attractive force on the test charge and energy would be released in moving it from outer space to that position. The energy spent or released in moving a test charge from outer space to point A is the electric potential energy (also known as the electrostatic potential energy) at that point. For simplicity, this energy is often called the potential at point A. If a different arrangement of charges exists at point B (as in Figure 1.2), then moving a test charge from outer space to point B is associated with a different electric potential energy.

There is not a practical way of measuring the absolute electric potential energy at point A or at point B. However, it is possible to measure the electric potential energy difference between points A and B. A common strategy is to define some point in the system under study as a reference point. Then, the potential at any other point in the system is the electric potential energy difference between the point in question and the reference point. In this

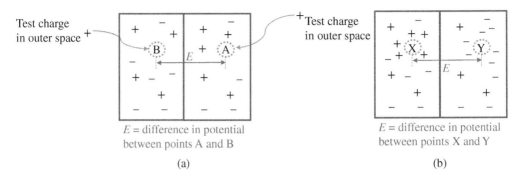

E = difference in potential
between points A and B

(a)

E = difference in potential
between points X and Y

(b)

FIGURE 1.2 a) The absolute electrical potential of a given point cannot be measured. However, the difference in potentials, E, between two points, A and B can be measured. In practice, some point within a system is defined as a point of reference so that the potential at any other point can be defined relative to reference point. b) The electric potential for a positive charge around X is much more positive than at Y driving a positive charge from X toward Y or driving a negative charge from Y to X.

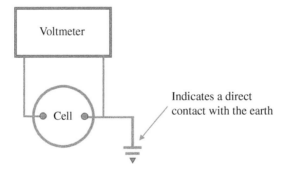

FIGURE 1.3 Electronic circuits form a continuous loop including all components. Usually some point in the circuit is linked to a conductor that has direct contact with the earth. That point becomes a reference point and is treated as though its potential is 0 V.

approach, no absolute electric potential energies need be evaluated. In the field of electronics, the reference point is often the electric potential energy of a conductor in direct contact with the earth (Figure 1.3). One might say that the electric potential of electrons in the ground is zero, but that is really just a statement about their relative energy; it does not represent an absolute value. (Furthermore, the ground is really just a local benchmark, because small variations in the electrical potential can be found at different places around the earth. Fortunately, a local reference is adequate for most practical situations.).

In electrochemical experiments, the reference potential is established by the use of a reference electrode. It is common practice to refer to a potential or voltage at an electrode when in actuality the value being discussed is the electric potential difference between the electrode in question and the reference electrode being used in that experiment. In older literature, this potential is also called the electromotive force or EMF as it is the energy available to drive charges from one point to the other and do work. Properties of reference electrodes are discussed in chapter 2 (section 2.3.4.3).

1.2.1. Volt Defined

In the thought experiment described earlier, a single particle was used as a test charge. The standard definition for electric potential is the energy required to move a unit of positive charge to the position in question. The standard unit of charge in physics is the coulomb and the standard unit of energy is the joule. Electrical energy is measured in volts. One volt represents one joule per coulomb of charge moved.

$$1\,V = 1\,J/C \tag{1.1}$$

The volt is the unit of electric potential energy per unit charge that one normally uses with simple meters in the laboratory. So, whenever someone refers to a voltage at some part of their system, they are describing the electric potential difference between that point and some reference point (usually the ground). The voltage is the number of joules released or spent in moving a coulomb of charge from the reference point to the point in question.

It is important to remember that the potential is the electrical work done per unit charge. However, a coulomb is a rather large amount of charge compared to the charge on a single electron. If moving a coulomb of charge from point A to point B costs $1.00\,J$ of energy, then how much energy is required to move a single positively charged particle between the same points? First, one can calculate charge in coulombs/electron using Faraday's constant, F, the number of coulombs per mole of electrons, $9.6485 \times 10^4\,C/mol$, together with Avogadro's number.

$$\frac{9.6485 \times 10^4\,C/mol}{6.022 \times 10^{23}\,particles/mol} = 1.6022 \times 10^{-19}\,C/particle \tag{1.2}$$

This value is the charge on an electron and is known as the elementary charge. One can calculate the energy that would be required to move a single electron through a voltage difference of $1\,V$ by multiplying the elementary charge by $1\,V$ in the units of $1\,J/C$:

$$(1.00\,J/C)(1.6022 \times 10^{-19}\,C/particle) = 1.6022 \times 10^{-19}\,J/particle \tag{1.3}$$

The result of this calculation is frequently useful and provides the definition for a separate unit of electric potential energy per elementary charge, namely, the electron-volt, eV.

$$1\,eV = 1.6022 \times 10^{-19}\,J/electron = 1.6022 \times 10^{-19}\,V \tag{1.4}$$

1.2.2. Current Defined

Current describes the movement of charge. It is the measure of the rate of change in charge moving past a specific observation point.

$$Current = i = \frac{\partial Q}{\partial t} \tag{1.5}$$

where Q is the charge in coulombs and time, t, is in seconds. Current has the units of coulombs per second or amperes.

$$1\,A = 1\,C/s \tag{1.6}$$

If one were to make the analogy of an electrical circuit with a river, then the current in amperes or coulombs per second parallels the volume flow rate of the river in gallons per second. The potential energy difference or voltage that is associated with any given component (such as an electrode in an electrochemical experiment) in the electrical circuit, is analogous to the energy available to do work per gallon of water as it drops over a waterfall. Current describes the rate of charge moving (amount per unit of time) and potential is a measure of the energy per unit charge in moving between two points.

1.2.3. Oxidation and Reduction

The exchange of electrons between two chemical species is generally known as an oxidation/reduction process or redox reaction. In a redox reaction occurring in a homogeneous solution, one reactant gains electrons while the other reactant loses. It is often useful to consider a redox reaction from the perspective of one of the reactants. Consider a redox reaction between cerium and iron ions in aqueous solution:

$$Ce^{4+} + Fe^{2+} \leftrightarrows Ce^{3+} + Fe^{3+} \tag{1.7}$$

The net reaction equation does not show any electrons as either reactants or products. However, it is useful to separate the two reactants into "half reactions" where electrons do appear.

$$Ce^{4+} + e^- \leftrightarrows Ce^{3+} \tag{1.8}$$

$$Fe^{2+} \leftrightarrows e^- + Fe^{3+} \tag{1.9}$$

A process in which a chemical species accepts one or more electrons is known as a reduction reaction; a process in which a species loses electrons is an oxidation reaction. Writing the half reactions indicates that the Fe^{2+} ion is being oxidized and the Ce^{4+} ion is being reduced. It also clearly states how many electrons are being transferred per mole of a given reactant.

1.2.4. Current and Faraday's Law

In many instrumental electrochemical methods, an electrode surface – usually a metal or carbon conductor – exchanges electrons with the analyte. In those cases, the electrode is treated as an inert source or sink for the electron exchange. The electrode does not appear in the reaction equation. Instead, the electrode reaction appears to be the same as the half reaction for the species being converted. A major benefit of using electrodes in place of a reagent in solution is the fact that the current passing through the electrode can be measured unlike the situation when two chemical species in solution exchange electrons directly. Furthermore, the current is a measure of the amount of analyte reacting.

Whenever electrons are transferred between the analyte and an electrode, the current can be integrated with respect to time in order to obtain the charge, Q, transferred.

$$Q = \int i\, dt \approx i_{average} * \Delta t \tag{1.10}$$

This charge is related to the moles of oxidized or reduced species by Faraday's law:

$$Q = nFN \tag{1.11}$$

where n = number of moles of electrons transferred per mole of reactant, N is the number of moles of the same reactant that undergo conversion, and F is Faraday's constant, 9.6485×10^4 C/mol of electrons.

1.2.5. Potential, Work, and Gibbs' Free Energy Change

If charge is moved, the amount of work done is proportional to the difference in voltage. Because the voltage difference, ΔV, is the energy spent per unit charge, the total work done in moving the charge, Q, is

$$\text{Electrical work} = \left(\frac{\text{energy spent}}{\text{charge}} \right) (\text{number of charges}) = \Delta V * Q \approx \Delta V * (i_{average} * \Delta t) \tag{1.12}$$

This is analogous to carrying a piano up a flight of stairs. The potential energy difference is fixed by the height of the stairs. To move two pianos requires twice the amount of work.

There are a couple of other conventions worth mentioning here. In electrochemical contexts, E is used instead of ΔV to represent the electrochemical potential energy difference. It is also common to equate electrical work and the Gibb's free energy change, ΔG. The relationship between potential and ΔG is usually expressed in terms of the energy per mole of reactant:

$$\Delta G = -nFE \tag{1.13}$$

where n is the number of moles of electrons/mol of reactant, F is Faraday's constant in coulombs/mol of electrons, and E is the potential difference in volts or joules/coulomb. A dimension analysis indicates that ΔG in Eq. (1.13) has the units of joules per mole of reactant. To find the total energy spent/released, or the total work done, one needs to multiply Eq. (1.13) by N, the number of moles of reactant being converted. Also, note that it is a matter of convention that favorable electrochemical processes are assigned positive potentials. Thus, the sign in Eq. (1.13) yields a negative ΔG for a positive value of E for a favorable process.

Another convention is to define the direction of a current as the direction that the positive charges move. This is the case, despite the fact that electrons are usually the major charge carriers and are moving in the opposite direction. That means that a current flows from a point of a higher potential to a point of lower potential; the electrons move in the opposite direction.

1.2.6. Methods Based on Voltage Measurement Versus Current Measurement

Potentiometry is a category of electroanalytical techniques that involves measuring the potential energy difference that develops at the boundary between a sensor and the sample solution as a function of the analyte concentration in the sample solution in which current does not flow. Alternatively, one can use an external power source, such as a battery, to impose a voltage to an electrode surface. This strategy drives an electron transfer reaction between an electrode and analyte in solution at select voltages. Current is measured in these experiments, and it may be proportional to analyte concentration under the right conditions. Voltammetry is a category of methods that measure the current in response to applying a range of voltages to the electrode/solution interface. The term "voltammetry" implies that the voltage is scanned in some manner. If the voltage is held at a constant value while measuring the current, the technique is called amperometry.

1.3. ELECTROCHEMICAL CELLS

1.3.1. Electrodes

Electroanalytical experiments are built around electrochemical cells (see Figure 1.4). There are some common features to electrochemical cells used for both potentiometry and voltammetry. The signal-generating event occurs at an electrode surface or, more precisely, at the boundary between the sample solution and an electrode surface. In voltammetry experiments, this electrode is often called the working electrode and is made from a metal that is not easily corroded, such as gold or platinum, or a highly conducting form of carbon. In potentiometry experiments, the signal is a voltage that develops at the

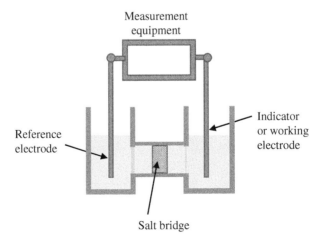

FIGURE 1.4 Basic arrangement of an electrochemical cell. Two electrodes are required to complete a circuit for the movement of charge. Each electrode is isolated in its own solution (or "half-cell"). A salt bridge keeps the two solutions from mixing but allows some ions to cross in order to complete the electrical circuit. The measurement equipment may be as simple as a voltmeter in a potentiometry experiment or, in the case of a voltammetry experiment, it may include a power source and a current meter.

indicator electrode. The indicator electrode may be as simple as a metal conductor in some experiments, but it is often a more elaborate device. Ideally, no current passes during a potentiometric measurement. The potential being measured is sometimes called the open circuit potential or the rest potential to emphasize the fact that passing a significant amount of current during the measurement can distort the signal. Potentiometric devices are discussed in depth in Chapters 2 and 3.

One electrode is not enough. Measuring current or voltage at an electrode requires that the device be incorporated into an electrical circuit (see Figure 1.4). The external equipment may be as simple as a voltmeter in a potentiometry experiment or a combination of a current meter and a voltage control unit (called a potentiostat) in the case of a voltammetry experiment. The circuit provides a path for charge to move from the external measuring device into the electrochemical cell and back out again to the meter in a complete loop. For example, consider a voltammetry experiment. If the external equipment pushes electrons into the working electrode to drive a reduction reaction (where some chemical species in solution accepts electrons from the electrode), then there must be a mechanism that can return electrons from the cell to the outside circuit to complete the cycle. A second electrode is introduced to provide a path for electrons to return to the meter. This second electrode is known as a reference electrode.

Occasionally, the components of an electrochemical cell are summarized in a schematic diagram written on a single line, such as this:

$$Cu/Cu^{2+}(5\,mM),\ KNO_3\,(0.1\,M)//KCl\,(0.1\,M)/AgCl/Ag$$
$$\text{Anode} \qquad\qquad\qquad \text{Cathode}$$

A single slanted line, $/$, indicates a phase boundary and a double slanted line, $//$, indicates a salt bridge separating the two half-cells. A potential may develop at any of those boundaries. The salt bridge may be as simple as a porous glass frit filled with a salt solution. It has two boundaries, one facing each of the two half-cell solution compartments. Components separated by a comma are together in the same solution. Electrode materials are specified at the beginning and the end of the line. The electrode where an oxidation process occurs appears on the left and is known as the anode. The electrode for the half-cell where a reduction process occurs appears on the far right. In this case, a copper electrode is placed in a solution of copper(II) ions together with an electrolyte solution of 0.1 M potassium nitrate. A salt bridge separates the first solution from a potassium chloride solution. In contact with the KCl solution, is a silver wire that has a coating of silver chloride. This particular diagram indicates that the two half-cell reactions are

$$\text{Anode}: Cu \rightleftarrows Cu^{2+} + 2e^-$$

$$\text{Cathode}: AgCl + e^- \rightleftarrows Ag + Cl^-$$

This type of diagram is more common in energy storage (batteries) and power generation systems, such as fuel cells. In many electroanalytical experiments, an external power source is used to apply a voltage to the system to drive a reaction of interest. In those cases, the applied voltage frequently changes in a manner so that the roles of the electrodes are reversed one or more times during the experiment. Therefore, such a diagram is only partly informative. A detailed description of the experimental conditions is more appropriate.

In a voltammetry experiment, it is necessary to apply a voltage to the working electrode in order to drive the reaction there. For that reason, it is very important to know how much energy is being applied to the working electrode. Reference electrodes are constructed so that the voltage at their surfaces remains constant. Consequently, any voltage applied to the cell from the outside is completely focused on the interface between the working electrode and the sample solution. A stable reference electrode is also essential for potentiometric experiments where the chemical composition of the solution induces a potential at the indicator electrode. The steadfastness of the reference electrode potential assures the experimenter that voltage changes in the cell represent potential changes of the same magnitude at the indicator electrode. Electrochemists say that the reference electrode is nonpolarizable; it is able to transfer whatever current is needed without budging from its reference potential. How reference electrodes maintain their nonpolarizability is discussed in Chapter 2.

The solution conditions in the immediate environment of the reference electrode are very carefully controlled in order to ensure that the reference electrode potential remains fixed. These conditions are often incompatible with sample solutions. Therefore, the reference solution is frequently isolated from the sample using a salt bridge. This salt bridge is usually a porous ceramic or polymer plug that provides ultrafine pores for the movement of ions but prevents significant mixing between the bulk solutions on opposite sides of the bridge (see Figure 1.4).

In many ways, a potentiometry experiment is simpler than a voltammetry experiment. A pH measurement with a glass electrode is a potentiometric experiment. The chemical composition of the sample solution surrounding the indicator electrode establishes an electrical potential energy difference across the boundary between the indicator electrode and the sample solution. The potential that the voltmeter reads is often called the cell potential, E_{cell}. It is common to think about the cell as an assembly of two "half-cells." Usually, the electron-transfer reaction taking place at the reference electrode constitutes one half reaction and the process occurring at the other electrode is the "indicator" half reaction. The measured cell potential represents the difference between the reference and indicator electrode potentials.

$$E_{cell} = E_{indicator} - E_{reference} \quad \text{or} \quad E_{indicator} = E_{cell} + E_{reference} \tag{1.14}$$

In practice $E_{reference}$ is a well-known constant so that any changes in the measured voltage for the cell can be interpreted as changes at the indicator electrode.

1.3.2. Cell Resistance

Voltammetry experiments, where currents are measured, require ions in the solution to carry charge between electrodes. Even though water ionizes to a small degree (for pure water $[H^+] = [OH^-] = 10^{-7}$ M), the conductance of water is usually too small for the purposes of most voltammetry experiments. Instead, one usually adds a pure salt to the sample solution. The salt is often referred to as the supporting electrolyte. Because the pH of a solution influences the electrode reaction in many cases, often an acid/base buffer is also included in the supporting electrolyte. The supporting electrolyte keeps the electrical resistance down. Lower resistance helps minimize voltage errors due to "ohmic losses" in voltammetry experiments. In a voltammetry experiment, current passes through the cell. The solution that carries that current has a finite electrical resistance. Energy, in the form of voltage, is lost overcoming the resistance according to Ohm's law. This loss represents an error in the measurement of the true voltage. The energy lost in volts, V, is given by

$$V = iR \tag{1.15}$$

where i is the current driven through the solution resistance, R. The actual voltage that reaches the electrode, V_{actual} is

$$V_{actual} = V_{applied} - iR_{cell} \tag{1.16}$$

In typical voltammetry experiments, the resistance is on the order of $100\,\Omega$. Consequently, errors on the order of $1\,mV$ or bigger occur when the current reaches $10^{-5}\,A$ ($=10^{-3}\,V/100\,\Omega$) or more. The energy lost in overcoming the solution resistance is energy that is not applied to the working electrode. Whenever the product, iR, is greater than a few millivolts, the assumption that all of the energy applied to the cell is focused onto the working electrode/solution interface no longer holds and the data are suspect.

1.3.3. Supporting Electrolyte

Supporting electrolyte is also important in potentiometry experiments, even though the current is virtually zero in those experiments. The reason for that is that all potentiometric indicator electrodes respond to the activity of an analyte, not just its concentration. The activity of an ion is a function of the ionic strength of the solution. Recall that the ionic strength, μ, is a measure of the concentration of charge:

$$\mu = \frac{1}{2} \sum c_i z_i^2 \tag{1.17}$$

where c_i is the molar concentration of an ion with charge, z_i, summed over all ions. In addition to the effect on activity of the analyte, the mismatch between the sample solution and reference solution in concentration and type of ions making up the supporting electrolyte contributes to an error called the liquid junction potential. (That phenomenon is addressed latter in this chapter.) Consequently, it is important to control the ionic strength. This is often done by the addition of a solution of a high concentration of electrolyte, known as an ionic strength adjustment buffer. Whenever that is not practical, an effort is made to keep the ionic strength constant among all the sample and calibration standards. (Ion activities and activity coefficients are discussed in Appendix A.)

A special voltmeter is used to monitor the potential between the electrodes. In order to function, any electronic meter requires some current to flow. As will be discussed later, drawing a significant level of current through the sensor distorts the voltage signal being measured. Therefore, the goal is to minimize the amount of current that is drawn. The type of voltmeter that is typically used in a pH meter or other potentiometric apparatus can measure the voltage while preventing a significant amount of current from flowing in the circuit. This attribute is what makes the meter special. Typical voltmeters sold in hardware stores draw current levels of around 10^{-6} A. For many electrochemical applications, a level above 10^{-12} A can be a significant amount of current. Voltmeters used in potentiometry are designed to draw tiny currents (10^{-12} A or smaller) during operation. They are said to have a high input impedance because they impede the flow of current into the meter.

1.4. THE ELECTRIFIED INTERFACE OR ELECTRICAL DOUBLE LAYER

Instrumental methods of electrochemical analysis depend upon chemical events at boundaries between two different phases. In potentiometric experiments, interesting processes give rise to a separation of charge at the boundary between the sample and sensor; in voltammetric experiments an outside power source applies a voltage to the working electrode creating a separation of charge that drives interesting processes there. It is common for charges to appear at many different phase boundaries in nature, for example, at the surface of biological cell walls, on the surface of water droplets or solid aerosols, and at the surface of wet materials such as ceramics, clays, sediments, and soils. The same electrochemical principles that are involved in electrochemical analysis drive lots of natural phenomena as well. One of the most important concepts that is universal is the boundary between two phases where charges accumulate. It is called the electrified interface or the electrical double layer.

1.4.1. Structure of the Double Layer

There are numerous electrochemical sensors that selectively respond to a specific chemical species of interest. For example, fluoride is routinely monitored in municipal drinking water by fluoride selective electrodes. Lithium ion can be determined in the blood or urine of a patient being treated for depression by lithium-containing medications using a lithium ion selective electrode. These devices are popular because of their simplicity of use and their reliability. The increasing interest in monitoring select chemical species in clinical, environmental, industrial settings and, more recently, in private homes and for personal health monitoring is likely to encourage the development and implementation of even more sensors of this type.

The heart of all electrochemical sensing devices is the boundary between the sensor and the test solution. It is there that a charge separation develops. Because of its importance, it is very useful to take a closer look at the structure of the boundary. Consider, for example, a metal wire dipping into a salt solution. Assume, for the sake of discussion, that an excess of negative charge (i.e. electrons) appears on the wire. Electromagnetic theory predicts that the excess charge will appear at the surface of the metal. The arrangement of charges on the solution side is a bit more complicated. The excess electrons will naturally attract cations from solution. In the mid-nineteenth century, the German scientist Herman Helmholtz imagined that all of the cations necessary to balance the charge on the metal surface migrate into position at a small distance from the surface forming a plane of charge [5]. It is now known that the cations do not actually come into contact with the

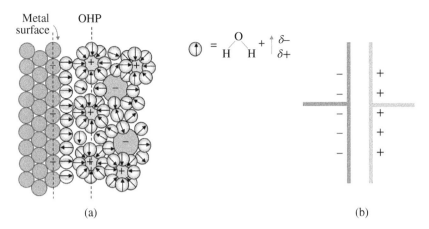

FIGURE 1.5   The Helmholtz model of the electrified boundary between a metal surface (dark spheres) with a net negative charge and an aqueous salt solution. (a) Cations are attracted to the surface forming a net positive layer to balance the negative charge in the metal. Water molecules occupy the first layer on the metal surface. They also surround ions in solution aligning their dipoles according to the type of charge on the ion. (Arrows point toward the oxygen atoms.) The charges on the solution side define a layer called the outer Helmholtz plane (OHP) [5]. (b) The double layer of charge behaves like a capacitor producing an electrical potential energy difference between the two layers whose magnitude is proportional to the charge.

metal surface because a monolayer of water molecules cling directly to the surface and are not easily displaced. Furthermore, individual cations are surrounded by a sphere of water molecules, known as the hydration sphere, that are also tightly bound. As a consequence, the cations approach the electrode surface no closer than about a distance equal to the length of two water molecules (about 5–6 Å total). Figure 1.5 shows cations with their hydration spheres parked in a line outside a layer of water molecules attached to the electrode surface. The centers of these cations represent a layer now known as the OHP. In the Helmholtz model, the charge on the OHP is equivalent in magnitude to the charge on the metal. This model closely resembles a simple capacitor.

$$Q_{\text{metal}} = -Q_{\text{OHP}} \tag{1.18}$$

In cases where the solid surface has a net positive charge, it attracts an excess of anions to balance the charge and the OHP is occupied by an excess of anions. In some cases, individual anions are able to come into direct contact with the metal surface. This phenomenon is called contact adsorption. Whether or not contact adsorption occurs depends upon the net free energy for three separate steps in the overall adsorption process. Two of the steps are obviously endothermic. Removing water molecules from the electrode surface to make room for the anion and removing part of the hydration sphere around the ion both cost energy. Therefore, only interactions between the ion and the electrode surface that lead to strong bonds make the adsorption process favorable. The electrostatic attractions between oppositely charged ions and the electrode are not decisive by themselves. Contact adsorption relies on London dispersion forces, overlap of electron orbitals, and image forces. An image force is similar to the mechanism known as London dispersion forces where a

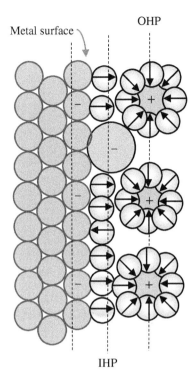

FIGURE 1.6 The inner Helmholtz plane (IHP) is a plane parallel to the electrode drawn through the center of ions or neutral molecules adsorbed directly in contact with the electrode surface. London and image forces as well as electron overlap between the metal and ion are responsible for the net attraction. These forces can result in binding even when charge considerations oppose adsorption [5].

momentary dipole resulting from the instantaneous arrangement of charge density around a molecule induces a rearrangement of electron density in a neighboring molecule creating a momentary dipole that results in dipole–dipole attraction. Unlike London dispersion forces, the image force is created by a permanent dipole or a charge on the ion inducing a dipole or local excess of charge in the electrode that leads to attraction at that location. The plane that includes the center of ions that are contact-adsorbed to the electrode surface is often called the inner Helmholtz plane (IHP) (see Figure 1.6). An interesting consequence of these additional forces is that some anions can remain attached to the electrode surface even when the electrode is also negatively charged. To account for contact adsorption, the equation for the charge balance becomes

$$Q_{\text{metal}} = -(Q_{\text{IHP}} + Q_{\text{OHP}}) \tag{1.19}$$

The model of Helmholtz suggests that all of the counter ions necessary to balance the charge on the electrode are held rigidly near the electrode surface. However, the experimental evidence suggests that thermal forces are great enough to dislocate counter ions to some degree. These observations inspired work by Louis Gouy [6] and by David Chapman [7] that led to a different model [5]. They proposed that the counter ions are distributed

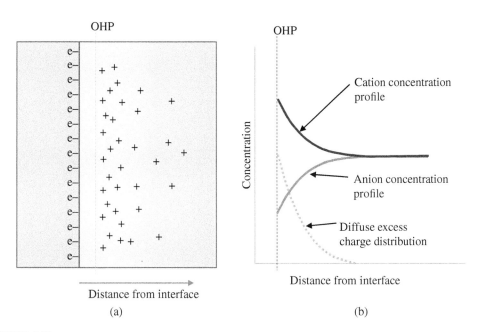

FIGURE 1.7 (a) Gouy–Chapman model of electrical double layer showing a diffuse region of excess charge on the solution side. (Solvent molecules are not shown for clarity.) (b) The excess concentration drops off exponentially with distance from the electrode surface.

in a nonuniform manner with a high concentration of counter ions near the electrode that falls off exponentially with distance from the electrode surface (Figure 1.7).

Unfortunately, this diffuse–charge model seemed to overcompensate. Experiments indicate that only a fraction of the charge appears to be disrupted by thermal agitation of the surrounding solution, whereas the diffuse charge model implied that it all was susceptible to thermal motion. In 1924, Otto Stern proposed a new model that was a synthesis of the earlier two [8]. Stern proposed that part of the compensating charge on the solution side is held tightly in the IHP plus the OHP and the remaining fraction of charge is contained in a diffuse zone of freely moving counter ions with a concentration that decreases with distance from the electrode surface. The IHP and OHP are sometimes collectively called the Stern layer. This layer is considered a stagnant zone of solution that clings to the surface. Figure 1.8 shows the arrangement of ions at a negatively charged electrode according to the Stern model. The charge balance equation for the double layer becomes

$$Q_{\text{metal}} = -(Q_{\text{IHP}} + Q_{\text{OHP}} + Q_{\text{diffuse}}) = -(Q_{\text{Stern}} + Q_{\text{diffuse}}) \tag{1.20}$$

Notice how the electrical potential changes as a function of distance from the electrode surface. The Stern model has several implications. One is that at higher electrolyte concentrations, the diffuse region of excess counter ions becomes more compact. The distance from the surface to the outer edge of the diffuse region is known as the Debye length, λ_{D}. The Debye length is inversely proportional to the square root of the electrolyte concentration. A good benchmark to keep in mind is that the Debye length is about 100 Å (10 nm) for

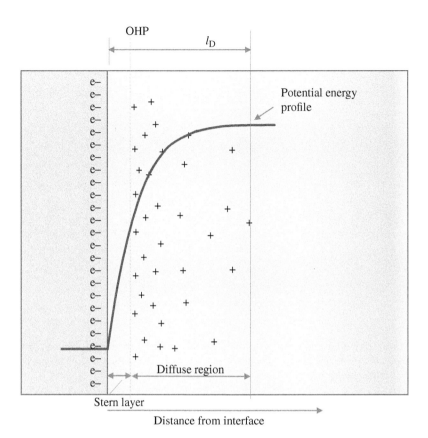

FIGURE 1.8 Stern model of the electrical double layer. Charges and solvent at the OHP cling to the solid surface. Not all of the charge on the surface of the electrode is compensated by the excess of cations at the outer Helmholtz plane (OHP). A fraction of the counter charge is represented in a diffuse region just beyond the OHP. (Only the charges for the excess cations – the number of cations greater than the local number of anions – are pictured for clarity.)

a sodium chloride solution of 0.01 M [9]. Most electroanalytical measurements are made at electrolyte concentrations of 0.01 M or higher. This profile for the potential has an important significance for voltammetry experiments where molecules must be transported to the electrode surface in order to exchange electrons with the surface. In most cases, the molecules can approach no closer than the OHP. Consequently, the electric potential that they experience is that of the OHP rather than the true potential of the electrode surface. As will be discussed in Chapter 5, this has important implications for the rate of electron transfer and the magnitude of the corresponding current in voltammetry experiments.

It seems appropriate to pause here and note an application of electrochemical principles to phenomena outside of the field of electroanalysis. Naturally occurring phase boundaries involving electrified surfaces often occur in a variety of environments. Here is just one example in which the structure of the electrical double layer is especially relevant. River waters frequently carry tiny soil particles that remain suspended in the water because of the negative charges on the mineral surface (see Figure 1.9). These surface charges arise from the crystal structure in which some Al^{3+} and Si^{4+} ions are replaced

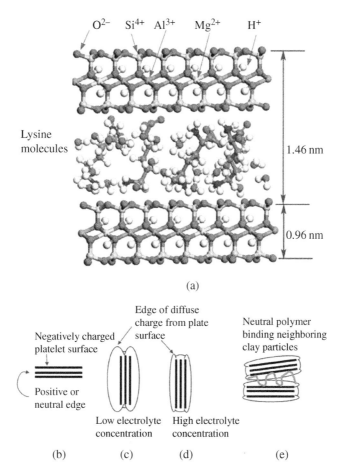

FIGURE 1.9 The electrical double layer plays an important role in the suspension or sedimentation of tiny particles, such as clay platelets. (a) Montmorillonite clay mineral structure with lysine bridging plates. Source: Adapted with permission from Zhu et al. [17]. Copyright 2019, Elsevier. (b) The surfaces of clay platelets are negatively charged as a result of lower valence cations replacing Al^{3+} and Si^{4+} ions in the crystal lattice. (c) At low ionic concentrations, the diffuse charge region of the sides extends several nanometers out into solution effectively pushing neighboring particles apart. (d) At high electrolyte levels, the field from the diffuse part of the electric double layer compacts allowing clay particles to approach more closely [12]. (e) Neutral polymers, such as naturally occurring polysaccharides, can adsorb at multiple points to neighboring particles leading to aggregation. Aggregation processes are known as coagulation and flocculation.

by lower valence cations, such as Mg^{2+}. Cations from the surrounding solution form an electrical double layer with the surface of each clay platelet. In the river, where the electrolyte concentration is low, these charged particles have electrostatic fields that extend far enough into solution to keep neighboring particles from approaching each other closely. However, whenever a river flows into the sea, the ionic strength of the mixture increases suddenly (the sea is about 0.5 M in NaCl) decreasing the distance that the electrostatic field reaches from the surface of a particle. Under these conditions, collisions bring the particles close enough for attractive interactions, such as van der Waals forces and hydrogen bonding, to overcome the repulsion of the charges. The particles cling to each other and can settle out faster as a result.

Seawater and soil particles are complicated mixtures and multiple mechanisms for binding particles together have been described. In some cases, calcium and other di- or tri-valent cations can bridge between adjacent particles [10]. Another mechanism has been exploited in water treatment and industrial applications of clay materials. Water soluble, neutral polymers, such as polysaccharide chains, are added to clay suspensions in industrial processes to bridge between particles by hydrogen bonding to oxygen atoms in the clay surface [11]. Polymers that attach at several points but still loop out into the solution appear to work best. Presumably, the loops in the chains extend far enough to reach across the electric double layer of neighboring particles. The compression of the electrical double layer is a key part of the mechanism in both industrial and natural processes. As these particles agglomerate at the mouth of rivers, they settle out of solution-carrying nutrients, and sometimes pollutants, into the sediments. This process has very important implications for the ecology of estuaries and the biological productivity of marine environments [12].

1.4.2. The Relationship Between Double Layer Charge and the Potential at the Electrode Interface

One can gain a lot of insight about electrochemical processes from approximating the behavior of an electrified interface with that of a capacitor. For example, it is interesting to think about how many charges are involved in creating a voltage across the boundary between two phases. An estimate can be made by modeling the electrical double layer as a simple capacitor where the solid metal constitutes one plate of the capacitor and the solution at the OHP (plus the diffuse region) serves as the second plate (Figure 1.10). (A similar argument can be made for other types of phase boundaries, such as between two ion-containing liquids.)

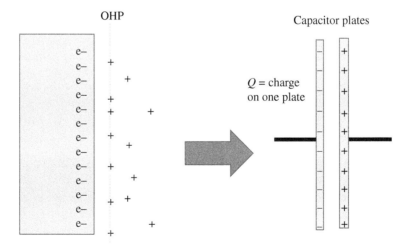

FIGURE 1.10 The electrical double layer can be modeled as a capacitor where the charge Q is the charge on one plate.

Consider how much charge would be required to create an interface potential differ-ence of 1.0 V. For the sake of discussion, consider a metal surface that has a net positive charge. The magnitude of charge, Q, that a capacitor accumulates on each side of the inter-face for a given voltage, V, separating the two plates is given by Eq. (1.21) [5].

$$Q = CV \qquad (1.21)$$

where the proportionality constant, C, is the capacitance of the dielectric medium separat-ing the plates. (Here is the reason that the model is an over-simplification. The double layer actually has a capacitance that varies with both the electrolyte concentration and with the double layer potential. However, this model does give results that set some upper limits. That is useful.) Empirically, the capacitance of the interface between a metal electrode in an aqueous salt solution is typically in the range of 10–40 µF/cm^2 [9]. For convenience, a value at the midpoint of that range will be used, i.e. 25 µF/cm^2 or 25×10^{-6} C/(V cm^2), to estimate the charge in this example. (Notice that the units on the value for the capacitance indicate that total charge depends on the electrode area.)

$$Q = CV = (25 \times 10^{-6}\ \text{C/(V cm}^2)) (1\ \text{V}) = 25 \times 10^{-6}\ \text{C/cm}^2 \qquad (1.22)$$

Because there are 9.6485×10^4 C/mole of charge:

$$Q, \text{in moles of charge} = \frac{25 \times 10^{-6}\ \text{C/cm}^2}{9.6485 \times 10^4\ \text{C/mol}} \approx 2.5 \times 10^{-10}\ \text{moles of charge/cm}^2 \qquad (1.23)$$

The calculation indicates that 2.5×10^{-10} mol of anions will be required to charge the double layer to a potential of 1.0 V. Thus, on a mole basis, the number of ions required to charge the solution side of the interface is tiny. For comparison, consider the number of moles of anions present next to a square electrode 1 cm on a side. Consider the chloride ions in a volume of 1 cm^3 solution of 0.1 M NaCl.

$$\text{Moles chloride ions} = 0.1\ \text{mol/l}(1 \times 10^{-3}\ \text{l}) = 1 \times 10^{-4}\ \text{mol} \qquad (1.24)$$

$$\frac{2.5 \times 10^{-10}\ \text{moles of charge in double layer}}{10^{-4}\ \text{mol Cl}^-\ \text{ions in 1 ml}} = 2.5 \times 10^{-6} = 0.000\ 25\% \qquad (1.25)$$

Charging the electrode to 1.0 V would require less than 0.0003% of the chloride ions from the surrounding milliliter of solution to be recruited into the double layer. Clearly, that amount represents a negligible loss to the Cl$^-$ concentration in the neighboring solution.

For every potential difference that appears across the double layer, there is a cor-responding arrangement of charge. If the number of charges changes, the double layer potential changes. Likewise, if one is applying a voltage to the interface, then one must move electrons and ions to establish any new arrangement of charge. Because the move-ment of charge constitutes a current, then a current will exist until the new arrangement of charge is established. This phenomenon is called the double layer charging current. It can be a problem in some voltammetry experiments, because the signal current may be much smaller than the double layer charging current. In applied potential techniques one is usually interested in measuring the current that is related to the amount of analyte that is being oxidized or reduced at the electrode interface. The signal current associated with the oxidation or reduction of a chemical species is called a Faradaic current because the charge exchanged between the electrode and the electroactive species in solution is proportional to the number of moles of analyte that is oxidized or reduced according to Faraday's law ($Q = \int i \, dt = nFN$). The double layer charging current is non-Faradaic; it represents a background component that one must remove from the signal in order to perform quantitative analyses. Methods for circumventing the double layer charging current are described in Chapter 5 on controlled potential techniques.

1.5. CONDUCTANCE

Electrical conductance is a measure of the ability to carry current. Resistance is defined as the reciprocal of conductance. It is easily measurable. Because the measurement of the resistance of a solution depends on the area of the electrodes and the distance separating them, the standard method uses two square platinum plates, 1 cm on each edge separated by 1 cm of solution (see Figure 1.11).

Of course, the interface between the solution and each plate develops an electrical double layer. As a consequence, the electrochemical cell behaves as a circuit with two capacitors in addition to the solution resistance. The resistance is measured using a special meter that applies an oscillating voltage to the electrodes and measures the current response. The resistance component has to be extracted from the response. The resulting resistance is called the specific resistance of the solution, ρ, and has the units of Ω cm. The electrical resistance for any other arrangement of electrodes is proportional to the length, ℓ, of solution between the electrodes of area, A.

$$R = \rho \frac{\ell}{A} \tag{1.26}$$

where ρ is the proportionality constant. Because conductance is inversely related to resistance one can define the conductance, G in Siemens, as follows:

$$G = \kappa \frac{A}{\ell} \tag{1.27}$$

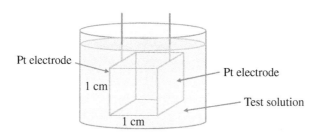

FIGURE 1.11   Conductance cell.

where κ is the electrical conductivity of the solution in units of Ω^{-1} cm^{-1} or S cm^{-1}. Although the standard method defines the shape and separation for electrodes, commercial instruments often have a different geometry and correct for differences by applying a calibration factor.

The solution resistance and conductance also varies with temperature [13].

$$G = \frac{1}{R_{\text{soln}}} = \kappa(1 + r(T - 25)) \tag{1.28}$$

where $T =$ the solution temperature in °C and r is a temperature coefficient in Siemens/degree for the solution. The temperature coefficient needs to be evaluated for different electrolyte solutions, but a representative value is $r = 0.0191$ for a 0.01 M KCl solution [13].

The conductance of a solution also depends on the type of ions that make up the electrolyte. The important point here is that ions move at different speeds. Ions move by diffusion, the process that is conceptualized as a random walk of individual particles, but under the influence of an electric field, they also migrate in the direction of the oppositely charged electrode. The velocity of an ion caused by an electric field is sometimes called the drift velocity or the migration velocity. It is proportional to the strength of the electric field, ε, driving the current.

$$v = u\varepsilon \tag{1.29}$$

where the electric field, ε, has the units of V/cm. It is the voltage difference between the electrodes divided by the distance between them. v is the drift velocity of the ion in cm/s and the proportionality constant, u, is the ion mobility. The units for the ion mobility are cm^2/(s V). The reason that the mobilities vary among ions is the fact that collisions with solvent molecules and other particles cause drag. Drag is related to the size of the ion. In this context, the size of the ion includes the sheath of solvent molecules that the ion drags with it, its solvent sphere. The bigger the solvated ion, the greater the viscous drag force opposing the ion's movement. All ions are slowed down by the viscosity of the solution. Because the viscosity decreases with temperature, the conductance increases with temperature as indicated in Eq. (1.28). The ion mobility is also proportional to the charge on the ion. Also, the bigger the charge, the greater the tug that the electric field exerts on the ion. Each ion carries a fraction of the total current in proportion to its mobility and its contribution to the total number of charges in solution. An important consequence of the variation in ion mobilities is the fact that the current is shared unevenly among the ions.

There are two practical applications of conductance or conductivity measurements that are of interest to analytical chemists. The first application is a semiquantitative estimate of ion concentration. One can calibrate conductivity measurements for an accurate quantitative determination of a specific salt, if that is the only source of ions in the sample. However, such a situation constitutes a special case. In the area of environmental science and agriculture, conductivity measurements are often made as a general indicator of water purity. Conductivity is also a parameter that is monitored at the outlet of a water purifying system commonly used for analytical and biochemistry laboratories. The quality of the water is often described in terms of the specific resistance. Theoretically, the specific resistance of water with no ions other than those from the dissociation of water is 18.3 MΩ.

There is one common application where conductivity provides good data for quantitative determinations of specific ions. Conductivity detectors are very popular for ion chromatography. In this case, the conductivity cell has been miniaturized so that it can operate on volumes on the microliter scale. Ions eluting from the separation column flow through the electrochemical device and cause a surge in conductance in proportion to their concentration. A full discussion of how a conductivity detector works in liquid chromatography can be found elsewhere [14].

1.6. MASS TRANSPORT BY CONVECTION AND DIFFUSION

The movement of ions and molecules in solution is important in many different aspects of electrochemical analysis. The term "mass transport" is often used to mean that reactant material is being driven by some force to the surface of an electrode. The rate at which reactant material is brought to the electrode surface influences the sensitivity of methods in many cases. The two most common mass transport mechanisms are convection and diffusion. In the first case, the bulk solution is mechanically stirred or pushed past an electrode such as in a flowing stream. The term "hydrodynamic system" is also used to mean a flowing or stirred solution that continuously brings material to the electrode.

The other mechanism for mass transport that is exploited in electroanalysis is called diffusion. Diffusion moves material by the force of a concentration gradient. This mechanism is subtler and deserves some discussion here. Imagine two solutions separated by a square window, 1 cm on each edge (see Figure 1.12). Molecules move very rapidly at room temperature, but they are frequently colliding with each other and the solvent. Consequently, the path of any individual molecule changes direction many times per second. The molecule appears to be moving randomly. How fast it moves depends on its solvated radius. Imagine also that one can count the molecules that pass through the window in each direction. The net excess going one direction or the other per second is called the flux for that molecule. A flux has the same dimensions as the product of a concentration and a velocity. The normal units are $mol/(cm^2\ s)$ (equivalent to $mol/cm^3 \times cm/s$). When the concentration for some molecule, M, on both sides of the window is equal, the number going from left to right through the window matches the number going from right to left each second. Consequently, the flux is zero.

Now, imagine starting the experiment over with a concentration of M at a value of C_M on the left side of the window and a concentration of 0 on the right side of the window.

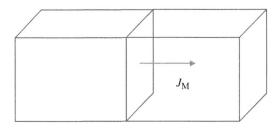

Flux is the number of $mol/(cm^2\ s)$ crossing a plane perpendicular to direction of net movement

J_M

FIGURE 1.12 The definition of flux is the net number of moles of molecules per second crossing a plane of solution with an area of $1\ cm^2$.

Because there are no molecules on the right side initially, none move through the window from right to left. However, many are going from left to right initially. Consequently, the flux is not zero initially. For ease of discussion, let the direction of left to right represent movement along the x-axis in the positive direction. Intuitively, the flux is never going to go from right to left as long as the C_M on the left is greater than it is on the right. It will never be greater on the right. At best, the concentration on the right will reach a value that equals that on the left, but only at equilibrium. Because there are more molecules to consider on the left side, the probability is always greater for a molecule to move from left to right through the window than in the other direction until equilibrium is reached. It also seems intuitive that the bigger the discrepancy in the concentrations on the two sides, the greater the excess in the number of molecules going in one direction. The following equation is a more elegant statement of these ideas. It is known as Fick's first law of diffusion.

$$J_M = -D_M \frac{dC_M}{dx} \tag{1.30}$$

where J_M is the flux of molecule, M, in mol/(cm^2 s), C_M is the concentration of M. The proportionality constant, D_M, is called the diffusion coefficient in cm^2/s. The gradient in concentration is the driving force for moving molecules across the plane perpendicular to the direction of motion. By convention a decreasing concentration in the x-direction is represented by a negative gradient, that is $dC/dx < 0$ in that case. The negative sign in front of the diffusion coefficient arises in order to make the flux positive for a concentration that decreases in the direction of increasing x. (It is just a convention.) The diffusion coefficient is related to the ion mobility described earlier by the Einstein–Smoluchowski equation [15]:

$$u_i = \frac{|z_i| F D_i}{RT} \tag{1.31}$$

As with ion mobility, the diffusion coefficient decreases with the solvated radius of an ion. The diffusion coefficients for a few ions in water are given in Table 1.2. Note that the

TABLE 1.2 Diffusion coefficients

Ions in water		
Ion	Diffusion coefficient[a] (cm^2/s)	Hydrated radius[b] (Å)
OH$^-$	52.73×10^{-6}	3.5
Na$^+$	13.34×10^{-6}	4.5
K$^+$	19.57×10^{-6}	3
SO$_4^{2-}$	10.65×10^{-6}	4
Ca^{2+}	7.92×10^{-6}	6
Cl$^-$	20.32×10^{-6}	3
Mg^{2+}	7.06×10^{-6}	8
H$^+$	96.6×10^{-6c}	9
NO$_3^-$	19.5×10^{-6c}	3

[a]From Samsonl et al. [18]. Copyright 2003. Used with permission.
[b]From Kielland [19]. Copyright 1937, American Chemical Society. Used with permission.
[c]Calculated from the electric mobility given by Bakker [15].

diffusion coefficients for OH^- and H^+ are rather large despite the fact that their hydrated radii are also large. That is the case because these ions do not move through water as an individual particle, but rather by exchange of hydrogen ions with molecules of water in their solvation sphere [5].

In a number of different electrochemical methods discussed in later chapters (see section 5.3), a reaction can change the local concentration of a reactant so that the concentration varies in adjacent regions of solution. This difference in concentration drives a net movement of material in the direction of higher to lower concentration.

1.7. LIQUID JUNCTION POTENTIALS

Because accuracy in measuring the voltage in potentiometry and in controlling the applied voltage in voltammetry experiments is so important, it is worth noting at this stage a common mechanism that can introduce errors in the cell potential measurement. Most electrochemical measurements involve the use of salt bridges to isolate reference electrodes from the sample solution. The contact between the sample solution and the salt bridge introduces another opportunity for a separation of charge to develop as shown in Figure 1.13. The mechanism for the build-up of a potential is driven by a difference in the mobility of the ions. For example, consider a salt bridge that contains 1 M NaCl as an electrolyte contacting a sample solution with 0.1 M NaCl.

Intuitively, one would expect that a difference in concentration of the ions at the boundary would lead to the movement of ions from the higher concentration toward the medium with the lower concentration. The movement of ions can be modeled mathematically using the Nernst–Planck equation [15]:

$$J_A = -D_A \frac{\partial C_A}{\partial x} - D_A C_A \frac{z_A F}{RT} \frac{\partial \phi}{\partial x} \tag{1.32}$$

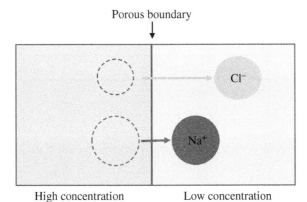

FIGURE 1.13 A salt bridge is a liquid junction between two solutions that allows a small exchange of ions but prevents the two solutions from mixing. A difference in concentration drives ions from the side of high concentration to the low concentration side. Smaller, faster ions move more charge of one sign across the boundary than the other. The result is a net separation of charge at the boundary causing a voltage difference known as a liquid junction potential.

where C_A is the concentration of solute A in mol/cm^3, D_A is the diffusion coefficient of that solute in cm^2/s, and J_A is the flux or moles crossing a plane perpendicular to the direction of movement (here, in the x-direction) per square centimeters per second, $\frac{\partial \phi}{\partial x}$ is the potential gradient, F is Faraday's constant, and T is the absolute temperature. (Notice that the flux, in mol/(s cm^2) has the same dimensions as the product of a concentration and a velocity, mol/cm^3 × cm/s.) Initially, both the sodium and chloride ions have the same concentration gradient, and there is no electric potential gradient (and the second term on the right is zero). The faster ion is the one with the larger diffusion coefficient. The diffusion coefficients in water of Na$^+$ and Cl$^-$ are 13.3×10^{-6} and 20.3×10^{-6} cm^2/s, respectively. That suggests that the chloride moves about 50% faster than the sodium ion. Consequently, a negative charge builds up on the lower concentration (sample solution) side and an equal amount of positive charge accumulates on the salt bridge side of the boundary due to the excess of sodium lagging behind. As a result, this process creates a potential energy difference across the boundary. However, the process quickly reaches a steady-state potential. The potential gradient across the boundary (the second term in Eq. (1.32)) is no longer zero. The potential gradient accelerates the movement of cations and decelerates the movement of anions so that the cation flux and anion flux become equivalent. This mechanism leads to the liquid junction potential. This junction potential is a part of the measured cell potential and it changes with solution conditions.

$$E_{cell} = E_{measured} = E_{indicator} + E_{junction} - E_{reference} \qquad (1.33)$$

The junction potential can introduce errors as big as 50 mV or more [16]. A useful equation for calculating the magnitude of this junction potential was presented by Henderson and is discussed in Appendix C.

Figure 1.14 can guide one's thinking about junction potentials in order to determine the effect on the measured cell potential. The reference point is the reference electrode. Going from the potential of the reference electrode through the electrochemical cell to the

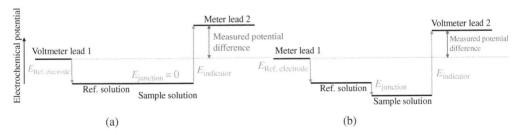

(a)   (b)

FIGURE 1.14 The measured potential between the indicator and reference electrodes includes all of the transitions in potential in between, represented by vertical arrows in the diagram. If anions are moving across the salt bridge faster than cations in the direction of the reference solution toward the sample solution, then $E_{junction}$ is negative; the vector for the potential for the junction is downward. (The sample solution potential is lower than that of the reference solution.) In this case, the $E_{indicator}$ is positive, so it is represented by an upward vector going from the solution potential toward the second meter lead. The cell potential is the value measured between the two voltmeter leads. If the junction potential had the opposite sign, then the vector for the junction would be upward and the measured potential would be larger than shown here. Both situations yield values that are different than the ideal (when $E_{junction} = 0$). (a) Ideal case and (b) real case; $E_{junction} \neq 0$.

other side, the path crosses the salt bridge. The argument above indicated that the sample solution will be at a lower potential (more negative) than the reference solution. The path from there, across the indicator electrode interface, starts from this solution potential which appears lower than it would have in the absence of a junction potential. The measured cell potential is indicated as a vector sum for all the transitions in potential between the meter leads. Ideally, the measured voltage is merely a difference between the reference and indicator electrode potentials, but the junction potential must also be included. The diagram indicates that the junction potential would be a negative number in Eq. (1.33). That is, the junction potential leads to an error that makes the measured potential appear more negative (or less positive) in this situation.

Of course, one way to avoid this error is to keep the conditions (other than the analyte concentration) of the sample solution and standards as similar as possible. This concern is another reason for using an ionic strength buffer. A further precaution that minimizes the junction potential is to use an electrolyte for both the salt bridge and test solutions in which the cation and anion have very similar diffusion coefficients, such as KCl. Potassium ions have a diffusion coefficient in water of 19.6×10^{-6} cm^2/s which is very similar to that of chloride ions (20.3×10^{-6} cm^2/s).

There is another consideration in setting up a salt bridge. Despite carefully matched diffusion coefficients, another mechanism can give rise to a junction potential. Glass or ceramic material are often used to make porous frits for salt bridges. These materials usually have a net negative charge on their surfaces. For glass with pores on the order of 5–10 nm in diameter, the negative charge can effectively screen anions from crossing the boundary leading to a junction potential on the order of 50 mV even with KCl as a supporting electrolyte (see Figure 1.15). Consequently, salt bridges made from material with larger (micron size) pores are preferable [16]. The trade-off is a greater leakage of ions from the salt bridge into the sample.

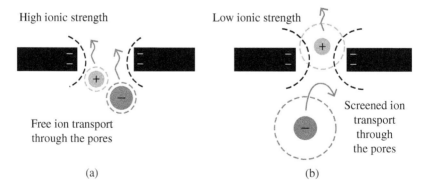

FIGURE 1.15 Microscopic pores in a glass frit used as a salt bridge. The electric field (shown by dashed lines) from negative charges on the glass surface extends a few nanometers into the surrounding solution. This distance is called the Debye length of the electrostatic field. The Debye length is inversely dependent on the ionic strength. (a) At high ionic strength, the anion can pass through the pore without its field interacting with that of the glass. (b) At low ionic strength, the field of the anion and the charge on the glass repel each other decreasing the migration of anions through the pore. Source: Adapted with permission from Mousavi et al. [16]. Copyright 2016, American Chemical Society.

PROBLEMS

1.1 A bourbon distillery treats its discharge water (that remains behind after the whisky distills off) electrochemically to remove copper by plating it out on an electrode. After 42 days of continuous operation the electrode was replaced. It was weighed before sending it off to a copper recycling service and was found to have recovered 450 g of copper. Calculate the average current during its use assuming that 100% of the current was used in the reduction of Cu^{2+} to Cu metal.

1.2 Consider a carbon electrode with a circular shape and a diameter of 3.0 mm dipping into a 0.1 M NaCl solution with a double layer capacitance of $20 \, \mu F/cm^2$. If the electrode receives a pulse of 1.0×10^{12} electrons, what will the change in voltage be at the electrode solution interface?

1.3 The diffusion coefficient for NO_3^- ion is bigger than the diffusion coefficient for Na^+ ion. Consider measuring the electrochemical cell potential in which a salt bridge is used between the reference solution containing 3 M $NaNO_3$ and a sample containing 0.1 M $NaNO_3$ supporting electrolyte solution. Explain whether there will be a junction potential and, if so, whether the cell potential with the junction potential will appear more positive or more negative than without it.

1.4 Explain why ohmic loss is more likely to cause a serious error in a voltammetry experiment than it is for a potentiometric experiment.

1.5 If 250 μA of current flows when a potentiostat applies −0.351 V to an electrochemical cell with a resistance of 152 Ω, what is the ohmic loss in voltage?

1.6 Explain two different mechanisms that could cause the potential of a pH electrode to shift upon the addition of 3 g of KCl to 100 ml of a solution of 0.1 M HCl.

1.7 How many moles of electrons would be required to change the voltage on a Pt circular disk electrode with a 2.0 mm diameter from −0.100 to −0.500 V in an electrolyte solution of 0.1 M KCl given the electrode solution capacitance of $24 \, \mu F/cm^2$?

1.8 The average thermal energy (or the average kinetic energy) in three dimensions for a molecule is often give as $3/2 \, kT$ where k is Boltzmann's constant and T is the absolute temperature. How does the average thermal energy of a molecule at 25 °C compare with 1 eV?

1.9 How does the energy of a blue photon at 400 nm compare to 1 eV?

1.10 How does the dissociation energy of the carbon–carbon bond in an ethane molecule compare with 1 eV?

REFERENCES

1. National Research Council (1987). *New Horizons in Electrochemical Science and Technology*. Washington, DC: The National Academies Press.

2. Pilla, A.A. (1974). *Bioelectrochem. Bioenerg.* 1 (1): 227.

3. Xiao, T., Wu, F., Hao, J. et al. (2017). *Anal. Chem.* 89 (1): 300–313.

4. Beni, V., Nilsson, D., Arven, P. et al. (2015). *ECS J. Solid State Sci. Technol.* 4 (10): S3001–S3005.

5. Bockris, J.O.'.M. and Reddy, A.K.N. (1970). *Modern Electrochemistry*, vol. 2. New York, NY: Plenum.

6. Gouy, G. (1910). *Compt. Rend.* 149: 654.

7. Chapman, D.L. (1913). *Phil. Mag.* 25: 475.

8. Stern, O. (1924). *Z. Electrochem.* 30: 508.

9. Bard, A.J. and Faulkner, L.R. (2001). *Electrochemical Methods*, 2e. New York, NY: Wiley.

10. Tang, J., Zhang, Y., and Bao, S. (2016). *Minerals* 6: 93. https://doi.org/10.3390/min6030093.

11. Fuller, L.G., Goh, T.B., Oscarson, D.W., and Biliadaris, C.G. (1995). *Clays Clay Miner.* 43 (5): 533–539. Grim, R.E. (1962). *Applied Clay Mineralogy*. New York: McGraw-Hill.

12. Furukawa, Y., Watkins, J.L., Kim, J. et al. (2009). *Geochem. Trans.* 10 (2) https://doi.org/10.1186/1467-4866-10-2.

13. Carlson, G.. Specific conductance as an output unit for conductivity readings. Technical Support, In-Situ Inc. https://in-situ.com/wp-content/uploads/2015/01/Specific-Conductance-as-an-Output-Unit-for-Conductivity-Readings-Tech-Note.pdf.

14. Vitha, M.F. (2017). *Chromatography, Principles and Instrumentation*. Hoboken, NJ: Wiley.

15. Bakker, E. (2014) Fundamentals of Electroanalysis: 1. Potentials and Transport. ebook. https://itunes.apple.com/us/book/fundamentals-electroanalysis-1-potentials-transport/id933624613?mt=11.

16. Mousavi, M.P.S., Saba, S.A., Anderson, E.L. et al. (2016). *Anal. Chem.* 88: 8706–8713.

17. Zhu, S., Xia, M., Chu, Y. et al. (2019). *Appl. Clay Sci.* 169: 40–47.

18. Samson, E., Marchand, J., and Snyder, K.A. (2003). *Mater. Struct.* 36: 156–165.

19. Kielland, J. (1937). *J. Am. Chem. Soc.* 59: 1675.

POTENTIOMETRY OF OXIDATION–REDUCTION PROCESSES

2

2.1. OVERVIEW

Potentiometry of redox solutions attempts to measure the open circuit potential of a system. The open circuit potential means that the measurement is made without drawing significant current. The result is often called the redox solution potential. The measurement is usually performed with an indicator electrode, such as a Pt wire. Electron exchange between the inert electrode and a species in solution raises or lowers the potential energy of electrons in the wire to match the energy of electrons in the reduced species in solution. The reaction in question is generalized as follows:

$$Ox + ne^- \rightleftarrows Red$$

The relationship between the solution potential and the activity of the redox species in solution is given by the Nernst equation:

$$E_{soln} = E^o - \frac{RT}{nF} \ln \left\{ \frac{a_{Red}}{a_{Ox}} \right\}$$

The standard potential, E^o, in the Nernst equation represents the solution potential that is observed at standard conditions, namely, unit activity for all reactants and products, 1.0 atm pressure, and 298 K. The Nernst equation can be applied to concentrations, as shown here:

$$E_{soln} = E^{o\prime} - \frac{RT}{nF} \ln \left\{ \frac{C_{Red}}{C_{Ox}} \right\}$$

where the activity coefficients have been combined with the standard E^o to yield a formal potential, $E^{o\prime}$ as long as the conditions that effect the activity coefficients, such as ionic strength, are constant. Often, hydrogen ion, hydroxide ion or ligands appear in the redox reaction, and the corresponding terms are incorporated into the formal potential as well. Knowledgeable workers are careful to state the conditions that apply to the formal potential.

The most commonly used reference electrodes are based on the reduction of silver chloride:

$$AgCl + e^- \rightleftarrows Ag + Cl^-$$

Electroanalytical Chemistry: Principles, Best Practices, and Case Studies, First Edition. Gary A. Mabbott.
© 2020 John Wiley & Sons, Inc. Published 2020 by John Wiley & Sons, Inc.

or mercurous chloride (calomel):

$$Hg_2Cl_2 + 2e^- \rightleftarrows 2Hg + 2Cl^-.$$

Variations in chloride activity for these reference electrodes lead to variations in the corresponding reference electrode potential. The ultimate reference electrode is the standard hydrogen electrode (SHE) also known as the normal hydrogen electrode (NHE):

$$2H^+ + 2e^- \rightleftarrows H_2$$

for 1 M hydrogen ion activity and 1 atm hydrogen gas. Although it is used only in special circumstances, it has a standard potential of 0.000 V, and most workers report their most important experimental results in terms of the potential that one would observe versus SHE as well as the potential actually measured with their own reference electrode. The scale conversion is

$$E_{\text{cell vs SHE}} = E_{\text{cell vs Ref1}} + E_{\text{Ref 1 vs SHE}}.$$

Redox titrations are still used when high precision and accuracy are both critical to the analysis of a redox active analyte. The course of the titration can be monitored potentiometrically with a platinum indicator electrode. Before the equivalence point, the potential of the solution is controlled by the half reaction representing the sample species; after the equivalence point, the titrant species governs the reaction solution potential. The equivalence point potential is approximately equal to the weighted average of the formal potentials for the sample and titrant species, $E_s^{o'}$ and $E_t^{o'}$, respectively.

$$E_{\text{eq.pt.}} = \frac{n_s E_s^{o'} + n_t E_t^{o'}}{n_s + n_s}$$

Another common application of oxidation–reduction potential (ORP) measurement is known as E_H or ORP. It is an environmental parameter that is used to assess the redox power of both natural water systems and soil environments as well as industrial fermentation tanks and disinfecting solutions. It correlates well with the preferred environment of various microbial communities and predicts the form of various redox species that are likely to be present. Because natural systems are rarely at thermodynamic equilibrium and because some important redox species do not transfer electrons readily with metal electrodes, the measured potentials cannot be applied to the Nernst equation for the quantitative estimate of concentrations of redox species. However, ORP measurements have proven to be a useful indicator of the general redox status of an aqueous system dominated by microbial activity.

The following sections are topics that are appropriate to a college instrumental analysis course and are essential to one's understanding of the materials in later chapters of this book. Section 2.2 describes measuring open circuit potentials, Section 2.3 introduces the meaning of a solution potential, and Section 2.3.1 explains how electron transfer between species in solution and a metal surface leads to a potential at the interface. Although the Nernst equation is often discussed in general chemistry courses, Section 2.3.2 describes the relationship in terms of activities and Section 2.3.3 defines formal potential. The next few sections build the foundation to understanding reference electrodes. Section 2.3.4 introduces the concept of an active metal indicator electrode, Section 2.3.4.1 indicates how a metal responds to the corresponding metal cation in solution, and Section 2.3.4.2 shows how the metal cation activity can be controlled by a species that reacts with it. These ideas form the basis of practical reference electrodes that are discussed in Section 2.3.4.3. The SHE is discussed in Sections 2.3.4.4 and 2.3.4.5 demonstrates how one can convert potentials measured in an experiment with one reference electrode to a value that one would expect using a reference electrode of a different type. These topics should be studied before moving on to other chapters.

Redox titrations are covered in Section 2.3.5. The basic theory and methodology of redox titrations are often covered in introductory courses in analytical chemistry (quantitative analysis). Section 2.3.6 introduces ORP as an environmental characteristic. This material is not ordinarily covered in a standard course in instrumental analysis despite its interesting applications to environmental chemistry and microbial systems.

2.2. MEASURING "OPEN CIRCUIT" POTENTIALS

Potentiometry is the term given to a class of electroanalytical methods based on measuring the voltage between two electrodes. The voltage that is measured is assumed to be a result of the electrode reaching an equilibrium with an analyte in solution. Therefore, a quantitative relationship between the activity of the analyte and the electrode voltage can be derived from thermodynamic principles. It is assumed that the measurement is performed without disturbing the equilibrium arrangement of ions in the electrical double layer. However, if the measurement process pushes current across the double layer, then the chance of disrupting the arrangement of charge and, consequently, moving the potential away from the thermodynamic value increases. In order to avoid that problem, early measurement equipment used cumbersome "null point" strategies to balance the cell potential against an adjustable reference voltage source (as shown in Figure 2.1). Adjustments were made until the current (as measured by a very sensitive meter known as a galvanometer) between the cell and the reference source was zero. At that point, the adjustable voltage matched the cell voltage. A mechanical device attached to the adjustable source indicated the percentage of the full-scale battery voltage that corresponded to the new setting [1, 2]. The adjusted potential was calculated and assumed to be equivalent to the cell potential that developed before it was placed in an electrical circuit. This was called the "open circuit potential" to emphasize the idea that the measured value was equivalent to the cell potential when no current was flowing.

Many useful sensors, such as pH electrodes, are potentiometric devices that use selective membranes to develop signals that follow the concentration of a single type of ion over many orders of magnitude. These devices are known as ion selective electrodes and have a tremendous number of applications in the chemical sciences as well as in clinical, industrial, and environmental analyses. These sensors depend on charge transfer by ions at the sensor/sample solution interface. No electron transfer is involved at the double layer that generates the signal of interest. Because of their different mechanisms of operation

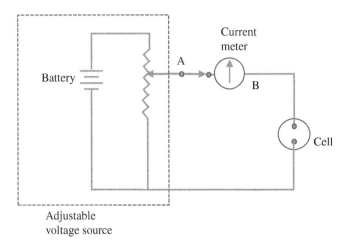

FIGURE 2.1 Null point method for measuring a voltage. The arrow pointing to the left at A represents a sliding contact that can be moved along a resistor to provide a variable voltage level at point A. The voltage at point A is adjusted until it matches the voltage at point B. When the meter indicates that current passing through it is zero, then there is no difference in the voltages at A and B (hence, the name, "null point"). At that point, the adjusted voltage at point A can be calculated as a percentage of the full-scale battery voltage. That percentage is equal to the percentage of the full range of the mechanical travel represented by the sliding contact at position A. The apparatus is equipped with a mechanism to read that position to at least three significant figures.

(ion transfer versus electron transfer) and their popular use, Chapter 3 is devoted to ion selective, charge transfer devices.

2.3. SOLUTION REDOX POTENTIAL

A second major application of potentiometry is the measurement of the redox potential of a solution. The redox potential is the driving force for electron transfer between a solution species and the indicator electrode surface. It also correlates with the strongest oxidizing agent available in the solution. That correlation makes it an important environmental parameter in natural water systems. It is sometimes called the oxidation–reduction potential or ORP in the literature. While pH is a measure of acidity, ORP is a measure of the relative oxidizing strength of a solution. In addition to pH, the redox potential is an important indicator of the biochemical status of a lake, sediment, or ground water. Chemical cycling in the natural world is largely determined by microbial activity. Microorganisms living in the water need some sort of oxidant, such as molecular oxygen, in order to metabolize food. In the absence of molecular oxygen, species such as iron(III) complexes, nitrate ions, and sulfate ions can serve as oxidants. Potentiometric measurement of the redox status of a solution is indicative of the oxidizing species that is available to microorganisms and, consequently, what class of microorganisms (aerobic versus anaerobic) will actively control the chemistry of the system. The predominant biological species will, in turn, determine what chemical species (nitrate, nitrite, ammonia, sulfate, sulfide, etc.) will be found as by-products of microbial metabolism in that environment [3]. This application to environmental analysis is described later in this chapter. In addition to measuring the redox potential of natural water systems, the redox potential of a solution can be used to follow the course of a titration or the status of a reaction in an industrial or lab setting [4]. This chapter focuses on the redox potential that develops at metal electrodes in both types of applications.

In studying any instrumental method of analysis, it is important to have a model for the underlying phenomena that give rise to the analytical signal. In this case, that understanding will lead to a quantitative relationship between the activity of the analyte and the potential that develops at the interface between the analyte solution and the surface of the sensor. It also provides insight into some of the constraints and limitations of potentiometric methods.

2.3.1. The Development of a Charge Separation

How does a voltage develop at the surface of an electrode? There are several different mechanisms, depending on the material that is used as the sensor. Perhaps, the simplest potentiometric electrode is one in which a metal surface, such as a platinum wire, exchanges electrons with the oxidized and reduced forms of some solution component. The state of the platinum does not change in this mechanism; the platinum is just a source or sink for electrons. It is regarded as inert (Figure 2.2). (The fact that "noble" metals like platinum and gold do not corrode easily is the basis for this assumption that they do nothing more than exchange electrons with electroactive species. That is a simplistic model [5]. Chapter 5 will address consequences when that assumption breaks down.)

Consider a simple mixture of Fe^{3+} and Fe^{2+} ions in a solution of nitric acid in contact with the metal surface. Some Fe^{3+} ions can take electrons from the platinum wire to form more Fe^{2+} ions.

$$Fe^{3+} + e^-_{(Pt)} \rightleftarrows Fe^{2+} \tag{2.1}$$

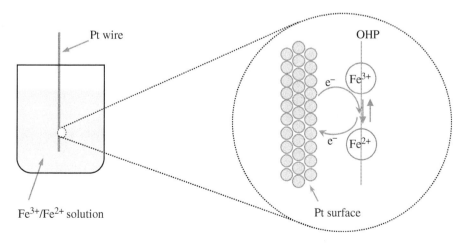

FIGURE 2.2 Electron transfer at a Pt electrode for Fe(III)/Fe(II) oxidation–reduction reaction.

It is also possible that the net movement of electrons is from the iron(II) into the wire accompanied by the formation of more of the oxidized species. Which direction the electrons will go depends upon the free energy of an electron in the platinum compared to the free energy of an electron in an iron(II) ion in solution. If the free energy for the electron is higher in the wire, then there is a net movement corresponding to the direction of an electron jumping from the wire to an iron(III) ion. Because the electron has a negative charge, a positive charge is left on the electrode surface when the electron leaves. Consequently, the metal becomes slightly more attractive to electrons. In other words, the free energy for an electron on the metal surface is slightly lower than before the electron left. As more electrons leave the electrode surface, the free energy for an electron on the platinum surface continues to fall until it matches the free energy of an electron on an iron(II) ion in solution. At this point, the rate of electrons moving in one direction matches the rate of electrons jumping in the opposite direction. It is assumed that an equilibrium is quickly reached for this exchange process and that the total amount of charge exchanged is tiny compared to the amount of Fe(II) and Fe(III) available in the surrounding solution. Typically, the number of electrons exchanged in this sort of situation is on the order of 100 pmol, or less, for electrodes with areas in the range of $1 \, cm^2$ [6]. Consequently, by placing the electrode into the solution is it unlikely that it will disturb the original concentrations of the redox species (except when the volume of the sample and the concentration of electroactive species are both very low). Also, note that this exchange of electrons takes place without any external circuit driving the process.

Consider the situation in which a positive charge builds up on the metal surface. An equivalent amount of charge carried by anions in solution is attracted to the interface. (This process forms the electrical double layer described in Chapter 1.) In this example, nitrate ions are the predominant anions in solution. Consequently, nitrate ions migrate to the surface to balance the net positive charge on the platinum surface. Of course, the initial conditions in solution determine whether the platinum surface takes on a net positive charge or a net negative charge, but the potential energy difference across the interface between the metal surface and the solution is related to the original ratio of Fe^{2+} and Fe^{3+} concentrations in solution. The potential that develops across the solution/metal interface in this case is called the redox potential of the solution.

2.3.2. The Nernst Equation

The quantitative relationship between the concentration of electroactive ions or molecules and the potential that develops at a metal surface was described in the late 1880s by Walter Nernst during his years as a postdoctoral researcher in the group of Friedrich Wilhelm Ostwald at the University of Leipzig, Germany [7]. The equation that Nernst developed was based on both experiment and theory. The equation's form evolved over the course of time. In the 1930s, Guggenheim demonstrated that the equation commonly used today can be derived from the concept of the electrochemical potential for ions [5, 8] (see Appendix E for the derivation). For a mixture of Fe^{2+} and Fe^{3+} the potential at the platinum/solution interface, E_{soln}, is

$$E_{soln} = E^{o}_{Fe^{3+}/Fe^{2+}} - \frac{RT}{nF} \ln \left\{ \frac{(a_{Fe^{2+}})}{(a_{Fe^{3+}})} \right\} \qquad (2.2)$$

In Eq. (2.2), R is the universal gas constant, 8.314 J/(K mol); T is the absolute temperature in kelvins; and n is the number of electrons in the balanced half reaction. From Eq. (2.1), $n = 1$ in this case. The activity, a_i, is the product of the concentration, c, and the activity coefficient, γ.

$$a_i = \gamma_i c_i \tag{2.3}$$

When the system is at the standard state, the activities of both Fe^{3+} and Fe^{2+} are 1 M and $E_{soln} = E^o{}_{Fe^{3+}/Fe^{2+}}$. The activity is the effective reaction concentration of the species. The presence of other ions in solution effects the behavior of the Fe^{3+} and Fe^{2+}. The activity coefficient, γ_i, accounts for this effect. Other ions in the neighborhood tend to screen the electric field of a reactant species decreasing its ability to interact with other species that would, otherwise, exchange an electron or a proton or react in some other way had the interfering ions not been there. Generally, the activity of an ion decreases as the concentration of the surrounding electrolyte increases. (A discussion of activity and activity coefficients appears in Appendix A at the end of this book.)

Equation (2.2) uses the ratio of activities for the products over the reactants as is the convention for defining equilibrium constants. Because the convention is to write a half reaction as a reduction process (electrons appear on the reactant side of the arrows), one can also write the ratio of activities for products (the reduced form) over reactants (the oxidized form) in the Nernst equation, if a negative sign is used in front of the logarithm term. In other words, each species on the reduced side of the equation appears in the numerator when a negative sign appears before the coefficient. A general procedure for formulating the equation for the redox potential for any balanced electron transfer reaction can be stated as follows: Consider the reduction of a moles of an oxidized species, Ox, by n electrons to form b moles of the reduced product, Red.

$$a\,Ox + ne \rightleftarrows b\,Red \tag{2.4}$$

The Nernst equation for the redox potential of a solution containing a mixture of the reactants and products can be written as

$$E_{soln} = E^o{}_{Ox/Red} - \frac{RT}{nF} \ln \frac{(a_{Red})^b}{(a_{Ox})^a} \tag{2.5}$$

Note that the number of electrons, n, appears in the denominator next to Faraday's constant. Here is another concrete example. 1,4-Benzoquinone is an organic molecule that is occasionally used as a mild oxidizing agent in synthesis work. It is also known as quinone (Q). It is reducible to 1,4-dihydroxybenzene or hydroquinone (H_2Q).

In applying the general approach described earlier, we include the activities of all the species on the reduced side of the equation in the numerator for the logarithm term and all of the species (accept the electrons) on the oxidized side of the equation in the denominator. The activity of each species is raised to a power equal to the stoichiometric coefficient for that species in the balanced half reaction.

$$E_{soln} = E^{o}{}_{Q/H_2Q} - \frac{RT}{2F} \ln \left\{ \frac{a_{H_2Q}}{(a_Q)(a_{H^+})^2} \right\} \tag{2.6}$$

Frequently, people express this equation for a temperature of 298 K (25 °C) and replace the natural logarithm with the common log ($\ln\{x\} = 2.303 \log\{x\}$. The coefficient of the logarithm term becomes $2.303 RT/nF = 0.059\,16/n$ V at 298 K.) The equation for the reduction of quinone then can be written as follows:

$$E_{soln} = E^{o}{}_{Q/H_2Q} - \frac{0.059\,16}{2} \log \left\{ \frac{a_{H_2Q}}{(a_Q)(a_{H^+})^2} \right\} \tag{2.7}$$

The activities of neutral molecules are essentially the same as their concentrations at ionic strengths below 0.1 M. Notice that the Nernst equation summarizes many different aspects of the behavior of redox systems such as the response to changes in activity of the reactants or products, changes in pH, temperature, and ionic strength (which affect activity coefficients). Consequently, it is a very important guide. Standard electrode potentials (E^o) are the redox potential for the half reaction when all species are at unit activity at a temperature of 298 K. Of course, the standard potential is also the observed redox potential of the solution whenever the ratio in the log term becomes unity, as would be the case when the activities of Fe^{3+} and Fe^{2+} are equal in Eq. (2.2). Tables for standard electrode potentials for many half reactions appear in the literature. A brief table of standard reduction potentials for some electroactive species appears in Appendix E in the back of this text. The significance of establishing standard electrode potentials for half reactions is similar to the importance of obtaining acid dissociation constants (K_as) for acids and bases in fields such as biochemistry. The E^o is a quantitative measure of the oxidizing power of the oxidized form of the redox pair. The more positive the value of E^o, the stronger is the oxidized form as an oxidizer; the more negative the standard potential, the stronger is the reduced form as a reducer. This is an important factor governing the action of redox enzymes in living systems, catalysts in many industrial processes, and electron-transfer reactions in many other contexts [7].

2.3.3. Formal Potential

It should be noted that E^o represents the redox potential that we would expect to measure when all reactants and products are present at *unit activity* at 25 °C. In dilute electrolyte solutions (say millimolar levels and lower) the activity coefficients approach a value of 1.0, and we can substitute concentrations for the activities in the Nernst equation with negligible error. In other situations, it may be convenient to use a fixed ionic strength in order to keep activity coefficients constant by adding a salt to the system. For example, one might choose to use 0.1 M KCl as a supporting electrolyte when other ionic components are 1 mM or less. In doing so, the higher KCl concentration dominates the ionic strength of the solution so that the activity coefficients (γ-values) are effectively constant throughout the experiment. This strategy also allows one to use concentrations instead of activities if one also evaluates the apparent E^o under the new conditions. This new constant (the measured potential for a mixture in which the concentration ratio of reduced and oxidized forms is unity) is known as the formal reduction potential, $E^{o\prime}$. In general terms for a half reaction:

$$a\text{Ox} + ne^- \rightleftarrows b\text{Red}$$

$$E_{\text{soln}} = E^o_{\text{Ox/Red}} - \frac{RT}{nF} \ln \frac{(a_{\text{Red}})^b}{(a_{\text{Ox}})^a} = E^o_{\text{Ox/Red}} - \frac{RT}{nF} \ln \frac{(\gamma_{\text{Red}} C_{\text{Red}})^b}{(\gamma_{\text{Ox}} C_{\text{Ox}})^a} \qquad (2.8)$$

Regrouping the activity coefficients with the standard potential:

$$E_{\text{soln}} = E^o_{\text{Ox/Red}} - \frac{RT}{nF} \ln \frac{(\gamma_{\text{Red}})^b}{(\gamma_{\text{Ox}})^a} - \frac{RT}{nF} \ln \frac{(C_{\text{Red}})^b}{(C_{\text{Ox}})^a} = E^{o\prime}_{\text{Ox/Red}} - \frac{RT}{nF} \ln \frac{(C_{\text{Red}})^b}{(C_{\text{Ox}})^a} \qquad (2.9)$$

where the formal reduction potential,

$$E^{o\prime} = E^o_{\text{Ox/Red}} - \frac{RT}{nF} \ln \frac{(\gamma_{\text{Red}})^b}{(\gamma_{\text{Ox}})^a} \qquad (2.10)$$

Obviously, the value of $E^{o\prime}$ depends on the specific conditions (the specific ionic strength and, therefore, the activity coefficients) that were used in this particular experiment.

In addition to the effect on activity coefficients, the solution conditions are important in other ways. For example, in the case of the nitric acid solution of iron(III) and iron(II) in Eq. (2.1), Fe^{3+} and Fe^{2+} were really abbreviations for the corresponding aquo-complexes, $Fe(H_2O)_6^{3+}$ and $Fe(H_2O)_6^{2+}$, respectively. Under other conditions, iron forms other complexes. For example, in a solution of 1 M HCl, iron(III) forms the following species: $FeCl(H_2O)_5^{2+}$, $FeCl_2(H_2O)_4^+$, $FeCl_3(H_2O)_3$, $FeCl_4(H_2O)_2^-$, and $HFeCl_4(H_2O)_4$. Usually, keeping track of the amount of iron(III) in all of the different species is not helpful to the task at hand. Often, the important concern is the total concentration of iron(III) in all of its forms. This total is referred to as the formal concentration of Fe^{3+}. When invoking a formal potential, the concentrations that appear in the Nernst equation are meant to be formal concentrations, unless specified otherwise. If one chooses to work with formal potentials, then it is very important to define for everyone else the conditions that are being used. The formal potential for the Fe^{3+}/Fe^{2+} redox couple is different in 1 M HCl (+0.700 V) compared to the standard potential (+0.771 V) as measured in 1 M $HClO_4$ where the iron is assumed to be all in the iron(III) and iron(II) aquo-complex forms.

There are other contexts for using formal potentials rather than standard electrode potentials. Most experiments are run under conditions far from standard state. For example, many half reactions involve H^+ or OH^- ions. Probably, the most common deviation in conditions is associated with the pH. Rarely do biochemists, for example,

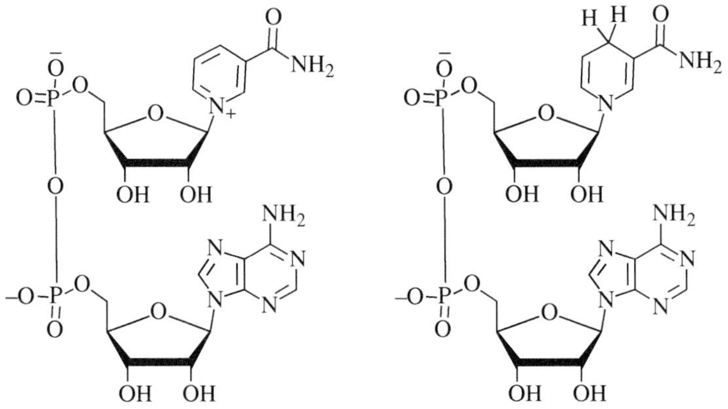

FIGURE 2.3 Structures of nicotinamide adenine dinucleotide in the oxidized form (NAD$^+$) and the reduced form (NADH). The site for redox activity is the nicotinamide ring at the top.

work at pH = 0, but whenever the half reaction includes H$^+$ ions, 1.0 M H$^+$ ion activity is standard state. For them, solution conditions of pH of 7.0 or 7.4 are more meaningful, because cellular system are buffered there. Consequently, in addition to fixing the ionic strength of the solution, they are likely to set the pH as well. In those cases, the formal potential may be defined to include the pH as well as the activity coefficients. For example, an important enzyme cofactor in many biochemical reactions is nicotinamide adenine dinucleotide, NAD$^+$ (Figure 2.3).

The reduction of NAD$^+$ at neutral pH is a two-electron, one proton process:

$$NAD^+ + 2e^- + H^+ \rightleftharpoons NADH \tag{2.11}$$

The Nernst equation for this system using the standard reduction potential is

$$E = E^\circ - \frac{0.059\,16}{2} \log \left(\frac{a_{NADH}}{(a_{NAD^+})(a_{H^+})} \right) \tag{2.12}$$

The formal potential is generally what one finds in the literature in searching for a standard electrode potential, or standard reduction potential, for a biochemically important redox agent. The formal potential, $E^{\circ\prime}$, for NAD$^+$ at pH 7.0 and 25 °C is −0.315 V [9]. This number represents the combination of the hydrogen ion activity with the E° (the standard electrode potential at pH 0):

$$E = E^\circ - \frac{0.059\,16}{2} \log \left(\frac{1}{(a_{H^+})} \right) - \frac{0.059\,16}{2} \log \left(\frac{a_{NADH}}{(a_{NAD^+})} \right) \tag{2.13}$$

or

$$E = E^\circ - 0.029\,58\text{pH} - \frac{0.059\,16}{2} \log \left(\frac{a_{NADH}}{a_{NAD^+}} \right) = E^{\circ\prime} - \frac{0.059\,16}{2} \log \left(\frac{a_{NADH}}{a_{NAD^+}} \right) \tag{2.14}$$

where

$$E^{o\prime} = E^o - \frac{0.059\,16}{n}\text{pH} = E^o - 0.029\,58\text{pH} \tag{2.15}$$

How would one calculate the formal potential for the $NAD^+/NADH$ redox pair at a different pH? For example, one might want to know the formal potential inside mitochondria (organelles where a lot of cell metabolism takes place) at pH 7.8. One approach would be to solve Eq. (2.15) for the standard electrode potential, E^o, and put that value back into Eq. (2.15) to calculate a new $E^{o\prime}$ for the new pH.

$$E^o = E^{o\prime}_{\text{orignal}} + 0.029\,58\text{pH}_{\text{original}} \tag{2.16}$$

$$E^{o\prime}_{\text{New}} = E^o - 0.029\,58\text{pH}_{\text{New}} = E^{o\prime}_{\text{orignal}} + 0.029\,58\text{pH}_{\text{orignal}} - 0.029\,58\text{pH}_{\text{New}} \tag{2.17}$$

Then, at pH = 7.8:

$$E^{o\prime}_{\text{New}} = E^{o\prime}_{\text{orignal}} + 0.029\,58\text{pH}_{\text{original}} - 0.029\,58\text{pH}_{\text{New}} = -0.315 + (0.029\,58)(7.0 - 7.8)$$

$$= -0.315 - (0.02958)(0.8) = -0.339 \tag{2.18}$$

Of course, working with a constant ionic strength would also enable the use of concentrations rather than activities, giving a slightly different value for the formal potential:

$$E = E^{o\prime\prime} - \frac{0.059\,16}{2}\log\left(\frac{[\text{NADH}]}{[\text{NAD}^+]}\right) \tag{2.19}$$

where

$$E^{o\prime\prime} = E^o - 0.029\,58\text{pH} - \frac{0.059\,16}{2}\log\left(\frac{\gamma_{\text{NADH}}}{\gamma_{\text{NAD}^+}}\right) \tag{2.20}$$

Once again, it is very important to state the formal conditions and whether the formal potential is meant to include the activity coefficients or not. Furthermore, the numerical value for the formal potential must be evaluated at the pH and ionic strength described for the experiment under consideration.

2.3.4. Active Metal Indicator Electrodes

In the previous section, a platinum wire (or other inert electrode material) was used to sense the redox potential established by an equilibrium between oxidized and reduced species, both of which were in solution. Neither metallic platinum nor platinum ions were a part of the electrochemical reaction equation. Because noble metals such as platinum and gold (and some forms of carbon) do not oxidize over a relatively wide voltage range, and they appear to exchange electrons easily with a wide range of electroactive species in solution, they have been used as indicator electrodes for redox-active species in solution. This section will examine surface reactions that develop potentials because of a transformation of the electrode material itself.

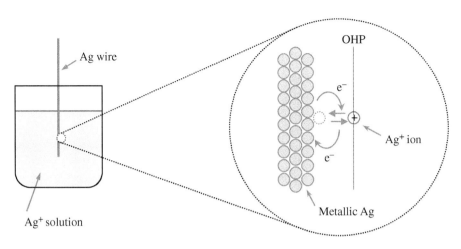

FIGURE 2.4   A metallic silver surface in electrochemical equilibrium with silver ions in solution.

2.3.4.1.   An Active Metal Electrode of the First Kind. Consider a piece of silver metal in contact with an aqueous solution of silver nitrate (Figure 2.4).

Whereas, it was assumed that platinum atoms on an electrode surface remain inert, silver atoms on a metallic silver surface are more active. That is, silver oxidizes more easily than platinum or gold. Tiny quantities of silver cations do leave the surface for the OHP and, in the process, each silver ion leaves behind an electron in the surface. The reverse process is also probable. It is possible for returning silver ions to find a place on the metal surface and become part of the metal by taking an electron out of the conduction band of the metal making the metal surface more positively charged. The process reaches an equilibrium that depends on the conditions. A net charge separation builds up between the metal and the OHP, leading to an electrical potential difference between the metal and solution. If one deliberately adds silver ions to the surrounding solution, the equilibrium shifts and that changes the potential on the surface. The electrode potential is controlled by this equilibrium reaction:

$$Ag^+ + e^- \;\rightleftarrows\; Ag \tag{2.21}$$

In writing the Nernst equation, metallic silver is the reduction product, so its activity should appear in the numerator of the logarithm term. Because metallic silver is a solid, its activity is unity by definition. So the net expression for the potential difference driving the reaction is written as follows:

$$E = E^o_{Ag^+/Ag} - \frac{RT}{F} \ln \left\{ \frac{1}{a_{Ag^+}} \right\} = E^o_{Ag^+/Ag} - (0.059\,16) \log \left\{ \frac{1}{a_{Ag^+}} \right\} \tag{2.22}$$

where $0.059\,16\,V$ is the value of $2.303(RT/F)$ at 298 K. For this electrode, E depends solely on the silver ion activity. This kind of electrode is, therefore, called an indicator electrode, meaning its potential directly indicates an activity of a species in solution (Figure 2.5).

It might appear that the silver wire could act as a selective sensor for Ag^+. That is true in limited situations. The problem is that metallic silver can also exchange electrons with

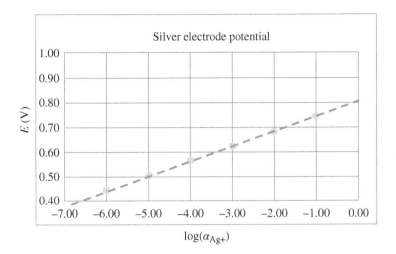

FIGURE 2.5 Plot of the half-cell potential at a silver metal electrode as a function of silver ion activity.

other electroactive solution species and behave in a manner similar to the platinum electrode described earlier. There is the possibility that more than one half reaction could be contributing to the potential at the silver electrode surface in complicated solutions. Under those circumstances, the relationship given in Eq. (2.22) may not hold. Consequently, a bare silver wire is not often used as a sensor for Ag^+ ions; there are better electrochemical devices for determining silver ion activity. Several other metal ions develop a redox potential with their corresponding metallic form. As with the Ag^+/Ag combination, these metals can be used to monitor the solution activity of the corresponding ion. However, this strategy is only for situations where other electroactive species that might exchange electrons with the metal are absent. Some examples are $Cu^+/Cu/$, Bi^+/Bi, and Ni^{2+}/Ni [10].

2.3.4.2. An Active Metal Electrode of the Second Kind. In the previous case, silver ions in solution are in equilibrium with the metal, and the potential is controlled by the activity of the free silver ions. An important second case arises if the solution surrounding the electrode contains anions that form a sparingly soluble salt with silver ions. For example, in the presence of chloride ions, silver precipitates to form crystalline silver chloride that has a solubility product of 1.78×10^{-10} (Figure 2.6) [11].

$$AgCl \;\rightleftarrows\; Ag^+ + Cl^- \tag{2.23}$$

In the presence of a solution of chloride ions, the silver ion concentration is controlled by the chloride activity:

$$K_{sp} = (a_{Ag^+})(a_{Cl^-}) = 1.78 \times 10^{-10} \tag{2.24}$$

or

$$(a_{Ag^+}) = \frac{K_{sp}}{(a_{Cl^-})} \tag{2.25}$$

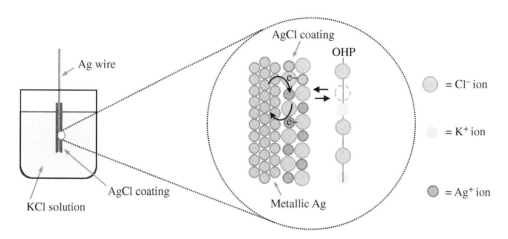

FIGURE 2.6 AgCl/Ag electrode. A coating of silver chloride provides silver ions close to the silver metal surface. Electron transfer is very favorable permitting the rapid conversion between Ag^+ and Ag metal. A chloride ion moves between the outer Helmholtz plane (OHP) and the AgCl crystal as necessary to balance the charge. The boundary potential between the solid AgCl and the OHP depends on the chloride ion activity in solution.

Substituting for the silver ion activity in the Nernst equation for silver ion reduction to silver metal gives:

$$E = E^o_{Ag^+/Ag} - \frac{RT}{F} \ln \left\{ \frac{1}{a_{Ag^+}} \right\} = E^o_{Ag^+/Ag} - \frac{RT}{F} \ln \left\{ \frac{(a_{Cl^-})}{K_{sp}} \right\} \tag{2.26}$$

Once again, the potential relies on the activity of only one ion, namely chloride in this case. Consequently, it is an indicator of the chloride activity. This type of indicator electrode is called an electrode of the second kind in older literature. With an active metal electrode of the first kind, the potential-controlling ion is the same element as the metal making up the electrode. With an active metal electrode of the second kind, the potential-controlling ion reacts with the ion made from the metal controlling the metal ion activity and, therefore, the potential indirectly. A convenient device can be made by depositing a coating of $AgCl_{(s)}$ directly onto the surface of the silver metal. The silver ions from the salt can be reduced directly to metallic silver:

$$AgCl + e^-_{Ag} \ \rightleftarrows \ Ag + Cl^- \tag{2.27}$$

From this reaction equation, the Nernst equation becomes:

$$E = E^o_{Ag/AgCl} - \frac{RT}{F} \ln\{a_{Cl^-}\} \tag{2.28}$$

The half reaction for the reduction of AgCl is related to the half reaction of Ag^+ as can be seen in Figure 2.7. There are two paths for the reduction AgCl. First of all, the salt can be reduced in a single step as indicated by path 1. Alternatively, the salt can dissolve to form

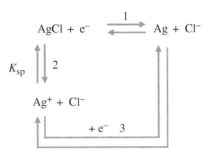

FIGURE 2.7 The reduction of AgCl to metallic Ag by two paths.

silver ions and chloride ions as indicated by step 2 followed by the reduction of the silver ions in solution by step 3. In both cases, the end result is metallic silver and chloride ions. Hess' law applies here. That is, the free energy change going from AgCl to the products by way of path 1 is equal to the free energy change for the path following step 2 and step 3. Considering this equivalence at standard state:

$$\Delta G_1^o = \Delta G_2^o + \Delta G_3^o \tag{2.29}$$

$$-nFE_{Ag/AgCl}^o = -RT \ln(K_{sp}) - nFE_{Ag^+/Ag}^o \tag{2.30}$$

Rearranging Eq. (2.30), one can solve for $E_{Ag/AgCl}^o$.

$$E_{Ag/AgCl}^o = E_{Ag^+/Ag}^o + \frac{RT}{nF} \ln(K_{sp}) \tag{2.31}$$

Consequently, the standard potential for the reduction of AgCl in Eq. (2.28) is a combination of two terms, the standard reduction potential for silver ions and the solubility product for silver chloride. Equation (2.31) provides a means of evaluating any one of these terms given data for the other two. Therefore, the standard electrode potential for the reduction of $AgCl_{(s)}$ is given by

$$E_{Ag/AgCl}^o = E_{Ag^+/Ag}^o + \frac{RT}{F} \ln\{K_{sp}\} = 0.779 + (0.059\,16) \log(1.78 \times 10^{-10}) = +0.222\,V \tag{2.32}$$

An interesting outcome of this arrangement is a device that responds to the logarithm of the chloride ion activity in solution.

A reliable way of measuring $E_{Ag/AgCl}^o$ would be to plot the potential at a silver/silver chloride electrode versus the logarithm of the chloride ion activity as in Figure 2.8. As Eq. (2.28) indicates, this plot will form a line that goes through $E = E_{Ag/AgCl}^o$ at a chloride activity of 1 M. This approach is recommended, because multiple measurements are required and the uncertainty in the resulting value is smaller than it would be using a single solution of 1 M chloride activity to measure $E_{Ag/AgCl}^o$ directly.

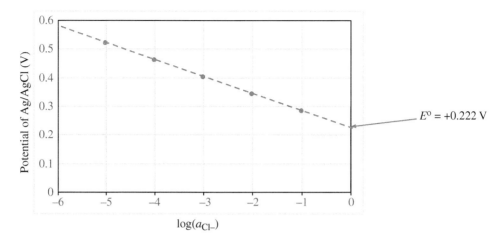

FIGURE 2.8   Plot of the potential of a silver/silver chloride electrode versus $\log(a_{Cl^-})$. The value of the standard potential, $E^o_{Ag/AgCl}$, for this half reaction is equal to the voltage at the y-intercept.

2.3.4.3.   Reference Electrodes.   An electrode of the second kind provides a very convenient method for creating a reference electrode. A silver chloride-coated silver wire is an easy-to-prepare and effective example of this type of electrode. This device is often called a silver/silver chloride electrode (or, sometimes, merely a silver chloride electrode). The surrounding chloride activity can easily be fixed and the electrode potential is set. The half-cell potential for a silver/silver chloride electrode with 1 M KCl is +0.222 V. A popular alternative is to use a saturated KCl solution (~4.17 M at 25 °C) so that there is no variation in the chloride activity, if any evaporation occurs. This saturated KCl, silver/silver chloride reference has a reference potential of +0.199 V. All of the species involved in the electron transfer are a part of the solid coating or the underlying silver. The electron transfer in either direction is very rapid. Chloride ions are a part of the overall reaction, but they are not directly involved in electron transfer with the silver. They mainly act to confine the silver ions to the coating. At high concentrations of chloride (≥0.1 M), this electrode robustly maintains the potential predicted by Eq. (2.28). If a silver chloride electrode is placed into an electrochemical cell along with an inert indicator electrode, such as a platinum wire and an outside voltage is applied to the cell, the silver chloride electrode will not budge from its rest potential as predicted by Eq. (2.28). All of the energy from the outside power source will be transferred to the double layer at the platinum electrode. This behavior is ideal for a reference electrode in an electrochemical measurement system. The key attribute is that the reference remains at its fixed potential regardless of current being forced through the cell in one direction or the other. Such an electrode is said to be "nonpolarizable." In practice, such ideal behavior is limited to low levels of current (less than a few microamps). In voltammetric experiments, an auxiliary electrode is usually added to carry the current for the reference electrode as a precaution. (Three-electrode systems are discussed in more detail in Chapters 5 and 7.)

Of course, applying a voltage to an electrochemical cell is a voltammetric experiment. An auxiliary electrode is not common in potentiometric work. In a potentiometric

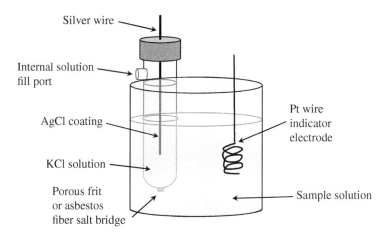

FIGURE 2.9 A complete potentiometric cell with a silver/silver chloride reference electrode and a Pt indicator electrode. This arrangement is commonly used for direct measurement of the redox potential of a sample solution or for a redox titration of an electroactive analyte, such as in the determination of vitamin C.

experiment, the cell current is nominally zero. (In practice it is finite, but very small, typically <1 pA.) Figure 2.9 shows an electrochemical cell for a simple potentiometric experiment, such as for a titration of vitamin C with I_3^-, a mild oxidizing agent. A platinum wire serves as an indicator electrode. Because a silver chloride electrode is nonpolarizable, changes in potential at the indicator electrode do not move the potential at the reference electrode in a potentiometric experiment either. Consequently, any change observed in the cell potential can be interpreted as a change in the potential of the indicator electrode. Therefore, as described in Chapter 1:

$$E_{cell} = E_{measured} = E_{indicator} + E_{junction} - E_{reference} \qquad (2.33)$$

or

$$E_{indicator} = E_{measured} - E_{junction} + E_{reference} \qquad (2.34)$$

The indicator potential, $E_{indicator}$, from Eq. (2.34) is related to the analyte concentration through the Nernst equation ($E_{indicator}$ is the same as the E_{soln} in Eq. (2.5)) for a redox system.

The silver/silver chloride reference electrode is the most commonly used reference in modern practice, but two other reference electrodes deserve special mention. The first of these is based on the reduction of a mercury(I) chloride salt, also known as calomel (see Figure 2.10). The half reaction is

$$Hg_2Cl_2 + 2e^- \; \rightleftarrows \; 2Hg + 2Cl^- \qquad (2.35)$$

$$E_{calomel} = E^o_{Hg_2Cl_2/Hg} - \frac{RT}{2F} \ln \{a_{Cl^-}\}^2 = E^o_{Hg_2Cl_2/Hg} - \frac{RT}{F} \ln\{a_{Cl^-}\} \qquad (2.36)$$

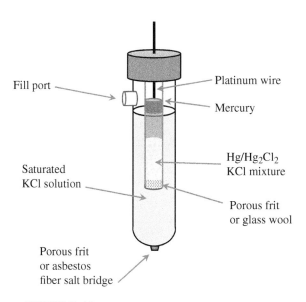

FIGURE 2.10 Saturated calomel electrode (SCE).

The standard potential, E°, for the calomel electrode (with a 1 M KCl solution) is +0.282 V and the saturated calomel electrode or SCE (a calomel electrode containing a saturated KCl solution) has a half-cell potential of +0.242 V [10].

2.3.4.4. The Standard Hydrogen Electrode. The ultimate reference electrode is based on the half reaction for the reduction of hydrogen ions to molecular hydrogen:

$$2H^+ + 2e^- \ \rightleftarrows \ H_2 \tag{2.37}$$

The corresponding Nernst equation for this half-cell is

$$E_{\text{NHE}} = E^o_{H^+/H_2} - \frac{RT}{2F} \ln \left\{ \frac{P_{H_2}}{(a_{H^+})^2} \right\} \tag{2.38}$$

In Eq. (2.38), the logarithm term is zero for a hydrogen atmosphere of 1.0 atm and a hydrogen ion activity of 1.00 M. By convention, $E^o_{H^+/H_2} = 0$. This is only a reference point for the purpose of establishing a relative potential energy scale; the absolute value for any single electrode potential is not known. One cannot know the absolute electrode potential because any experiment designed to measure electrode potentials requires two electrodes. The question of what is the potential of a single electrode is similar to asking "what is the distance to Chicago?" [10]. The obvious concern is from where? The answer only makes sense, if one specifies the starting point for measuring the distance to Chicago. Likewise, there must be a point of reference for measuring the potential at an electrode. An electrode potential can only be measured in an electrochemical cell with a second electrode as a reference. The ultimate reference has been arbitrarily chosen (by the International Union of Pure and Applied Chemists) to be the half-cell for the reduction of hydrogen ions to hydrogen gas at standard conditions. The apparatus in Figure 2.11 consists of a piece of platinum

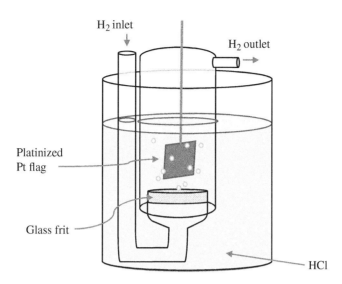

FIGURE 2.11 The standard hydrogen electrode (SHE). The half-cell is based on the reduction of hydrogen ions to hydrogen gas. The standard conditions of unit activity hydrogen ions and 1.0 atm pressure of hydrogen gas are an ideal. It is not practical for direct measurements.

metal dipping into a solution of HCl (1.0 M H^+ activity for standard state) and exposed to a stream of hydrogen gas at 1 atm [12]. In order to enhance the adsorption of hydrogen and the kinetics of electron transfer, the electrode surface is "platinized" by reducing chloroplatinate from solution to form a film of finely divided platinum particles on the Pt plate. The electrochemical potential of this half-cell is taken to be 0.000 V and standard electrode potentials are reported in comparison to this reference in the literature by convention. This electrode is generally known as the SHE or sometimes, the NHE (for Normal Hydrogen Electrode). However, NHE is technically a misnomer [13]. The term was originally introduced when "normal" referred to a *concentration* of 1 M. Because the high ionic strength of a 1 M solution of HCl lowers the activity coefficient, unit molarity is significantly different from unit activity. That being said, NHE and SHE are used synonymously now to mean the hydrogen electrode at unit hydrogen ion activity.

Data for $E°$-values for other half reactions are reported in the literature as though they had been measured with this electrode as the reference. The standard reduction potentials listed in the literature correspond to the cell potential for the half-cell of interest (at standard state) together with the SHE as a reference electrode and, since the SHE has a half-cell potential of 0.000 V, the measured cell potential is numerically equal to the half-cell potential for the second electrode. However, the direct application of a hydrogen electrode using a hydrogen activity of 1.0 M is not really practical [11]. The aforementioned moderate hydrogen ion concentrations (>0.01 M) in the salt bridges develop junction potentials that are difficult to compensate for. Instead, the hydrogen half-cell, as shown in Figure 2.12, is mainly used to calibrate a calomel electrode that, subsequently, is used as a secondary standard reference electrode for work with other half-cells. The cell potentials are plotted versus the log of H^+ activity and extrapolated to $a_{H^+} = 1.0$ (where the hydrogen reference electrode has a potential of 0.000 V by definition) to find the cell potential that is equivalent

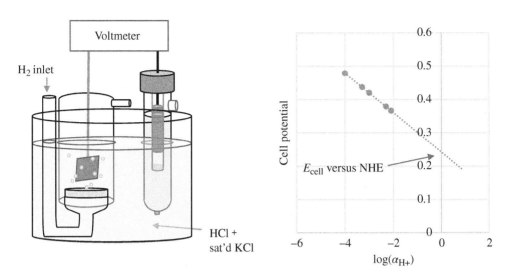

FIGURE 2.12 In calibrating a calomel electrode, the electrode is introduced directly into the outer beaker of the hydrogen half-cell (Figure 2.11). In that case, there is only one salt bridge separating the calomel electrode solution and the hydrogen half-cell. That salt bridge contains saturated KCl (or about 4.17 M KCl at 25 °C). KCl salt is added to saturate the solution in the hydrogen half-cell as well. The high concentration of KCl accounts for most of the ion traffic across the salt bridge and with an equal concentration on both sides, no junction potential develops. Above a level of 0.01 M HCl on the hydrogen electrode side, a hydrogen ion gradient is created through the salt bridge that introduces a voltage error. In order to avoid this error, cell measurements are made for a variety of hydrogen activities below 0.01 M. Then the data is graphed and extrapolated to a hydrogen activity level of 1.0 M. Because the potential of the SHE is defined as 0.000 at those conditions, this extrapolated cell potential is taken as the true reference potential of that calomel electrode on the SHE scale [12].

to the half-cell potential of the secondary reference electrode on the NHE scale. This calibration is done mainly for thermodynamic studies or cases where potential measurements of high precision and accuracy are needed.

2.3.4.5. Comparing Reference Electrodes. Most workers use calomel or silver/silver chloride reference electrodes for their experiments. That raises the question of how one can compare the results between experiments when the reference electrodes are different. The convention is to convert the measured cell potential to the SHE scale. That is, one calculates the potential that would have been observed had the reference electrode been the SHE instead (using Eq. (2.39)):

$$E_{\text{cell vs SHE}} = E_{\text{cell vs Ref1}} + E_{\text{Ref 1 vs SHE}} \tag{2.39}$$

For example, Figure 2.13 indicates the half-cell potentials for a few common reference electrodes in a ladder diagram. The values of the various reference electrode potentials are listed with respect to the SHE on the left. Also, included in the diagram is the standard electrode potential for the ferricyanide/ferrocyanide ($Fe(CN)_6^{3-}/Fe(CN)_6^{4-}$) redox couple on the same scale. If a group of workers were to measure the solution potential for

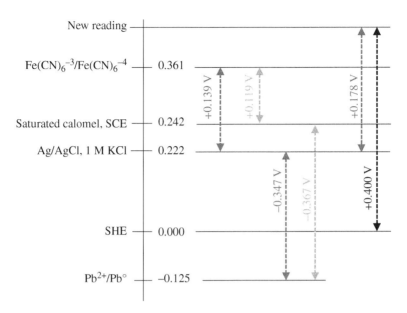

FIGURE 2.13 Diagramming half-cell potentials on an energy ladder using the SHE electrode as the zero point helps in converting measured cell potentials using one reference half-cell to the scale for a different reference electrode. For example, a solution with an equimolar mixture of $Fe(CN)_6^{3-}/Fe(CN)_6^{4-}$ would have a cell potential of +0.139 V using a Ag/AgCl (1 M KCl) reference electrode or a value of +0.119 V for an SCE reference. A half-cell with 1 M Pb^{2+} and a metallic lead electrode would produce a cell potential of −0.347 V with a Ag/AgCl (1 M KCl) reference electrode or a value of −0.367 V combined with an SCE reference electrode. Consider a new mixture that produces a cell potential of +0.178 V versus a Ag/AgCl (1 M KCl) reference electrode. It would have a value of +0.400 V on the SHE scale [14].

a mixture of $Fe(CN)_6^{3-}$ and $Fe(CN)_6^{4-}$, ions at the same activity using a platinum electrode and a Ag/AgCl (1.0 M KCl) reference electrode, they would read a cell potential of +0.139 V. The best way for them to report their results would be not only to present the value that they measured with their apparatus but also to calculate the value that one would expect to see had the SHE electrode been used as a reference, namely +0.361 V ($=E_{cell\ vs\ Ref\ 1} + E_{Ref\ 1\ vs\ SHE} = 0.139 + 0.222$). Reporting the value on the SHE scale is a convenience for other workers who might want to predict what they would see upon repeating the first group's work, but using their own reference electrode. For example, if the second team wanted to know what cell potential that they would see for the solution with the same ferricyanide/ferrocyanide composition measured against a calomel reference electrode, they would merely subtract their reference electrode potential from the value reported against the SHE, namely, $0.361 - 0.242 = +0.119$ V. Diagramming the relationships for half-cell potentials on the SHE scale, as in Figure 2.13, can be more helpful than trying to remember equations for converting between reference electrodes.

Also shown in Figure 2.13, is the standard potential for the reduction of Pb^{2+} ions to metallic lead. This process has an E° that is more negative than the potential for the SHE. Consequently, the reduction of Pb^{2+} would appear at even more negative voltages in a cell using either a Ag/AgCl or a calomel reference electrode.

TABLE 2.1   Temperature dependence of two common reference electrodes, calomel (saturated KCl) = SCE and Ag/AgCl (1 M KCl)

Temperature (°C)	E_{SCE} (V)	$E_{AgCl/Ag}$ (V)
0		0.236 55
5		0.234 13
10		0.231 42
15	0.250 8	0.228 57
20	0.247 6	0.225 57
25	0.244 4	0.222 34
30	0.241 7	0.219 04
35	0.239 1	0.215 65

Source: Heinemann 1989 [10]. Reprinted with permission of John Wiley & Sons.

Potentials for reference electrodes (and, therefore, cell potentials) are subject to changes in temperature. Some temperature effects are readily apparent, as it is with the coefficient for the logarithm term in the Nernst equation (see Eq. (2.5)). However, other temperature effects are subtler. The fact that the reference potential for the Ag/AgCl electrode depends on the K_{sp} of AgCl introduces another temperature effect, because equilibrium constants are temperature-dependent. Also, activity coefficients and junction potentials are sensitive to temperature. Consequently, careful work requires consistent temperature control for measuring all sample and standard solutions. Table 2.1 shows the variation in the reference electrode potential with temperature for the SCE and for the silver/silver chloride electrode in 1.0 M KCl [10].

2.3.5.   Redox Titrations

Although titrations are old technology, they are still in use in many important analyses today. For example, redox titrations are used to determine the level of moisture in pharmaceuticals because it is an important parameter in determining the shelf life of many medications. Titrations are also applied to monitoring the level of chlorine that is used to control bacteria in water for cooling towers [15, 16]. Titrations are particularly attractive whenever high precision and accuracy are needed. Typically, instrumental analyses (not based on titration) have uncertainties of about 1%. Using titration, a skilled analyst can usually produce data with a relative standard deviation of 0.1–0.3% depending on the conditions. Another advantage of titration techniques is that they can be performed with inexpensive equipment. Even a potentiometric titration can be performed with a pH meter with reference and indicator electrodes. Furthermore, autotitration equipment can remove much of the tedium and increase the productivity of the method.

Of course, the goal is to find the precise point on the volume axis that corresponds to the equivalence point where the moles of the sample and titrant species are in the exact ratio for a complete reaction according to the stoichiometry defined in the overall balanced reaction equation. The equivalence potential is not necessarily at the center of the steeply rising portion of the titration curve. However, the equivalence point potential can be calculated. An example case will help to make this clear. Consider titrating a sample of iron in the Fe^{2+} form with permanganate, MnO_4^-, titrant in an acid solution. (The acid helps prevent the precipitation of metal hydroxides and keeps the formal potential for the permanganate highly positive which favors the complete reaction between MnO_4^- and Fe^{2+}.)

$$Fe^{2+} \rightleftharpoons Fe^{3+} + e^- \tag{2.40}$$

$$MnO_4^- + 8H^+ + 5e^- \rightleftharpoons Mn^{2+} + 4H_2O \tag{2.41}$$

The net reaction equation is a combination of Eqs. (2.40, 2.41). In order to balance the number of electrons, each term in Eq. (2.40) must be multiplied by 5 before summing the two lines. The resulting equation gives the overall stoichiometry needed for the analysis:

$$5Fe^{2+} + MnO_4^- + 8H^+ \rightleftharpoons 5Fe^{3+} + Mn^{2+} + 4H_2O \tag{2.42}$$

The potential of the redox titration solution comes to an equilibrium after the addition of each aliquot of titrant. Before the equivalence point, the potential of the solution is in the neighborhood of the formal potential for the sample redox half reaction, namely $E^{o\prime}_{Fe^{3+}/Fe^{2+}}$. Halfway to the equivalence point, the potential at the indicator electrode is exactly equal to the formal potential of the sample species. After the equivalence point, the potential of the solution is in the neighborhood of the formal potential for the titrant species' half reaction, namely $E^{o\prime}_{MnO_4^-/Mn^{2+}}$. Because the half reaction for the permanganate titrant involves hydrogen ion, the formal potential also includes the pH.

$$E_{titrant} = E^o_{MnO_4^-/Mn^{2+}} - \frac{0.059\,16}{5} \log \left\{ \frac{a_{Mn^{2+}}}{a_{MnO_4^-}(a_{H^+})^8} \right\}$$

$$= E^{o\prime}_{MnO_4^-/Mn^{2+}} - \frac{0.059\,16}{5} \log \left\{ \frac{[Mn^{2+}]}{[MnO_4^-]} \right\} \tag{2.43}$$

where

$$E^{o\prime}_{MnO_4^-/Mn^{2+}} = E^o_{MnO_4^-/Mn^{2+}} - \frac{0.059\,16}{5} \log \left\{ \frac{\gamma_{Mn^{2+}}}{\gamma_{MnO_4^-}} \right\} - \frac{0.059\,16}{5} \log \left\{ \frac{1}{(a_{H^+})^8} \right\}$$

$$= E^o_{MnO_4^-/Mn^{2+}} - \frac{0.059\,16}{5} \log \left\{ \frac{\gamma_{Mn^{2+}}}{\gamma_{MnO_4^-}} \right\} - \frac{0.059\,16}{5}(8)pH \tag{2.44}$$

At the equivalence point, the potential of the indicator electrode will be equal to the potential of the reaction solution, $E_{eq.pt}$. It will also be equal to $E_{titrant}$ as given in Eq. (2.43) and to E_{sample} as given by the Nernst equation for the iron half reaction in Eq. (2.45):

$$E_{eq.pt} = E_{sample} = E^{o\prime}_{Fe^{3+}/Fe^{2+}} - 0.059\,16 \log \left\{ \frac{[Fe^{2+}]}{[Fe^{3+}]} \right\} \tag{2.45}$$

In principle, either Eq. (2.44) or Eq. (2.45) could be used to calculate the equivalence point potential. However, the concentrations of the reactant species (MnO_4^- and Fe^{2+}) are unknown. Fortunately, a substitution can be made. At the equivalence point, there are 5 mol of Fe^{3+} produced for every mole of Mn^{2+} formed.

$$5[Fe^{3+}] = [Mn^{2+}] \tag{2.46}$$

It is also true that at the equivalence point, any starting material remaining unreacted will also be equivalent. That is, the iron and manganese terms on the left side of Eq. (2.42) are equal to each other.

$$5[Fe^{2+}] = [MnO_4^-] \tag{2.47}$$

Substituting for the concentration of Mn^{2+} and MnO_4^- in Eq. (2.43) yields:

$$E_{\text{eq.pt}} = E^{o\prime}_{MnO_4^-/Mn^{2+}} - \frac{0.059\,16}{5} \log \left\{ \frac{5[Fe^{3+}]}{5[Fe^{2+}]} \right\} \tag{2.48}$$

By multiplying Eq. (2.48) by 5, the logarithm terms can be combined when it is added with Eq. (2.45):

$$6E_{\text{eq.pt}} = 5E^{o\prime}_{MnO_4^-/Mn^{2+}} - 0.059\,16 \log \left\{ \frac{5[Fe^{3+}]}{5[Fe^{2+}]} \right\} + E^{o\prime}_{Fe^{3+}/Fe^{2+}} - 0.059\,16 \log \left\{ \frac{[Fe^{2+}]}{[Fe^{3+}]} \right\}$$

$$= 5E^{o\prime}_{MnO_4^-/Mn^{2+}} - 0.059\,16 \log \left\{ \frac{5[Fe^{3+}]\,[Fe^{2+}]}{5[Fe^{2+}]\,[Fe^{3+}]} \right\} + E^{o\prime}_{Fe^{3+}/Fe^{2+}} \tag{2.49}$$

It is apparent that the argument of the logarithm term is unity in Eq. (2.49), consequently,

$$6E_{\text{eq.pt}} = 5E^{o\prime}_{MnO_4^-/Mn^{2+}} + E^{o\prime}_{Fe^{3+}/Fe^{2+}} \tag{2.50}$$

Solving for the equivalent point potential yields:

$$E_{\text{eq.pt}} = (5E^{o\prime}_{MnO_4^-/Mn^{2+}} + E^{o\prime}_{Fe^{3+}/Fe^{2+}})/6 \tag{2.51}$$

In general, the equivalence point potential is a weighted average of the formal potentials for the titrant and sample species:

$$E_{\text{eq.pt}} = (n_T E^{o\prime}_T + n_S E^{o\prime}_S)/(n_T + n_S) \tag{2.52}$$

where n_T indicates the number of moles of electrons in the balanced half reaction for the titrant species and n_S is the number of moles of electrons in the balanced half reaction for the sample species.

The information linking the shape of the titration curve and the titrant and sample species is summarized in Figure 2.14. It is an advantage that the equivalence point can be calculated in advance. However, because it is often difficult to calculate the junction potential and precise formal potentials due to high ionic strength levels that make activity coefficients uncertain and the changing pH of the reaction mixture, the best practice is

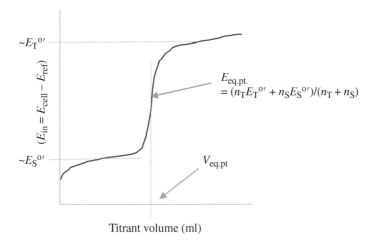

FIGURE 2.14   General shape of redox titration as monitored by potentiometry. The equivalence point potential is a weighted average of the formal potentials for the sample species half reaction and the titrant species half reaction.

to titrate a well-characterized sample with an analyte concentration similar to the anticipated level of the unknown samples and empirically determine the apparent equivalence point potential for the experimental conditions. Fortunately, when the formal potentials of the sample and titrant species are far apart, then the slope of the curve near the equivalence point is very steep and an error of a few millivolts in the equivalence potential will correspond to a relatively small error in the equivalence point volume.

2.3.6.   Oxidation–Reduction Potential (ORP) or E_H

In addition to its function as a laboratory tool for such tasks as monitoring the course of redox titrations, potentiometry with a platinum indicator electrode has applications in a variety of fields, such as environmental science, soil science, agronomy, industrial fermentation, and disinfection methods. Rivers, lakes, oceans, sediment, soil, fermentation vats, and similar aqueous environments are very complex systems. Most of these environments involve the activity of microorganisms. Microorganisms respond to their surroundings and can catalyze major chemical changes in their neighborhood. Just as pH is an important characteristic of a system, the local redox potential affects the chemistry and biochemistry of these complex systems.

The measurement of the apparent redox potential of a natural water system has some useful diagnostic value, but it must be used with caution. Many different redox components are present together in these settings, but they are not in equilibrium with each other. Under such circumstances, the potential developed at the platinum electrode surface is a result of a combination of electrochemical processes, leading to a response called a "mixed potential" [3].

What is a mixed potential? Consider a simplified representation of an environmental sample in which a solution contains a mixture of the oxidize and reduced forms of an organic redox couple, represented by M_{Ox} and M_{Red}. The electrode reactions can be

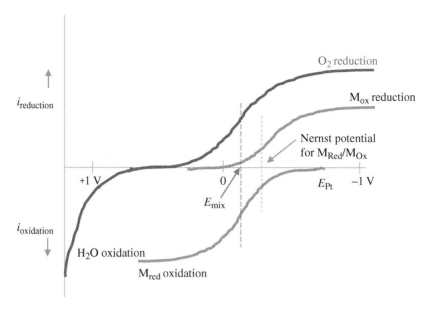

FIGURE 2.15 The platinum indicator electrode attains a potential where the net current is zero. For a simple mixture of M_{Red} and M_{Ox}, that potential agrees with the Nernst equation for that half reaction. In a complicated mixture, the current from each electroactive species contributes to the overall current. Consequently, the point of zero current (the potential measured at the platinum electrode) is not necessarily in agreement with the Nernst potential for a given half reaction and the corresponding activities of that redox couple [3].

summarized as follows:

$$M_{Ox} + ne^- + nH^+ \rightleftarrows MH_{n,Red} \tag{2.53}$$

Recall that no current flows at the measured potential in a potentiometric experiment. The platinum electrode develops a potential where the current for all reduction processes is equal to the current for all reactions going in the opposite direction. If M_{Ox} were present by itself, the current that it would produce at the platinum electrode as a function of voltage, in a stirred solution, would appear as the curve for M_{Ox} reduction in Figure 2.15. The current that would be observed for the oxidation of M_{Red} by itself also has been plotted on the same graph. The potential that the platinum electrode would attain in a mixture of both M_{Ox} and M_{Red} would be the point where the reduction current for M_{Ox} and the oxidation current for M_{Red} sum to zero. The potential where the sum of these two currents reaches zero is shown with a vertical dashed line on the right. It is approximately $-0.3\,V$ in this example. Now consider adding O_2 to the solution. (At $25\,°C$ the solubility of molecular oxygen is about $250\,\mu M$ in water.) The half reaction for the reduction of O_2 to water is

$$O_2 + 4e^- + 4H^+ \rightleftarrows 2H_2O \tag{2.54}$$

The oxidation of water is kinetically unfavorable at a platinum electrode, which means that it occurs only at fairly positive voltages. Consequently, the reaction going from right to left in Eq. (2.54) can be ignored in this discussion. The current as a function of voltage for the reduction of molecular oxygen by itself is also shown in Figure 2.15. Now combining the current for the reduction of molecular oxygen with the oxidation and reduction currents for M_{Red} and M_{Ox}, shifts the potential for the point of zero current to more positive values as marked by E_{mix}. The measured potential is not at the value predicted by the Nernst equation for M_{Red}/M_{Ox}.

The important point here is that one cannot apply the electrode potential reading to the Nernst equation for a given half reaction and expect to calculate a valid ratio of activities for the oxidized and reduced forms for a particular redox pair. Furthermore, there are also some other criticisms to make. Not all species that act as oxidizers or reducers in biochemical reactions exchange electrons easily with the platinum electrode. (They do not contribute to the measured potential.) In addition, there are experimental challenges to keeping the electrode surface and salt bridge clean and properly functioning [17]. These are serious limitations, but all is not lost. The measurement does have value as an operational parameter (a qualitative indicator) [17]. A general correlation has been observed between this measurement and the type of oxidizing agents that may be available for microbes to use for respiration. Taken together with pH measurements, the redox potential also indicates the stable species of various elements, the type of microbes that will be favored and the chemical changes that might be expected to occur under those conditions. In order to emphasize the operational nature of this measurement, some scientists use a special term for the redox potential in such complex systems. The electrode potential is called "E_H" in these cases. The "H" is a reminder that the potential is expressed on the normal hydrogen (SHE) scale [3].

2.3.7. Environmental Applications of Redox Measurements

2.3.7.1.   Soil E_H and pH. Even though a natural system is not at equilibrium, the pH and E_H of a soil or aquatic system is indicative of the type of microbial activity occurring there. Microorganisms control the chemistry and influence all other plants and organisms in the neighborhood. Both E_H and pH are important. They are both governing factors for and consequences of microbe-catalyzed chemistry in the soil. For example, iron is an essential nutrient at trace levels because of its role in the catalytic site of many different enzymes. It is critical to all organisms for that reason. However, most forms of iron found in oxygen-rich environments are insoluble. Because iron is so important to nitrogen-fixing bacteria, they produce and release special chelating agents, called siderophores, into the soil which bind and mobilize iron that the bacteria subsequently absorbs. In addition, phosphate, a nutrient that is needed at much higher levels than iron for growth, tends to be sequestered by adsorption on minerals, especially iron oxides and hydroxides.

Consequently, conditions that favor the dissolution of these solids also increase the availability of phosphate for plants and microbes. It has been observed that root tips of plants exude organic acids. This action lowers the local pH and mobilizes phosphate. It also has been observed that the organic acids are food for bacteria whose decomposition of the organic material tends to lower the E_H in the region around the plant roots. Both of these factors tend to mobilize inorganic nutrients for the plant. The optimum E_H range for many plants appears to be around +0.45 to +0.4 V. This redox environment also happens to coincide with conditions where a mix of nitrate and ammonia are both present [18].

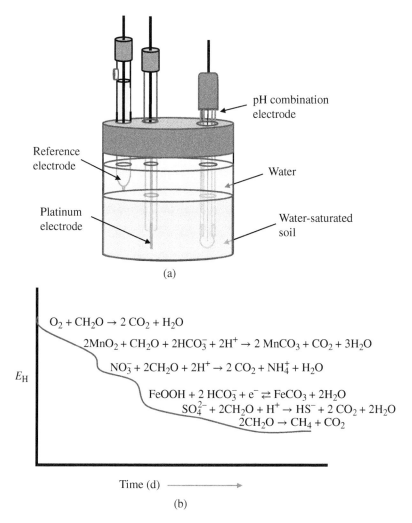

(a)

(b)

FIGURE 2.16 (a) Diagram of soil experiment. (b) The redox potential of a water saturated soil tends to drop with time following immersion because of the metabolic activity of the bacteria and the fact that molecular oxygen is slow to diffuse back into depleted regions. Microbes switch to different oxidizers for their respiration establishing a pattern of step-like transitions between levels of potential in rough approximation to the formal potential of the strongest oxidizer remaining.

In order to illustrate the correlations between the measured redox potential, E_H, and the chemistry of a complex natural system, consider a classical lab experiment in soil science. Imagine a sealed jar containing soil covered by water. Also, inside the container is a set of electrodes to monitor both pH and the redox potential (see Figure 2.16).

Bacteria living in the soil feed on organic matter present (sometimes represented by carbohydrates and abbreviated by CH_2O), oxidizing the organic material through a series of enzyme-controlled reactions ultimately to form CO_2 and water. When air is present, aerobic bacteria use O_2 as the final electron acceptor in the chain of reactions. The electrode potential starts out near the formal potential for the half reaction for the strongest available

TABLE 2.2 Redox reactions for common environmental oxidizers

	$E°$ (V)	$E°'$ (V) at pH 7
$O_2 + 4H^+ + 4e^- \rightleftarrows 2H_2O$	1.227	0.813
$NO_3^- + 2H^+ + 2e^- \rightleftarrows NO_2^- + H_2O$	0.837	0.422
$MnO_2 + HCO_3^-(10^{-3}\,M) + 3H^+ + 2e^- \rightleftarrows MnCO_3 + 2H_2O$	—	0.526^a
$NO_3^- + 10H^+ + 8e^- \rightleftarrows NH_4^+ + 3H_2O$	0.88	0.363_8
$NO_2^- + 8H^+ + 6e^- \rightleftarrows NH_4^+ + 2H_2O$	0.90	0.344
$FeOOH + 2HCO_3^-(10^{-3}\,M) + e^- \rightleftarrows FeCO_3 + 2H_2O$	—	-0.047^a
$SO_4^{-2} + 9H^+ + 8e^- \rightleftarrows HS^- + 4H_2O$	0.2514	−0.2218
$CO_2 + 8H^+ + 8e^- \rightleftarrows CH_4 + 2H_2O$	0.169	−0.2443

[a]These formal potentials are reported for the presence of 10^{-3} M bicarbonate concentration, because that is a more common condition for natural aquatic systems.
Source: Stumm and Morgan 1996 [3]. Adapted with permission of John Wiley & Sons.

oxidizer. At the start, the strongest oxidant is O_2, so the potential is in the neighborhood of the formal potential for the reduction of molecular oxygen to water (see Table 2.2).

The initial E_H is around +0.8 V for atmospheric O_2 levels and a neutral pH. Over the course of a few days, the E_H drops corresponding to a succession of microbial dynasties. Because the jar is sealed, the molecular oxygen is eventually depleted. When the oxygen is gone, aerobic bacteria (those that depend solely on O_2 for respiration) either die or go into a dormant phase, leaving facultative bacteria (those organisms that can utilize a variety of oxidizers for metabolism) to take control. Usually, the next most desirable oxidizer – the one that has the most positive formal potential and, therefore, provides the most energy – is the nitrate ion. If available, nitrate and nitrite can be reduced to ammonia. (The half reaction that pairs nitrate with nitrite is most often observed in aerobic conditions going from nitrite to nitrate following the production of nitrite by nitrogen-fixing bacteria, such as Nitrobacter [3].) The E_H during the time nitrate and/or nitrite are being utilized is roughly +0.3 V. Iron and manganese minerals can also act as oxidizers when present. When solid MnO_2 is present, its reduction occurs at about the same time as the reduction of nitrate. The iron(III) is utilized after that at lower E_H levels. The products of these reactions are usually solid $FeCO_3$ and $MnCO_3$ rather than free Fe^{2+} and Mn^{2+}. Carbonate is available because of the millimolar levels of carbon dioxide (that hydrolyzes to HCO_3^-) that are a result of earlier respiration. However, at low pH, these carbonate salts can dissolve, releasing Fe^{2+} and Mn^{2+} into the water column. As the E_H drops below −0.2 V, anaerobic bacteria begin to dominate. At this point, sulfate is the strongest oxidizer available and sulfide (or, more likely, HS^-) is the by-product. Interestingly, other bacteria can utilize carbon dioxide as an electron acceptor to breakdown organic compounds generating methane in the process.

There are some parallels to a titration experiment here in the sense that the reduced organic matter in the soil or water column consumes (with the help of microbes) all of the strongest oxidizer available before moving on to the next strongest oxidant. The E_H drops in a stepwise fashion and plateaus, while a particular oxidant is being consumed before the system goes on to the next. This general sequence of events occurs in nature when a soil is flooded, cutting it off from the supply of oxygen carried by the air. It is also observed in the sediment of deep lakes when the heat of the summer warms the water

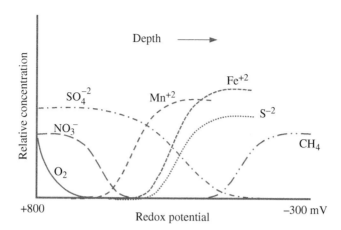

FIGURE 2.17 Soil chemistry changes with depth corresponding to the availability of oxidizing agents that are used by bacteria for respiration. E_H correlates well with these reactions and products. E_H values are calculated from the cell potential (with respect to the reference electrode). The diagram indicates the concentration profile for various common redox species. The redox potential decreases going deeper into the soil. A similar graph is expected for E_H measured at a single point in the upper level of the soil column and time after flooding of the soil replacing the distance axis. Source: Adapted with permission from Ref. [19].

near the lake's surface preventing mixing of air with the deep water and cutting the sediments off from oxygen. A similar pattern occurs on dry land. The oxygen that reaches the soil decreases with depth (see Figure 2.17). The succession of oxidizers follows what was described earlier where the time axis is replaced by distance below the surface.

Figure 2.18 is a diagram summarizing the different stable forms of iron as a function E_H and pH. This type of graph is often called a Pourbaix diagram. Moving down the diagram toward lower E_H, the conditions favor reduced forms of iron as would be expected in deeper regions of the soil. Depending on the pH and carbonate levels, the iron either forms solid iron(II) carbonate or soluble Fe^{2+}. Under very anaerobic conditions, the presence of sulfides from sulfate reduction traps the iron again as iron(II) sulfide.

Iron and manganese hydroxides adsorb many other species in addition to phosphate. Many metal ion pollutants readily adsorb to these surfaces. (Many classical methods of analysis exploit the fact that the precipitation of ferric hydroxide carries transition metal ions, and other cations, out of solution so that they can be concentrated from trace levels for analysis.) Mercury, chromium, arsenate, cobalt, nickel, cadmium, and lead ions are some metal ions with this tendency [3, 18, 21]. In ocean sediments, sulfate reduction can make iron and manganese sulfides important sinks for other trace metals due to adsorption. These minerals are also susceptible to dissolution and re-mobilization in the presence of oxygen. The mixing of fresh and marine water are places where such turnover can occur [3]. Rice fields are another area where low E_H levels are observed in the water-covered soil and where potential uptake of pollutants, such as arsenic, has been a concern [22]. In rice field studies, arsenic appeared to be less likely to be taken up by plants when the fields

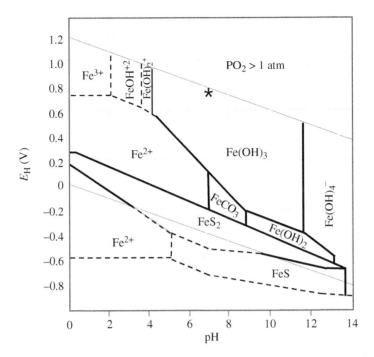

FIGURE 2.18 E_H–pH diagram shows the equilibrium between iron species in natural environments. The asterisk marks a fairly common set of conditions for an aerated soil. Progressing downward (either vertically or toward slightly lower pH), one sees the type of species normally encountered in deeper soils or in sediments cut off from air. Source: Adapted from Stumm and Morgan 1996 [3] and Elder [20].

were drained intermittently. The sequestering of the arsenic was attributed to the air exposure promoting oxidizing conditions where iron and aluminum oxides and hydroxides adsorbed the arsenate ions.

It is clear that the redox potential, E_H, is a valuable parameter for characterizing the local status of soil or natural water system. Even though a natural system is not at equilibrium, the E_H and pH indicate what sort of bacteria control the local chemistry, what oxidizing agents are available for microbes, what minerals will be stable, and whether certain pollutants might be mobilized as a result. That makes redox potential an important indicator in environmental science.

2.3.7.2. Applications to Fermentation Processes.

The fermentation industry represents a second area where monitoring redox potential can play a key role. Fermentation is the exploitation of the metabolic activity of a microorganism for the production of a desirable product. Yeast has been used for centuries to produce alcoholic beverages and to make bread, but industrial fermentation can also be used to produce other important chemical commodities. The extracellular redox environment can redirect the cell's metabolic profile to favor targeted products [23]. Although the biochemistry takes place inside the cells, the extracellular redox potential can be manipulated in order to activate desired enzymes and fuel selected pathways. Of course, DNA modification is another tool that has been used to expand the possibilities of fermentation products. Redox potential is influential, in part, because metabolism is based mainly on a series of oxidation/reduction pathways.

Biochemical reactions are thermodynamically favorable, but kinetically slow unless catalyzed by specific enzymes. Enzymes that control important redox reactions utilize cofactors that act as oxidizing or reducing agents. Two major cofactors are NADH and NADPH. (The structure of NADPH is similar to NADH. See Figure 2.19.) There are 129 enzymes that need NADH as a cofactor in driving 931 different redox reactions and 108 enzymes that require NADPH as cofactor in order to metabolize compounds in 1099 different redox reactions [23]. These two cofactors have similar two-electron half reactions with formal potentials (at pH 7) of -0.315 and -0.320 V, respectively [9].

$$NAD^+ + 2e^- + H^+ \rightleftarrows \ NADH \tag{2.55}$$

$$E = -0.315 - \frac{0.059\,16}{2} \log \left(\frac{a_{NADH}}{a_{NAD^+}} \right) \quad \text{at pH 7} \tag{2.56}$$

A large fraction of the enzymes controlling oxidation/reduction reactions in yeast have been shown to be sensitive to the extracellular redox potential. That is, controlling the external E_H can trigger changes in the activity of these enzymes inside the cell. One way of controlling the E_H of the growth medium is to seal the bioreactor from the air and add separate oxidizers or reducing agents via solution [23] (see Figure 2.20). Alternatively, gaseous O_2 can be bubbled through the mixture to raise the E_H or nitrogen can be added to remove O_2 thereby lowering the E_H. Knowing when to make an addition and how much to add depends on having a combination of E_H and reference electrodes sealed into the system.

An example of a commercially useful commodity produced by redox-controlled fermentation is 1,3-propanediol, an important starting material for making polymers, made by growing a bacterium on glycerol while keeping the medium between -0.2 and -0.4 V. The production of butanol, a biofuel, can be produced at a much higher level when controlling the growth medium of another bacterium at -0.29 V and adding nicotinic acid, a precursor for the formation of NADH which stimulates the appropriate pathways [23].

2.3.7.3. Applications to Sterilization.
Checking the potency of disinfecting regents represents another application where the monitoring of redox potential is beneficial. Strong oxidizing agents such as hypochlorite or ozone are commonly used to kill bacteria in food industry and healthcare facilities. For example, fruit and vegetable crops are frequently washed after harvesting in order to remove pesticides and kill pathogens, such as Salmonella [27]. The right level of hypochlorite is best determined by the E_H and pH of the solution rather than the total concentration of chlorine. Too much hypochlorite is actually less effective. Agronomists recommend recycling the rinse water and adding fresh hypochlorite as needed to maintain the redox potential between $+0.7$ and $+0.65$ V at a pH level of 6.5. Under those conditions, HOCl kills Salmonella and many other microbes within 30 seconds [27].

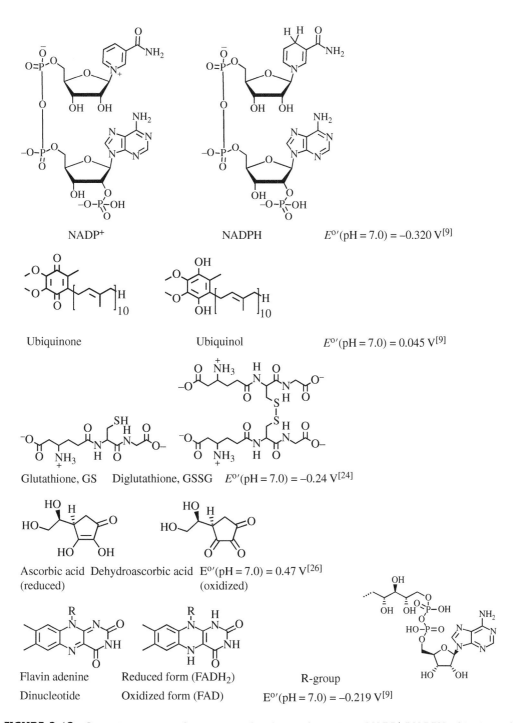

FIGURE 2.19 Some important redox agents in biochemical reactions. NADP⁺/NADPH, ubiquinone/ ubiquinol, and FAD are important enzyme cofactors that are recycled in many different metabolic processes. Ascorbic acid (vitamin C) is a cofactor for a few enzyme reactions, but is very important as a reducing agent that scavenges highly reactive oxidizing species that are often by-products of enzymatic and nonenzymatic reactions [24, 25]. Glutathione is also an important scavenger of reactive oxidizing species. It is present in cells at a relatively high concentration of 1–15 mM [26].

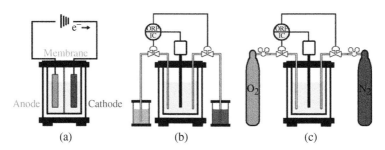

FIGURE 2.20 Methods for controlling the redox potential of growth media in fermentation baths. (a) Direct application of a voltage from an external power source using electrodes. (b) Controlled additions of separate oxidizing or reducing reagents. (c) The direct addition of O_2 or its removal using N_2. Source: Chen-Guang Liu et al., 2017 [23]. https://www.intechopen.com/books/fermentation-processes/fermentation-and-redox-potential. Licenced under CC BY 3.0.

PROBLEMS

2.1 Calculate the reference electrode potential for a Ag/AgCl electrode in a solution with a 0.10 M KCl activity.

2.2 Is H_2O_2 a more powerful oxidizing agent in an acid solution or a basic solution? Justify your answer with a quantitative argument.

2.3 If a null-point potential measurement device has a current meter capable of detecting a microamp of current, what is the maximum resistance that it can have in order to keep the error in the voltage (between point A and point B in Figure 2.1) ≤0.05 mV?

2.4 The limit to the precision of the potential measured by a null-point apparatus (as in Figure 2.1) is associated with the precision of the variable resistor used to adjust the voltage at point A. Precision variable resistors are commonly available with a rotary control that moves the sliding contact from the bottom position to the other extreme in 10 turns. Assuming that the dial on the rotary control can discriminate between settings that are 1° apart and assuming that the voltage on the battery is 1.5000 V, what is the uncertainty in volts in the voltage provided at point A?

2.5 What is the difference in mV in the voltage that one would observe for a Ag/AgCl reference electrode prepared using 0.50 M chloride concentration compared to 0.50 M chloride activity? (See Appendix A for help.)

2.6 (a) Write the Nernst equation for the half reaction for the reduction of permanganate ion to MnO_2.

$$MnO_4^- + 4H^+ + 3e^- \rightleftarrows MnO_2 + 2H_2O \quad E° = 1.70$$

(b) Calculate the formal potential for this half reaction at pH = 7.0.

2.7 Ferric ion is not very soluble at neutral pH. (The K_{sp} for $Fe(OH)_3$ is $\sim 3 \times 10^{-38}$.) Yet evidence of iron is often observed in water pumped directly from simple, hand-operated pumps in rural locations. (It is noticeable in the taste of the water and in the dark reddish-orange stain on the concrete support where water from the pump splashes.) Use the concept of E_H to describe the mechanism that solubilizes iron in the water deep in the ground water and leads to the rust color on the concrete base at the surface.

2.8   Consider the titration of samples solutions of $\sim 0.1\,M\,Sn^{2+}$ in $0.5\,M\,H_2SO_4$ solution using $0.1\,M\,Cr_2O_7^{2-}$ titrant in $0.5\,M\,H_2SO_4$.

(a) What would the equivalence point potential be (neglecting the effect of activity coefficients and any junction potential)?

(b) What would the equivalence point potential be if the sulfuric acid concentration were $0.05\,M$ and all other conditions were the same (neglecting the effect of activity coefficients and any junction potential)?

2.9   Temperature affects a reference electrode in several ways. Consider a drop in temperature from 25 to $20\,°C$ for a saturated calomel electrode (SCE). Explain the different mechanisms that lead to a change in the electrode's potential. (Can you describe four different influences?)

2.10   Consider the reduction of hydroxyl amine in an acid solution:

$$NH_3OH^+ + 2H^+ + 2e^- \; \rightleftarrows \; NH_4^+ + H_2O \quad E^o = 1.35\,V$$

(a) Write the Nernst equation for the half reaction.

(b) Calculate the formal potential for the same half reaction at a pH of 2 and assuming unit activity coefficients.

2.11   If a cell potential was measured to be $+0.150\,V$ using a saturated calomel reference electrode, what would the measured potential be if the reference electrode were replaced by a $Ag/AgCl$ electrode in a $0.10\,M\,KCl$ solution? (Assume that the junction potential is negligible in both cases.)

2.12   What potential would be expected for a cell with a platinum indicator electrode in an aqueous solution of $1.65\,mM\,Br_2$, $0.1\,M\,HCl$, and $0.753\,mM\,NaBrO_3$ versus a saturated calomel electrode (SCE) (neglecting activity coefficients and junction potentials)? See Appendix D for the appropriate standard electrode potential.

2.13   What would be expected for a cell with a platinum indicator electrode in an aqueous solution of $1.65\,mM\,Br_2$ and $0.753\,mM\,NaBrO_3$ versus a saturated calomel electrode (SCE) accounting for activity coefficients in a supporting electrolyte solution of $0.100\,M\,HCl$ (neglecting the junction potentials)?

2.14   Consider the salt bridge for a $Ag/AgCl$ reference electrode with a $1.0\,M\,KCl$ internal solution dipping into sample solution with a $1.0\,M\,NaCl$ supporting electrolyte solution. Explain the effect of the Na^+ ion, K^+ ion, and Cl^- ion on the junction potential.

2.15   Using the Henderson equation and table of diffusion coefficients in Appendix C, calculate the junction potential for a salt bridge with $3\,M\,KCl$ solution contacting a sample solution with $0.1\,M\,NaOH$.

REFERENCES

1. (1980). *The Beckman Handbook of Applied Electrochemisry*. Fullerton, CA: Beckman Instruments.

2. Francesco, S. and Bruno, L. (2009). *J. Chem. Educ.* 86: 246–250.

3. Stumm, W. and Morgan, J.J. (1996). *Aquatic Chemistry: Chemical Equilibria and Rates in Natural Waters*, 3e. New York, NY: Wiley.

4. Good sources for literature describing contemporary applications of redox titration procedures are manufacturer websites such as application library at Mettler Toledo. https://www.mt.com (accessed 9 August 2019).

5. Bockris, J.O.'.M. and Reddy, A.K.N. (1970). *Modern Electrochemistry*, vol. 1 & 2. New York, NY: Plenum.

6. Bard, A.J. and Faulkner, L.R. (2001). *Electrochemical Methods*, 2e. New York, NY: Wiley.

7. See for example,(a)Reckenfelderbäumer, N. and Krauth-Siegel, R.L. (2007). *Biochem. Biophys. Res. Commun.* 357: 804–808. (b) Arai, T., Yanagida, M., Konishi, Y. et al. (2008). *Electrochemistry* 76: 128–131.

8. Bakker, E. (2014) Fundamentals of electroanalysis: 1. Potentials and transport. ebook. https://itunes.apple.com/us/book/fundamentals-electroanalysis-1-potentials-transport/ id933624613?mt=11 (accessed 9 August 2019).

9. Voet, D., Voet, j.G., and Pratt, C.W. (1999). Chapter 13, Introduction to metabolism. In: *Fundamentals of Biochemistry*, 373. New York, NY: Wiley.

10. Heinemann, W. (1989). Introduction to electroanalytical chemistry. In: *Chemical Instrumentation: A Systematic Approach*, 3e (H.A. Strobel and W.R. Heinemann), 981. New York, NY: Wiley.

11. Blaedel, W.J. and Meloche, V.W. (1963). *Elementary Quantitative Analysis*, 2e. New York, NY: Harper and Row.

12. Biegler, T. and Woods, R. (1973). *J. Chem. Educ.* 50: 604–605.

13. Ramette, R.W. (1987). *J. Chem. Educ.* 64: 885.

14. Bard, A.J., Parsons, R., and Jordan, J. (1985). *Standard Potentials in Aqueous Solution*. New York, NY: International Union of Pure and Applied Chemistry, Marcel Dekker.

15. (a) Mettler Toledo.Introduction to Karl Fisher titration. Mettler Toledo KF Guide 1. https://www.mt.com/us/en/home/library/guides/laboratory-division/1/karl-fischer-titration-guide-principle.html (accessed 9 August 2019).

16. Monitoring chlorine in turbid cooling tower water. Hach application note. 2017. https://www.hach.com/titration-systems/automatic-titrators-instrument/family-downloads? productCategoryId=35547627765 (accessed 9 August 2019).

17. Carlsson, T. and Muurinen, A. (2008). Practical and Theoretical Basis for Redox Measurements in Compacted Bentonite a Literature Survey. *Working Report 2008-51, Posiva, Eurajoki, Finland*. http://www.iaea.org/inis/collection/NCLCollectionStore/_Public/43/066/43066524 .pdf (accessed 9 August 2019).

18. Husson, O. (2013). *Plant Soil* 362: 389–417.

19. Dorau, K. (2017). Monitoring of reducing conditions in soils and implications for biogeochemical processes. PhD dissertation. University of Koln. kups.ub.uni-koeln.de/7439/1/ Dissertation_Dorau.pdf. Figure created by Dorau from (a) Fiedler, S., Vepraskas, M.J., and Richardson, J.L. (2007). *Adv. Agron.* 94: 1–54. (b) Reddy, K.R. and DeLaune, R.D. (2008). *Biogeochemistry of Wetlands: Science and Applications*. Boca Raton, FL: CRC Press. (c) Strawn, D.G., Bohn, H.L., and O'Connor, G.A. (2015). *Soil Chemistry*, 4e. Hoboken, NJ: Wiley.

20. Shaposhnik, V. (2008). *J. Analyt. Chem.* 63 (2): 199–201.

21. Elder, J.F. *Metal Biogeochemistry in Surface-Water Systems, A Review of Principles and Concepts*, U.S. Geological Survey Circular 1013. https://pubs.usgs.gov/circ/1988/1013/report.pdf.

22. Honma, T., Ohba, H., Kaneko-Kadokura, A. et al. (2016). *Environ. Sci. Technol.* 50: 4178–4185.

23. Liu, C.-G., Qin, J.-C., and Lin, Y.-H. (2017). Fermentation and redox potential. In: *Fermentation Processes* (ed. A. Jozala). IntechOpen https://doi.org/10.5772/64640.

24. Bánhegyi, G., Benedetti, A., Margittai, E. et al. (2014). *Biochim. Biophys. Acta, Mol. Cell. Res.* 1843: 1909–1916.

25. Tur'yan, Y.I. and Kohen, R. (1995). *J. Electroanal. Chem.* 380: 273–277.

26. Pastore, A., Federici, G., Bertini, E., and Piemonte, F. (2003). *Clin. Chim. Acta* 333: 19–39.

27. Suslow, T.V. (2004). Oxidation–reduction potential for water disinfection monitoring, control, and documentation, University of California Davis. http://anrcatalog.ucdavis.edu/pdf/8149.pdf (accessed 9 August 2019).

POTENTIOMETRY OF ION SELECTIVE ELECTRODES

3

3.1. OVERVIEW

Potentiometric sensors play many important roles in modern chemical analysis. The glass pH electrode is a spectacular example of how valuable a reliable sensor can be. In modern science and technology, pH measurements are ubiquitous and relatively easy to perform. The measurement of hydrogen ion activity using a pH electrode is the most important example of a class of potentiometric sensing devices known as ion selective electrodes or ISEs. It is easy to see why the efficiency and trustworthiness of a pH sensor are so important. The knowledge of pH is critical to controlling industrial processes as diverse as brewing, fabric dyeing, the manufacture and quality control of pharmaceuticals, and virtually any other area where aqueous reactions are involved. It was the industrial need to measure the pH in fruit juice rapidly and reliably that led to the development of a commercial pH meter and the start of the first company to produce electronic chemical instrumentation [1]. ISEs are popular tools because they quickly provide important measurements and because they are reliable, portable, rugged, inexpensive, and accurate.

This chapter covers the fundamentals principles on which ISEs operate as well as practical considerations for their use. Before looking at the mechanisms and details of the phenomena that influence the behavior of an ISE, it is helpful to have a general understanding of the experimental set-up. Figure 3.1 is a block representation of the main components of an electrochemical cell arranged for an ISE measurement.

The signal that is measured is the voltage difference between the two reference electrodes. The most commonly used reference electrode is a silver chloride-coated silver wire. Its potential is set by the chloride activity in its immediate environment.

$$AgCl + e \rightleftharpoons Ag + Cl^- \tag{3.1}$$

The potential of each reference electrode is given by the corresponding Nernst equation:

$$E_{ref} = E^o_{Ag/AgCl} - \frac{RT}{nF} \ln(a_{Cl^-}) = E^o_{Ag/AgCl} - 0.05916 \log(a_{Cl^-}) \tag{3.2}$$

In Eq. (3.2), a_{Cl^-} is the chloride ion activity and $E^0_{Ag/AgCl}$ is the standard potential of the electrode or the potential when the chloride activity is one. Because these two reference electrodes are often similar, measuring a difference in their two potentials yields a

Electroanalytical Chemistry: Principles, Best Practices, and Case Studies, First Edition. Gary A. Mabbott.
© 2020 John Wiley & Sons, Inc. Published 2020 by John Wiley & Sons, Inc.

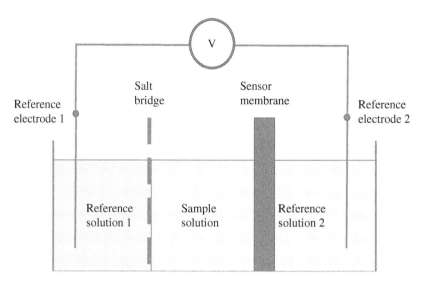

FIGURE 3.1 General diagram of an electrochemical cell for a measurement based on an ion selective electrode. The meter, V, measures the difference in voltage between the two reference electrodes that are situated on opposite sides of the sensing membrane.

value close to zero. Consequently, the meter reading reflects mainly the potential sources between the two electrodes, the biggest part of which is the potential across the sensor membrane. Even when the reference electrodes are not closely matched, their reference solution conditions are kept constant. In that way, their voltage difference is also a constant value and does not change as the sample solution changes.

The sensor membrane is the key component in generating a signal that is dependent on the analyte activity. A potential develops across the interface between the sample solution and the membrane because the analyte ion is preferentially adsorbed by (or partitions into) the membrane from the sample solution. *Note that this phenomenon is not a redox process.* Unlike at the silver/silver chloride electrode surfaces, there is no electron transfer between the two phases. Instead a small amount of charge carried by a small number of moles of analyte ions crossing the phase boundary produces an electrical potential energy difference at the membrane surface.

There are three general classes of membrane materials that have been used for sensors: glasses, ionic crystalline materials, and liquid membranes (usually made from an organic liquid or polymer solution). In the ideal case, the response of an ISE follows a relationship similar to the Nernst equation that was discussed for redox processes:

$$E_{meas} = E_{const} + \frac{RT}{z_i F} \ln (a_i)_{spl} \tag{3.3}$$

where E_{meas} represents the measured voltage, E_{const} is a constant for the system, z_i is the charge on the ion of interest, and a_i is the activity of the ion of interest in the sample solution (spl).

Ideally, the difference in potential measured between the two reference electrodes gives a potential that is proportional to the logarithm of the analyte ion activity. Any other voltage contributions (such as small differences in the reference electrode potentials) contribute to the constant term, E_{const}, and are independent of the activity of the analyte.

An understanding of the mechanisms that lead to the preference of the membrane for the selective exchange of the analyte ion with the sample solution is a goal of this chapter. It is also important to understand the mechanisms that lead to deviations from the ideal because they are encountered frequently, and they limit the accuracy of these measurements. Two of these problems are worth introducing here. They are also discussed in more detail in later sections. The first issue is the presence of interfering ions. Ideally, the membrane responds to a single type of ion, but in practice a small number of other ions are also capable of interacting with the membrane to a lesser degree and can lead to errors. For example, the glass pH electrode is also sensitive to sodium ion activity. The influence of the interfering sodium ion can be accounted for (in a derivation that appears in Appendix B) giving a variation of Eq. (3.3) known as the Nicolsky–Eisenman equation.

$$E_{memb} = E^o + \frac{RT}{zF} \ln \left(a_i + K_{ij}^{POT} \cdot a_j^{z/y} \right) = E^o + \frac{RT}{F} \ln \left(a_{H^+_{spl}} + K_{HNa}^{POT} \cdot a_{Na^+} \right) \qquad (3.4)$$

Equation (3.4) applies to ion selective devices in general, where z is the charge on the analyte and y is the charge on the interfering ion; a_i is the activity of the analyte; and a_j is the activity of the interferent. The term K_{HNa}^{POT} is the selectivity coefficient of the membrane for sodium compared to hydrogen ion. Selectivity coefficients are determined empirically for each type of membrane and pair of analyte and potential interfering ion and are generally small numbers. The smaller the selectivity coefficient, the less sensitive the ISE is to that particular interferent. Upon examining Eq. (3.4), one sees that sodium will not interfere with the determination of the hydrogen ion activity as long as $a_{H^+} \gg K_{HNa}^{POT} \cdot a_{Na^+}$. For example, for a glass pH electrode, the selectivity coefficient for sodium ion compared to hydrogen ion is K_{HNa}^{POT}, is $\sim 10^{-10}$. If one is making measurements in solutions where the Na^+ activity is 0.1 M, then the sodium ion will not interfere until the hydrogen ion activity drops to a level near $K_{HNa}^{POT} \cdot a_{Na^+} = 10^{-10} \cdot 0.1 = 10^{-11}$ M. The error becomes perceptible when hydrogen ion activity is within a factor of 10 of the product, $K_{HNa}^{POT} \cdot a_{Na^+}$, or 10^{-10} M (pH 10 or higher) in this case.

Figure 3.1 depicts the main components in an electrochemical cell for an ISE measurement in a conceptual way. The actual physical arrangement of parts is better portrayed by Figure 3.2. In a conventional ISE, the sensing membrane is sealed onto the bottom of a tube. Usually, the chamber inside the tube contains a silver chloride-coated silver wire immersed in a salt solution with a fixed chloride ion activity and a fixed activity for the analyte ion that the sensor is designed to measure. This internal reference electrode provides an electrical contact between the internal solution and the outside measurement circuit. This combination creates a handy probe that incorporates half of an electrochemical cell. It is common practice to refer to the combination of the sensor membrane and the internal reference as an "indicator electrode" or ISE. It is easy to insert this probe into a sample solution along with a second reference electrode that completes the electrochemical cell. This external reference electrode makes contact with the solution on the sample side of the sensor membrane. It usually uses a salt bridge to separate the sample solution from the reference solution. This arrangement helps one keep the chloride activity around the Ag/AgCl wire constant. Sometimes an additional chamber, isolated by another salt bridge is used in

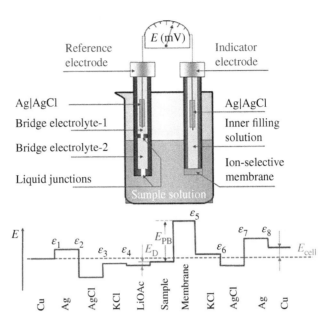

FIGURE 3.2 Diagram of an electrochemical cell using an ISE showing the internal reference electrode and reference filling solution and the external reference electrode. Source: Reproduced with permission from Lindner and Pendley [2]. Copyright 2013, Elsevier.

between the sample and the external reference electrode to help minimize contamination of the sample solution by ions from the reference electrode. This second electrolyte chamber with the extra salt bridge is called a "double junction" and is used when the intent of the analysis is the determination of chloride ions, for example. (A double junction is depicted for the external reference electrode in Figure 3.2.) In such a case, the extra chamber might be filled with a lithium acetate electrolyte solution or other electrolyte combination that will not interfere. In principle, the sample solution and this reference solution do not mix significantly, but in reality, some ions do diffuse across the salt bridge because of a concentration gradient. A tiny amount of leakage is accepted because some ion traffic is necessary in order to allow electrical communication between electrodes to complete the circuit.

In addition to the boundary between the sample solution and the sensor membrane, there are several other places where different phases or materials come into contact in order to make a complete circuit for charges (either electrons or ions) to travel between the outside measuring equipment through the electrochemical cell and out again to the meter. Every contact between different materials (or solution phases) provides an opportunity for an electrical potential difference to develop across that interface. The line diagram at the bottom of Figure 3.2 is meant to represent these transitions in electrical energy from one connection to the voltmeter through the cell to the other meter input. Fortunately, with a little care, each of these contributions to the measured potential can be kept constant so that they become part of E_{const} in Eq. (3.1). The two exceptions are the phase boundary potential, E_{PB}, at the sample solution/sensor membrane interface (it is one of two membrane/solution interfaces) and the diffusion (or junction) potential, E_D, where the salt bridge and sample solution meet. Of course, the first of these is the source of the desired signal and is a function of the analyte activity. The second is the junction potential as described in Chapter 1. It can cause an error in our estimate of the analyte activity in the sample solution. One of the approaches to minimizing junction potential errors involves adding salts to the sample solution (in the form of "ionic strength adjusting buffers") so that the rate of cation diffusion across the salt bridge matches that of the anions. These rates can be matched by eliminating concentration differences across the salt bridge for major ions or matching the diffusion coefficients of anions and cations. Rarely is the junction potential completely eliminated. The best situation that is normally achieved is a consistent junction potential between the reference solution and all sample and calibration solutions. In such cases, the junction potential combines with all of the other constant potential terms and is compensated for in the calibration process. Whenever sample conditions cannot be adjusted (such as in remote environmental measurements or continuous blood monitoring during surgery), a reliable means of evaluating the junction potential experimentally or estimating it mathematically is necessary. (See Appendix C for a discussion of the Henderson equation for estimating junction potentials.)

In addition to the glass pH electrode, insoluble ionic salts and liquid membranes have been developed as ion selective sensors. The analyte selectively adsorbs to crystalline lattices because the crystal structure has holes for the analyte to fill that have an optimum size and shape for strong interactions with ions of the opposite charge. In contrast, hydrophobic liquid membranes ordinarily resist ions and dissolve only hydrophobic organic salts with large nonpolar branches. However, these membranes can be coaxed to accept specific analyte ions when a neutral, nonpolar molecule with a high affinity for the analyte is added to the membrane matrix. These neutral attractants are known as ionophores.

The first half of this chapter represents a good foundation for understanding ion selective devices and are recommended for college courses on instrumental analysis. The mechanisms for ion selective devices are described in Sections 3.2–3.4. Section 3.5 discusses the use of calibration curves based on the Nicolsky–Eisenman equation (3.4). Section 3.9 should also be part of a course's reading assignment, because it addresses the important practical issues of ionic strength buffers and signal drift. Although ISEs are part of a mature field, some exciting developments have occurred in the last 20 years. The second half of the chapter describes several of these advances beginning with the insight that has pushed detection limits to the 10^{-11} M level for some ISEs (Section 3.6). Other more recent efforts (Section 3.7) to create robust ISEs without an internal reference electrode and the use of super-hydrophobic membranes show promise of making sensors that are smaller, more selective and more robust.

3.2. LIQUID MEMBRANE DEVICES

3.2.1. Selective Accumulation of Ions Inside an Organic Liquid

The separation of charge at a boundary between two different phases is a common phenomenon. For ion-containing materials, it happens whenever one type of ion is attracted across the boundary more favorably than ions of the opposite charge. For example, imagine a separatory funnel with a layer of dichloroethane and an aqueous solution of tetrabutylammonium nitrate (TBA^+, NO_3^-) (Figure 3.3) [3].

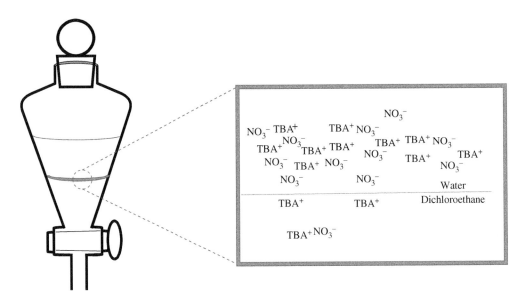

FIGURE 3.3 An aqueous solution of tetrabutyl ammonium nitrate ($TBA^+ NO_3^-$) in equilibrium with dichloroethane. The dichloroethane solvates TBA^+ well, but NO_3^- is more strongly solvated by water. Therefore, more TBA^+ crosses the phase boundary than does NO_3^- leading to a charge separation across the interface. Source: Bühlmann and Chen 2012 [3]. Reproduced with permission of John Wiley & Sons.

When the system is mixed and allowed to come to equilibrium, some of both ions will be found in the organic phase. Because of its alkyl side chains, the tetrabutylammonium ion is solvated by the dichloroethane more strongly than by water. That is evident from the negative Gibbs free energy change, $-21.8 \, kJ/mol$, for the process of TBA^+ going from water into the dichloroethane [3]. On the other hand, water solvates nitrate better than the organic solvent does and the ΔG is positive, $33.9 \, kJ/mol$, for nitrate going into the dichloroethane. Consequently, more cations than anions will cross the boundary into the organic phase. (Notice that the ΔG values for the transfer of the two ions do not need to be of opposite sign in order for a charge separation to result. If the free energies are numerically different, then some preference for one ion over the other will exist and some charge separation will result.) Therefore, there will be a small excess of positive charge in the organic solvent and an equal excess of negative charge on the aqueous side.

Of course, over the entire system, the charges balance. Taking the two solvents together, the system is electrically neutral. In addition, throughout the bulk of either solvent, the solution is also assumed to be electrically neutral. All of the excess charge appear at the interface of the two phases, more positive charge on the organic side, and, consequently, an excess of negative charge on the aqueous side in this case. This separation of charge creates a phase boundary potential.

This simple example does not make a very selective system. Some other interaction in addition to favorable solvation is necessary to make one ion preferred over all others. For liquid membrane devices, the selectivity depends on another solute in the organic phase that has a high affinity for the analyte. Most ISE liquid membranes contain an ionophore.

The term, ionophore, is Greek for "ion carrier". An ionophore is a molecule that is highly soluble in the membrane and binds strongly to the ion of interest. The complex of the analyte ion and ionophore must also be highly soluble in the membrane. Many ionophores are neutral molecules, but that is not an absolute requirement. (If the ionophore is charged, the system will be more complicated because a counter ion is required in order to get it to dissolve into the hydrophobic environment of the membrane.) Many ionophores for cationic analytes bind by forming multiple covalent coordinate bonds in which lone pairs of electrons on the ionophore form bonds with previously unoccupied orbitals on the analyte. As with chelation chemistry between metal ions and chelators, such as ethylenediaminetetraacetic acid (EDTA), the charge density of the metal ion, the solvation energy of the metal in the aqueous solution, steric factors associated with overlapping orbitals between the analyte and ionophore as well as the type of atoms participating in the new bond, the solvation energy associated with the complex within the membrane medium and the number of participating binding sites per ionophore, all influence the relative stability of the ionophore–analyte complex. These factors also lead to the preference of the membrane for one analyte over all others. In the end, it is the *relative stability* of the analyte–ionophore complex in the membrane (rather than the absolute stability) that makes the membrane selective as a sensor.

One of the most successful and widely used liquid membrane ISEs responds to K^+ ions. The membrane in this device contains an organic complexing agent that is selective for binding potassium ions. Valinomycin is a naturally occurring, neutral molecule that has a very large binding constant ($K_f \sim 10^6$) for K^+. Binding of valinomycin to potassium ion is at least 10 000 times greater than its affinity for binding Na^+ (the most common of possible interfering species in a blood sample), for example. The valinomycin increases the solubility of potassium ions in hydrophobic environments by surrounding the ion, insulating it from the nonpolar medium. The carbonyl oxygen atoms form ion/dipole bonds with the potassium ion, displacing the water molecules that normally hydrate it. Because of the fact that it selectively transports potassium ions across biological cell membranes, valinomycin is also a potent antibiotic.

Figure 3.4 shows how the ionophore complexes the potassium ion [4]. The size of the cavity in the folded ligand is the optimum size for an ion with the diameter of the potassium ion. For several decades, valinomycin has been used to study biochemical processes that produce electrochemical potential differences across biological membranes. When introduced to the medium around a biological cell, the hydrophobic valinomycin will associate itself with the lipid membrane of the cell wall. The valinomycin can pick up a potassium ion on one side of the membrane and carry it across to the opposite side. This process naturally drives K^+ ions across the membrane from a high concentration to the lower concentration solution. Because only potassium ions are being transported, this process builds up positive charge on the side receiving the net gain in potassium ions.

In commercial ISEs, the membrane is much thicker than a biological membrane. Furthermore, in an ISE, it is desirable to minimize the net movement of analyte from one side of the membrane to the other. Figure 3.5 shows a schematic diagram of a potassium ion electrode based on valinomycin and a liquid membrane. The sensor is constructed from a tube sealed at the bottom with the sensor membrane. The tube is filled with a reference solution of potassium ions at a well-defined activity (such as 0.1 M) and a fixed

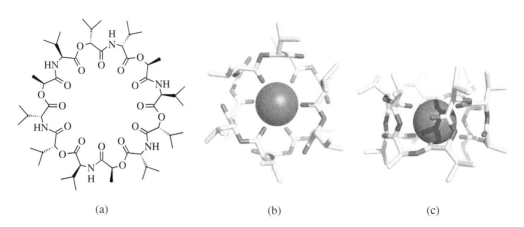

FIGURE 3.4   (a) The structure of valinomycin. Valinomycin is a neutral complexing agent that binds to K$^+$ ions. The oxygen atoms form ion–dipole bonds with potassium ions leaving hydrophobic side chains to interact with the nonpolar interior of lipid membranes. (b) 3-D model of the K$^+$-valinomycin complex. The size of the ion and the cavity created by the cage-like framework of the valinomycin match much better for K$^+$ than other common ions. (c) Side view. Source: Reprinted with permission from Palmer et al. [4].

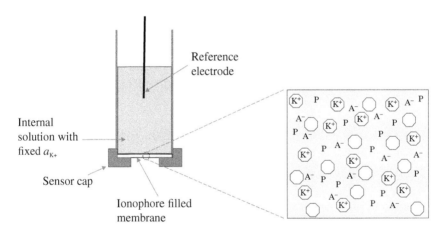

FIGURE 3.5   Conventional liquid membrane K$^+$ ISE. The membrane is usually a polymer film, such as polyvinyl chloride. A selective complexing agent or ionophore (open polygons in the figure) is added to the membrane along with a plasticizer (P) such as dioctyl phthalate. The plasticizer is added to make the medium more fluid so that the ionophore diffuses easily. A hydrophobic counter ion (A$^-$) is added at half the level on a mole basis as the ionophore. The counter ion, A$^-$, is also known as the ionic site. If the ionic site is added as a sodium salt, for example, then K$^+$ analyte ions eventually replace the sodium ions in the membrane and complex with the ionophore.

concentration of chloride ions that sets the potential for the silver/silver chloride reference electrode. The sensor membrane is made from a polymer film (often polyvinyl chloride) that acts as the mechanical support for the liquid membrane components. One of the components is a plasticizer. Plasticizers are small molecules, such as dioctyl phthalate, used in polymer materials to adjust the physical properties of the material, such as its flexibility.

In an ISE, a plasticizer also improves the solubility and movement of the ionophore in the film.

It was asserted at the beginning of this chapter that the response of an ISE would be of the form of Eq. (3.3). However, that relationship is a special case that evolves from the more general expression for the phase boundary potential. It holds only if certain conditions are met. It is enlightening to consider the derivation for the boundary potential here. Fortunately, the derivation is fairly straightforward using the concept of electrochemical potential for the ion of interest [5].

3.2.2. Theory of Membrane Potentials

Potassium ions cross the membrane/solution interface at each boundary. Assuming an equilibrium is established, we can derive the relationship between the potassium ion activity in the aqueous sample solution and the potential that develops across the membrane.

$$K^+_{(aq)} \rightleftharpoons K^+_{(memb)} \tag{3.5}$$

The electrochemical potential, $\overline{\mu}_{K^+,aq}$, for the potassium ion in the aqueous phase is

$$\overline{\mu}_{K^+,aq} = \mu^o_{K^+,aq} + RT\ln(a_{K^+,aq}) + (1)F\phi_{aq} \tag{3.6}$$

In Eq. (3.6), $\overline{\mu}_{K^+,aq}$ represents the electrochemical potential energy associated with the uncomplexed potassium ion in the aqueous solution, $\mu^o_{K^+,aq}$ is the chemical potential energy associated with creating a mole of potassium ions in the aqueous solution at standard temperature and pressure, and R is the ideal gas constant that has a value of 8.314 V per degree per Kelvin. T is the absolute temperature of the system, $a_{K^+,aq}$ is the activity of the potassium ion in the aqueous solution in contact with the membrane, F is Faraday's constant or 96 485 C/mol of charges and ϕ_{aq} is the electrical work done to move a coulomb of charge into the aqueous phase. The coefficient in front of $F\phi_{aq}$ is unity, because the charge on the potassium ion is +1.

On the membrane side of the boundary, the potassium ion electrochemical potential is

$$\overline{\mu}_{K^+,memb} = \mu^o_{K^+,memb} + RT\ln(a_{K^+,memb}) + (1)F\phi_{memb} \tag{3.7}$$

At equilibrium, the electrochemical potentials of the species in the two phases are equal to each other. Therefore, expressions (3.6) and (3.7) are equal to each other and can be used to solve for the difference in the electrical potential energy of the two solutions, $\phi_{memb} - \phi_{aq}$, which is also the phase boundary potential, E_{PB}.

$$\mu^o_{K^+,aq} + RT\ln(a_{K^+,aq}) + (1)F\phi_{aq} = \mu^o_{K^+,memb} + RT\ln(a_{K^+,memb}) + (1)F\phi_{memb} \tag{3.8}$$

$$F\phi_{memb} - F\phi_{aq} = \mu^o_{K^+,aq} - \mu^o_{K^+,memb} + RT\ln(a_{K^+,aq}) - RT\ln(a_{K^+,memb}) \tag{3.9}$$

$$\phi_{memb} - \phi_{aq} = \frac{\mu^o_{K^+,aq} - \mu^o_{K^+,memb}}{F} + \frac{RT}{F}\ln(a_{K^+,aq}) - \frac{RT}{F}\ln(a_{K^+,memb}) = E_{PB} \tag{3.10}$$

$$E_{PB} = E_{const} + \frac{RT}{F}\ln(a_{K^+,aq}) - \frac{RT}{F}\ln(a_{K^+,memb}) \tag{3.11}$$

The special conditions that render Eq. (3.11) equivalent to Eq. (3.3) are those in which $a_{K^+,memb}$ is a constant. In that case, the term containing $a_{K^+,memb}$ can be collected with the other constants to form a new constant, E'_{const}. Then,

$$E_{PB} = E'_{const} + \frac{RT}{F} \ln(a_{K^+,aq}) = E'_{const} + 0.05916 \ln (a_{K^+,aq})_{spl} \qquad (3.12)$$

An attractive strategy for holding $a_{K^+,memb}$ constant is to form a buffer for the analyte ion inside the membrane. A buffer for potassium ion can be created by combining a relatively high concentration of the potassium–valinomycin complex and free valinomycin together in the membrane. (Millimolar levels is sufficiently concentrated for this purpose.) Furthermore, in a manner that is analogous to an acid/base buffer where the highest buffering capacity is obtained when the mole ratio of acid to base is 1 : 1, the optimum conditions for the potassium ion buffer are equal molar amounts of complex and free valinomycin. Also, as with a pH buffer, its capacity to maintain those conditions increases with the combined amount of the components. A moderately high concentration of ionophore and complex is desirable. Because the complex carries a positive charge, in order to dissolve a sufficient quantity of the complex in the nonpolar membrane, there will need to be some lipophilic anion, A^-, present to balance the charge of the initial membrane mixture. These counter ions are often referred to as the "ionic sites" in the membrane [6]. Derivatives of tetraphenyl borate are commonly used for anionic sites, while tetraalkyl-ammonium cations with long carbon chains are used for cationic sites in anion ISEs [3]. Consequently, one might prepare a potassium ISE membrane by mixing valinomycin with half as many moles of the potassium salt of the ionic sites in an appropriate nonpolar solvent (that will swell the polymer film) along with a plasticizer (Figure 3.6).

Naturally, the quantity of ionic sites should be equivalent on a mole basis to the original number of moles of analyte ion in the membrane (or half the number of moles of ionophore). When the anionic site is not available as a salt of the analyte, another salt must be used. Then soaking the membrane for a few hours (exposing both sides) in a highly concentrated aqueous solution of the analyte ion is necessary to "pre-equilibrate" the membrane. This pretreatment allows for a simple cation exchange replacing the unwanted ion with the analyte ion in the membrane. The exchange for the analyte will be favored because of the selectivity of the ionophore.

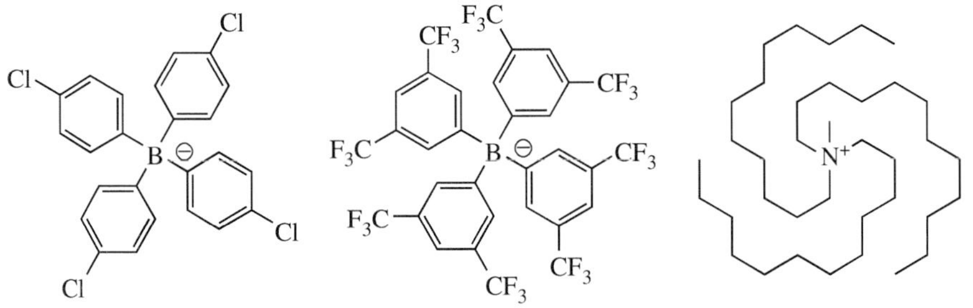

FIGURE 3.6 Examples of common hydrophobic counter ions used as ionic sites in liquid membrane ISEs. Derivatives of tetraphenyl borate are frequently used to counter cation analytes. For ISEs designed for anion analytes, quaternary amines with long hydrocarbon side chains are frequently used for ionic sites. Source: Bühlmann and Chen 2012 [3]. Reproduced with permission of John Wiley & Sons.

EXAMPLE 3.1

How much charge crosses the phase boundary? In the derivation of the phase boundary potential above, it was assumed that the activity of the K^+ ion inside the membrane remains constant. The combination of both free and complexed ionophore act as a buffer for the free K^+ ion activity there. However, that will only be the case, if the quantity of K^+ ion moving across the sample solution/membrane boundary during the measurement process is small compared to the original level of K^+ ions in the membrane. That raises an important question: How much K^+ actually crosses the solution/membrane interface in order to establish the phase boundary potential? A key concept that is helpful in answering that question is the fact that the phase boundary potential behaves in a manner similar to a capacitor in terms of storing charge. All of the excess charge appears at the boundary and the potential energy difference across the boundary is proportional to that net charge (on either side). Consider a fairly extreme case where the phase boundary potential is 1 V. The relationship between charge and voltage stored on a capacitor is

$$Q = CV \tag{3.13}$$

where Q is the excess charge on one side of the boundary and V is the voltage that results from the charge separation. The capacitance, C, is the capacitance of an aqueous double layer. For a solution of 0.1 M KCl, or other salt, C is typically around 10–40 $\mu F/cm^2$ or $(10–40)*(10^{-6}$ C/V)/cm^2 [7]. For the purpose of this calculation, we will use the higher value, 40 $\mu F/cm^2$.

Assuming an area for the ISE membrane surface exposed to a sample to be 1 cm^2. Then, the charge becomes

$$Q = (40 \times 10^{-6} \text{ C/(V cm}^2))(1 \text{ V})(1 \text{ cm}^2) = 4 \times 10^{-5} \text{ C} \tag{3.14}$$

Faraday's constant for the charge on a mole of electrons or singly charged ions is 96 485 C/mol of charge. So, the charge, Q, corresponds to

$$N_{ions,xs} = \frac{Q}{zF} = \frac{(4 \times 10^{-5} \text{ C})}{(1)(96\,485 \text{ C/mol})} \approx 4 \times 10^{-10} \text{ moles of ions} \tag{3.15}$$

That number of moles of excess charge represents the number of ions crossing the phase boundary. How does it compare to the number of moles of ionophore (or the number of moles of potassium ions originally doped into the membrane)? A reasonable concentration of ionophore is 1 mmol/kg of membrane. Because the density of the combined membrane materials is roughly 1 g/cm^3, the ionophore concentration in the membrane is also approximately 1 mM. Assuming that the membrane thickness is 100 μm, then the volume of the membrane associated with a surface area of 1 cm^2 is

$$Vol_{memb} = (1 \text{ cm}^2)(100 \text{ } \mu m)(10^{-4} \text{ cm/}\mu m) = 0.01 \text{ cm}^3 \tag{3.16}$$

Therefore, the number of moles of ionophore in the membrane is

$$N_{ionophore} = [\text{ionophore}](Vol_{memb}) = (1 \times 10^{-3} \text{ mol/l})(0.01 \text{ cm}^3)(10^{-3} \text{ l/cm}^3) = 1 \times 10^{-8} \text{ mol} \tag{3.17}$$

From the number of moles of excess ions, $N_{ions,xs}$, in Eq. (3.15) and the number of moles of ionophore, $N_{ionophore}$, from Eq. (3.17), the ratio of excess K^+ ions to moles of original ionophore is

$$\frac{N_{ions,xs}}{N_{ionophore}} = \frac{(4)(10^{-10})}{1 \times 10^{-8}} = 0.04 \text{ or } 4\% \tag{3.18}$$

Thus, only a small fraction of the original ionophore is needed to stabilize the excess potassium that crosses the phase boundary. Just as an acid/base buffer maintains the hydrogen ion activity upon a small addition of acid or base, the combination of free ionophore and K^+-ionophore complex keeps the K^+ ion activity within the membrane virtually constant. This calculation shows that a millimolar level of ionophore is sufficiently large to ensure that is the case.

3.2.3. Liquid Membrane Ionophores

Table 3.1 shows a short list of selective reagents and the ions that they bind in a number of successful liquid membrane ISEs. The selectivity coefficients listed in the table indicate the degree to which common ions interfere with the ability of the ISE to measure the target

TABLE 3.1 Selected ionophores used for ISEs

Ionophore	Analyte	K_{ij}^{POT}, selectivity coefficients for interfering ions	References
Valinomycin	K^+	$K_{KH}^{POT} = 4.0 \times 10^{-4}$, $K_{KNa}^{POT} = 6 \times 10^{-5}$, $K_{KMg}^{POT} = 2 \times 10^{-8}$, $K_{KCa}^{POT} = 1 \times 10^{-7}$	[8]
ETH 1001	Ca^{2+}	$K_{CaMg}^{POT} = 4.0 \times 10^{-5}$, $K_{CaLi}^{POT} = 4 \times 10^{-5}$, $K_{CaK}^{POT} = 4 \times 10^{-6}$, $K_{CaNa}^{POT} = 4 \times 10^{-6}$, $K_{CaNH_4}^{POT} = 1 \times 10^{-5}$	[9]
Nonactin	NH_4^+	$K_{NH_4K}^{POT} = 0.1$, $K_{NH_4Na}^{POT} = 2.5 \times 10^{-3}$	[10]

TABLE 3.1 (Continued)

Ionophore	Analyte	K_{ij}^{POT}, selectivity coefficients for interfering ions	References
NH$_4^+$–I R$_1$, R$_2$, R$_3$, R$_4$ = CH$_3$			
ETH 5435	Pb^{2+}	$K_{PbNa}^{POT} = 2.0 \times 10^{-5}$, $K_{PbCa}^{POT} = 2.5 \times 10^{-9}$, $K_{PbH}^{POT} = 3.2 \times 10^{-4}$, $K_{PbLi}^{POT} = 7.9.0 \times 10^{-6}$, $K_{PbMg}^{POT} = 4 \times 10^{-10}$	[11]
	Li$^+$	$K_{LiNa}^{POT} = 5 \times 10^{-4}$	[12]
where R = $-[CF_2]_9CF_3$	Ag$^+$	$K_{AgK}^{POT} = 2.5 \times 10^{-12}$, $K_{AgNa}^{POT} = 1.2 \times 10^{-13}$, $K_{AgCa}^{POT} = 2.0 \times 10^{-13}$, $K_{AgCu}^{POT} = 9.1 \times 10^{-14}$	[13]
Zn(II) tetraphenylporphyrin	CN$^-$	$K_{CN^-Cl^-}^{POT} = 2.0 \times 10^{-4}$, $K_{CN^-Cl^-}^{POT} = 2.0 \times 10^{-4}$, $K_{CN^-Cl^-}^{POT} = 2.0 \times 10^{-4}$	[14]

ion activity. In the table, the selectivity coefficient, K_{ij}^{POT}, represents the relative affinity of the membrane for the interfering ion, j, compared to that of the analyte ion, i. The smaller the selectivity coefficient, the less sensitive the ISE is to the interfering ion. A discussion of the influence of interfering ions on the detection limits for liquid membrane ISEs is taken up later in this chapter.

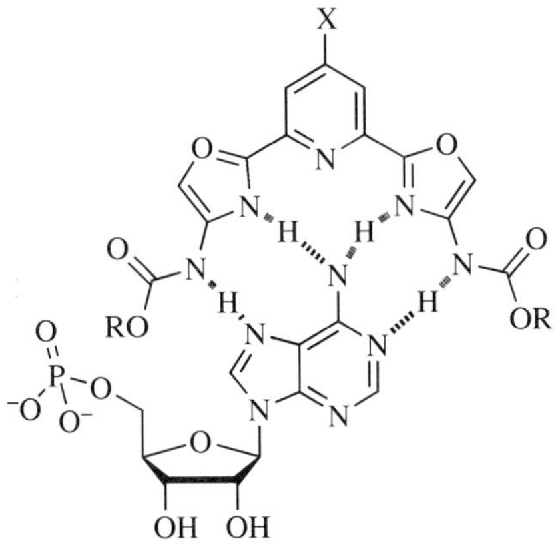

FIGURE 3.7 Ionophore that binds adenosine monophosphate. Source: Bühlmann and Chen 2012 [3]. Reproduced with permission of John Wiley & Sons.

The examples discussed here have focused on simple inorganic species as analytes. However, there are many organic ions, such as natural products or drugs and their metabolites that are important analytes for a host of clinical, environmental, and industrial applications. It should be possible to extend the application of ISEs to these analytes as well. The development of new sensors depends upon synthesizing ionophore molecules that optimize host–guest interactions in order to favor binding the analyte over potentially interfering ions. Figure 3.7 shows an example of an ionophore that binds adenosine monophosphate, AMP [3]. Relatively few ISEs for organic analytes have been demonstrated to date. However, the large number of important applications and the advantage of direct determination by an ISE for various analytes makes this a tremendously appealing and promising field for the future.

3.3. GLASS MEMBRANE SENSORS

3.3.1. History of the Development of a Glass Sensor of pH

Most commercial glasses are amorphous materials that are formed from melting silicon dioxide, in the form of sand, together with fluxes, such as sodium carbonate and potassium oxide. The purpose of the flux is to lower the melting temperature of the sand. Often other metal oxides, such as alumina or boric oxide, are also added to influence the physical properties of the glass. Ordinary glass is an electrical insulator. Unlike the atoms in metal alloys, the silicon atoms in glass form covalent bonds and do not have available valence electrons that can move freely in a conduction band as they do in metals. Consequently, most glasses do not conduct electricity. In the early twentieth century, glass makers experimented with various mixtures and discovered some materials that could conduct ions [15]. However, it should be emphasized that the level of current that the glass could conduct was still quite low even for the short distance across a thin glass membrane of $100\,\mu m$. Corresponding electrical resistance values were greater than 10^7–$10^9\,\Omega$. (Compare the resistance of a cube of copper 1 cm on an edge: $1.7\times10^{-6}\,\Omega$; the resistance of seawater is about $20\,\Omega$; that of

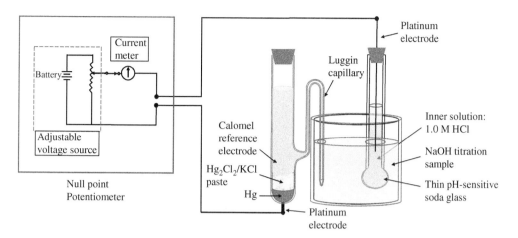

FIGURE 3.8　Early apparatus for pH titration. A calomel reference electrode with a salt bridge, known as a Luggin capillary, contacted the solution on the outside surface of the glass membrane where the titration took place. The thin glass bulb made from pH-sensitive, soda glass was filled with 1 M HCl and a platinum wire was used for completing the electrical circuit [17].

deionized water is about $1.8 \times 10^7\,\Omega$.) Nevertheless, the observation of some conductivity through these materials was evidence that some ions were able to migrate through the glass. A more interesting observation with this type of glass was reported by Cremer in 1906 [16]. He measured relatively large potential differences (on the order of 0.5 V) for a "low resistance" glass used as a barrier to separate acidic and basic solutions. By fixing the conditions of the solution on one side of the glass, one could measure the potential as a function of the acidity of the second solution on the other side. In 1909, Fritz Haber and Zygmunt Klemensiewicz demonstrated an acid/base titration using a pH-sensitive glass bulb sealed onto the end of a glass tube that held the reference solution [17] (see Figure 3.8).

A wave of research activity followed the report by Haber and Klemensiewicz. In the next two decades, thousands of publications appeared exploiting similar glass electrode systems. However, the glass electrode remained strictly a research tool because the process of measuring potentials in high resistance circuits was tedious and cumbersome with the conventional apparatus of the time. In the early 1930s, vacuum tube technology ushered in electronic amplifiers that could measure a voltage without drawing significant current. In 1935, Arnold Beckman commercialized a pH meter and glass pH electrode that became popular because together they gave accurate readings from a much more compact and easy-to-use apparatus [1].

3.3.2.　Glass Structure and Sensor Properties

How does a glass membrane develop a signal that is sensitive to pH? The answer lies in the structure of the glass. The atomic structure of glass includes negatively charged sites that bind H^+ ions. An early pH sensitive glass made by Corning has the composition of 22% Na_2O, 6% CaO, and 72% SiO_2 [15]. The term "glass" suggests that the material is amorphous; there is no long-range crystalline order to the arrangement of atoms. It is helpful

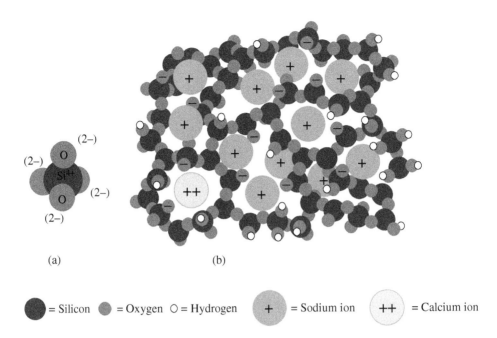

(a)   (b)

⬤ = Silicon   ⬤ = Oxygen   ○ = Hydrogen   ⊕ = Sodium ion   ⊕⊕ = Calcium ion

FIGURE 3.9   Glass structure. (a) Silicate tetrahedron. (b) Conceptual model for glass showing tangled strands made from chains of silicate units. Charges remaining on some silicate units are balanced by cations. Occasionally, hydrogen ions bond to anion sites. Source: Adapted with permission from Perley [18]. Copyright 1949, American Chemical Society.

to think of the framework as an assembly of silicate units that form tetrahedra. A silicon (4+) cation sits in the center of each tetrahedron (Figure 3.9a). An oxygen (2−) ion sits at each vertex of the tetrahedron. In a perfect network solid, each of these oxygen ions would also be a part of another adjacent tetrahedron as well. One can think of the two negative charges associated with each oxygen atom being divided between two tetrahedra. In that sense, each oxygen ion contributes half of its charge to each tetrahedron it participates in. As a result, each individual tetrahedron would be balanced with respect to charge (4+ for the silicon atom + [4 oxygen atoms] × [−1]). Although there are possibly small regions where this uniform Si–O network extends in all directions, the structure mostly resembles chains of tetrahedra, tangled together with frequent cross-links between chains (as in Figure 3.9b). Any oxygen atom that does not bridge between two silicon atoms confers a net negative charge on the one tetrahedron that it belongs to. These charges attract Na^+ or other cations as the original molten mixture cools in the glass-making process. Upon exposure to water, hydration of the surface of the glass occurs to a depth of about 0.1 µm. As water penetrates this thin zone, sodium cations can be displaced by H^+ ions.

$$Na^+_{(glass)} + H^+_{(solution)} \rightleftharpoons Na^+_{(solution)} + H^+_{(glass)} \tag{3.19}$$

The sites in the structure where this exchange occurs are the charged silicate tetrahedra. The equilibrium constant for this exchange strongly favors the hydrogen binding to the glass. George Eisenman rationalized this preference for H^+ as being the result of the higher field strength at the surface of the H^+ ion because of its smaller ionic radius [19]. The positive charge is spread out over a smaller surface of the hydrogen cation making the intensity of its field greater at the hydrogen ion surface than for the larger sodium ion surface. It is also helpful to think of this in terms of functional groups. Protonating a silicate anion creates a silanol group – a weak acid.

$$\equiv Si - O^-_{(membrane)} + H^+_{(aq)} \rightleftharpoons \equiv Si - OH_{(membrane)} \tag{3.20}$$

Within the hydrated zone of the glass wall, there can exist some H^+ ions in solution. These H^+ ions balance the total charge on the glass surface, but while in solution, these ions are on the other side of the electrical double layer formed at the membrane surface. They are part of the charge that creates an electrochemical potential energy difference at the interface. The extent to which the charge appears on the membrane depends on the degree of dissociation of the silanol groups on the surface. If a silanol group dissociates, the H^+ becomes solvated and becomes part of the solution side of the interface. It leaves behind a negative charge, increasing the negative charge on the membrane side of the interface. When a hydrogen ion bonds to an anion site in the glass, one or more molecules of water surrounding the ion are stripped away. At the same time, the charge separation decreases (Figure 3.10) because a charge on both sides of the interface has disappeared, one on the glass surface and one in solution at the outer Helmholtz plane (OHP). This mechanism accounts for the selectivity of the glass for H^+ ions as well as the variation in charge as a function of H^+ ion activity.

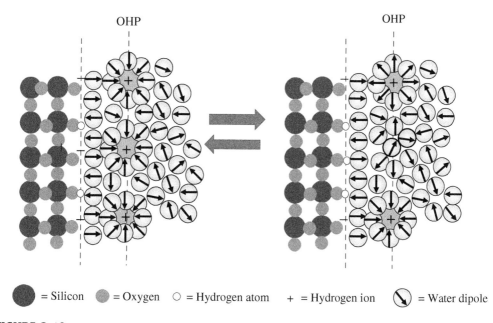

● = Silicon　● = Oxygen　○ = Hydrogen atom　+ = Hydrogen ion　◑ = Water dipole

FIGURE 3.10 Hydrogen ion transfer across the phase boundary from the outer Helmholtz plane (OHP) on the solution side to the glass surface. The negatively charged oxygen atom becomes a silanol (Si–O–H) group. With each ion that moves from the solution to the surface, the charge on both sides of the interface decreases by one.

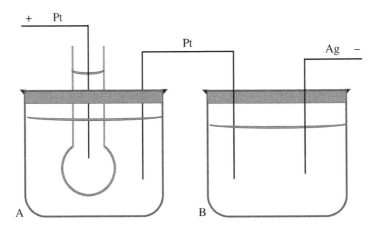

FIGURE 3.11 Haugaard's demonstration of sodium conductivity in glass. Beaker A contained a solution of 0.02 M HCl and a glass bulb made of pH sensitive glass. The inner tube was filled with the same solution. Platinum electrodes were used on both the inside and outside of the bulb. Beaker B contained 0.02 M silver nitrate solution. A large voltage was applied between the platinum electrode inside the glass bulb in beaker A and the silver electrode in beaker B. As current passed, silver ions were reduced and plated out at the silver electrode in proportion to the total charge transferred through the system. Source: Adapted with permission from Haugaard [20]. Copyright 1941, American Chemical Society.

Unlike the liquid membrane system in which K^+ ions can diffuse through the membrane, H^+ ions do not reach the other wall of the glass. Instead of hydrogen ions, sodium ions are responsible for the modest current. An elegant demonstration of sodium ion migration was reported by the Danish chemist, G. Haugaard in the 1940s [20]. He set up the apparatus depicted in Figure 3.11. A pH-sensitive glass bulb was filled with a dilute solution of HCl and placed into a beaker of the same solution. Haugaard placed a second container with a solution of silver nitrate next to the first one (Figure 3.11). A platinum electrode was used to bridge between the two beakers and a silver electrode dipping into the silver nitrate solution was connected to the negative side of the circuit. Then, over a three-week period, a current was forced through the bulb (and the solution in the two beakers) using a large applied voltage (220 V).

At the end of the experiment, solution from inside beaker A was removed and evaporated to dryness. HCl is volatile, so if it were the only electrolyte in solution, there would be no residue. However, after evaporation a solid residue of NaCl was found in beaker A where no sodium was present previously. Haugaard was able to collect a weighable mass of sodium chloride from the solution on the negative side of the glass bulb and determine the moles of sodium chloride after evaporation. In addition, for every charge passing through the system, an atom of silver was deposited on the silver wire.

$$Ag^+_{(aq)} + e \rightleftharpoons Ag_{(s)} \tag{3.21}$$

The mass of the silver wire was compared before and after the experiment. Haugaard demonstrated that the number of moles of sodium chloride accumulating in the solution outside the glass bulb was equal to the number of moles of silver gained by the silver wire.

The glass membrane was the only source for the sodium found in the outer solution. These observations were a strong support for the conclusion that the current was being carried through the glass bulb by sodium ions.

Empirically, in experiments in which glass membranes were used to monitor hydrogen ion activity, the voltage across the glass membrane was shown to follow a logarithmic dependence on the hydrogen ion activity. Both the solution/glass interface at the inside wall facing the reference solution and the sample solution/glass interface on the outer wall contribute to the membrane potential. As with conventional liquid membrane ISEs discussed earlier, the conditions for the reference side of the glass membrane are held constant, normally at pH 7.00, so that any measured potential changes can be assumed to be associated only with the outside interface between the sample solution and glass surface [15].

3.3.3. Selective Ion Exchange Model

The theoretical models to explain the Nernstian-like behavior of the glass membrane potential on the hydrogen ion activity evolved during the middle of the twentieth century. Nicolsky developed the idea that the hydrogen ions near the surface and the sodium ions in the hydrated layer of the glass undergo an ion exchange process. He derived an equation that correctly predicts a membrane potential that follows the log of the hydrogen ion activity [21] (see Appendix B for more details).

$$H^+_{(glass)} + Na^+_{(solution)} \rightleftharpoons H^+_{(solution)} + Na^+_{(glass)} \tag{3.22}$$

Eisenman expanded on this concept and included in the model, the influence on the membrane potential of the diffusion of the sodium ions through the glass. The resulting relationship is now known as the Nicolsky–Eisenman equation [22, 23] (Eq. (3.23)).

$$E_{memb} = E_{const} + \frac{RT}{F} \ln \left(a_{H^+_{spl}} + \sum_j K^{POT}_{Hj} \cdot a_j^{z_H/z_j} \right)$$

$$= E_{const} + (0.059\,16) \log \left(a_{H^+_{spl}} + \sum_j K^{POT}_{Hj} \cdot a_j^{z_H/z_j} \right) \tag{3.23}$$

In this expression, E_{const} represents an "off-set" voltage (a constant). Cations such as sodium can interfere. That is, under some conditions, they can compete with the hydrogen ion for the anionic sites on the glass surface and influence the membrane potential. In the equation, the interfering ion has an activity of a_j and a charge of z_j. Divalent ions do not show much exchange with H^+ in the glass, so z_j is usually 1. The term K^{POT}_{Hj} is known as the selectivity coefficient for ion j compared to the analyte ion, H^+. When $a_{H^+_{spl}} \gg \sum_j K^{POT}_{Hj} \cdot a_j^{z_H/z_j}$, then Eq. (3.23) simplifies to Eq. (3.3) and the membrane potential changes by 0.059 16 V (or 59.16 mV) for every unit step in pH. When $a_{H^+_{spl}} \ll \sum_j K^{POT}_{Hj} \cdot a_j^{z_H/z_j}$, then the membrane potential becomes insensitive to pH; it responds, instead, to the changes in the activity of the interfering ion. This effect

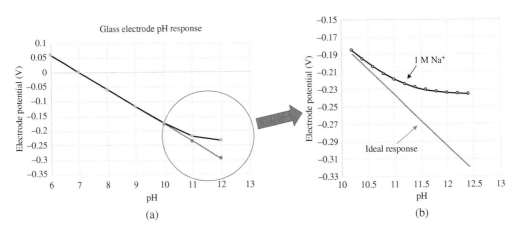

FIGURE 3.12 (a) Glass electrode response to pH deviates from the ideal in the presence of Na^+ ions. (b) Expanded plot shows deviation of the electrode response in the presence of 1 M Na^+. Under these conditions, the electrode is essentially unresponsive to H^+ at pH 12. Interferences from Li^+, K^+, and Ca^{2+} ions are smaller [15].

is observed at high pH and is sometimes referred to as the sodium error or alkaline error. Lithium and potassium can also compete with hydrogen ions for the negatively charged adsorption sites at high pH. Figure 3.12 shows the deviation in potential from the ideal response for a glass electrode as a function of pH. (The "ideal" in this case is $E = E_{const} - 0.059\,16pH$.) Figure 3.12b shows calculated plot for the electrode response in the presence of 1 M levels of Na^+ (based on $K^{POT}_{HNa} = 1 \times 10^{-11}$). The glass is essentially unresponsive to H^+ ion beyond pH 12 under these conditions. The selectivity coefficients of a common pH-sensitive glass (Corning 015) are in the order of 10^{-11} for Na^+, 10^{-12} for Li^+, 10^{-13} for K^+, and 10^{-14} for Ca^{2+} [15]. This information suggests that titrations at high pH would be subject to smaller errors in pH using KOH rather than NaOH as a titrant. Interestingly, the selectivity of the glass for H^+ ions over Na^+ ions is better (K^{POT}_{HNa} is smaller) for glasses made with Li_2O instead of Na_2O. Of course, the current-carrying ion in these glasses is lithium instead of sodium. Lithium is still an interfering ion for pH sensors made from lithium glasses, but it provides an attractive option since samples containing high lithium ion concentrations are much less common than high pH and high sodium conditions are [15]. Glass electrodes of this sort with less alkaline error are commercially available.

3.3.4. The Combination pH Electrode

Most pH glass electrodes sold commercially are combination electrode devices that include both reference electrodes in a single probe (see Figure 3.13). The pH sensitive glass is fused to an inner glass tube that contains a reference solution, typically 3 M KCl buffered at pH 7.00 [24]. This solution sets both the hydrogen activity that controls the inner glass/solution interface potential and also the chloride activity that fixes the silver/silver chloride electrode potential. A second glass tube surrounds the tube with the glass bulb. The two tubes are sealed at the bottom creating an annular cavity where

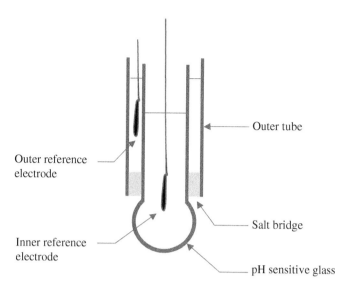

Outer tube

Outer reference
electrode

Salt bridge

Inner reference
electrode

pH sensitive glass

FIGURE 3.13 Combination pH glass electrode. Both reference electrodes are AgCl-coated silver wires. The inner reference solution is commonly 3 M KCl buffered at pH 7.00. The outer reference solution is typically 3 M KCl [24].

the second reference electrode and its solution are housed. This second chamber contains the same electrolyte as the inner reference electrode. It also contacts the sample solution through a salt bridge. Historically, the salt bridge was an asbestos wick or a cracked glass bead sealed into the wall of the outer glass tube. Modern salt bridges are usually a ring-shaped ceramic or a porous polymer forming a seal between the two glass tubes or a plug in the side wall of the outer tube. The salt bridge allows a tiny rate of diffusion of ions between the reference solution in the outer tube and the sample solution. Of course, the bridge must not allow a significant amount of mixing of the two solutions, since a change of the chloride activity in the reference solution could cause a drift in the potential and leakage of reference solution could cause contamination of the sample.

3.3.5. Gas-Sensing Electrodes

ISEs for single chemical species are very attractive. They are simple to use, portable, inexpensive, and have reasonably fast response times and are applicable over a wide range of concentrations. Although their development has been limited to certain ionic analytes, there have been imaginative strategies that have led to selective sensing devices for molecular analytes based on a pH electrode. One of these approaches is a gas sensing device. A carbon dioxide sensor can be made by trapping a thin layer of bicarbonate solution around the glass bulb of a combination pH electrode using a thin, gas-permeable polymer film such as Teflon™ (see Figure 3.14). Such a device was first introduced for blood CO_2 measurements by Richard Stow in 1954. Stow used deionized water for the medium inside the gas-permeable membrane. However, the potential for that device was prone to drift. The sensor was improved for clinical purposes by John Severinghaus and A. Freeman Bradley

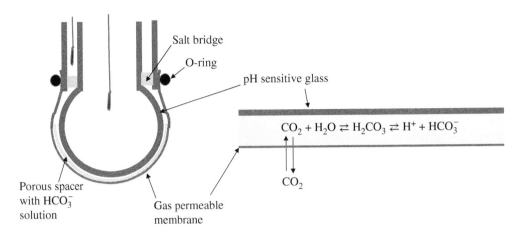

FIGURE 3.14 Carbon dioxide sensor based on a combination pH electrode known as the Severinghaus electrode. A gas permeable polymer membrane (such as silicone) traps a bicarbonate buffer solution around the glass bulb. Carbon dioxide can diffuse across the plastic film into the bicarbonate solution where it hydrolyzes to form carbonic acid and lowers the pH.

in 1958. Severinghaus added $NaHCO_3$ to the trapped solution and the system became much more stable [25–27].

Carbon dioxide crossing the polymer film changes the pH in the trapped solution because it forms carbonic acid when it dissolves in water:

$$CO_{2(g)} + H_2O_{(l)} \rightleftharpoons H_2CO_{3(aq)} \tag{3.24}$$

This equilibrium can be represented algebraically by

$$K_{eq} = \frac{1}{k_H} = \frac{a_{H_2CO_3}}{P_{CO_2}}$$

or

$$a_{H_2CO_3} = K_{eq} \cdot P_{CO_2} = \frac{P_{CO_2}}{k_H} \tag{3.25}$$

where k_H is the Henry's law constant for the process of CO_2 dissolving in water.

Since H_2CO_3 is a weak acid it can dissociate:

$$H_2CO_{3(aq)} \rightleftharpoons H^+_{(aq)} + HCO^-_{3(aq)} \tag{3.26}$$

This equilibrium influences the pH of the solution behind the polymer film and can be described algebraically:

$$K_{a1} = \frac{(a_{H^+})(a_{HCO_3^-})}{(a_{H_2CO_3})} \tag{3.27}$$

or

$$(a_{H^+}) = K_{a1}\left(\frac{a_{H_2CO_3}}{a_{HCO_3^-}}\right) = K_{a1}\left(\frac{K_{eq} \cdot P_{CO_2}}{a_{HCO_3^-}}\right) \tag{3.28}$$

The solution that is trapped between the gas-permeable membrane and the glass electrode is prepared with a high, fixed concentration of sodium bicarbonate so that the HCO_3^- activity, $a_{HCO_3^-}$, is essentially constant and the H^+ ion activity becomes proportional to the partial pressure of CO_2 in the sample solution. An inert porous material (such as a piece of fabric) is used to keep the membrane from collapsing against the glass.

$$(a_{H^+}) = K_{a1}\left(\frac{K_{eq} \cdot P_{CO_2}}{a_{HCO_3^-}}\right) = K_{a1}\left(\frac{P_{CO_2}}{k_H a_{HCO_3^-}}\right) = k' P_{CO_2} \tag{3.29}$$

So the potential at the glass electrode is proportional to the logarithm of the CO_2 level.

$$E_{electrode} = E_{const} + \frac{RT}{F}\ln(a_{H^+}) = E'_{const} + 0.059\,16\log(P_{CO_2})$$

$$= E''_{const} + 0.059\,16\log([CO_2]) \tag{3.30}$$

where E_{const}, E'_{const}, and E''_{const} are constants when $a_{HCO_3^-}$ is constant.

The practical range for this device spans approximately 160 mV corresponding to a partial pressure of CO_2 of about 0.015–0.6 atm or a solution concentration range of 5×10^{-4} to 2×10^{-2} M $H_2CO_{3(aq)}$. Figure 3.15 shows a calibration curve for a sensor immersed directly into carbon dioxide standard solutions [28]. A procedure recommended by vendors for determining the total carbon dioxide and carbonate species involves adding 5 ml of a buffer to 50 ml of sample (or standard) solution in order to adjust the pH to a value between 4.8 and 5.2 before the sensor is inserted into the solution [29]. At that pH, virtually all carbonate present converts to CO_2. Once the probe is placed in the sample, the solution is stirred to promote exchange of CO_2 between the sample solution and the solution in direct contact with the pH electrode. This approach gives a measure of the

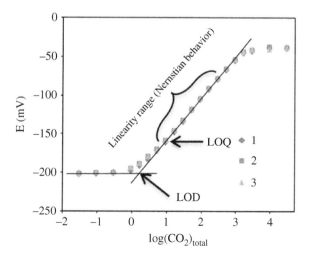

FIGURE 3.15 Carbon dioxide electrode response in mV as a function of dissolved carbon dioxide. Units for [CO_2] for this calibration curve were mg/l. Source: Reproduced with permission from Ricardo Hipólito-Nájera et al. [28]. Copyright 2017, American Chemical Society.

total carbonate, bicarbonate, and carbon dioxide for the sample. Typically, readings reach a stable value within a minute. At longer times, there is significant risk of carbon dioxide loss from the sample solution to the atmosphere. Direct comparison of signals from sample solutions to a standard calibration curve or the use of the method of standard additions are both applicable. Carbon dioxide sensors are also used *in situ*, such as in fermenters in breweries. Of course, in those cases, the measurement reflects the partial pressure of CO_2 in the sample solution rather than the total carbonate concentration. As such, the signal is dependent on the pH of the system being sampled.

EXAMPLE 3.2

Reproducibility is reported to be about 2 mV for replicate measurements for a gas-sensing electrode. What uncertainty in the CO_2 level does that correspond to? Consider two measurements that are 2 mV apart. For each measurement, assume that the electrode response follows a logarithmic dependence on the partial pressure of CO_2 and that the measurements are made directly in solutions at 298 K.

$$E_1 = E'_{const} + 0.059\,16\,\log(P_1) \quad \text{and} \quad E_2 = E'_{const} + 0.059\,16\,\log(P_2) \tag{3.31}$$

If the difference in the two readings is 2 mV,

$$E_1 - E_2 = 0.002 = E'_{const} + 0.059\,16\,\log(P_1) - E'_{const} - 0.059\,16\,\log(P_2) \tag{3.32}$$

$$0.002 = 0.059\,16\,\log\left(\frac{P_1}{P_2}\right) \tag{3.33}$$

$$\frac{0.002}{0.059\,16} = \log\left(\frac{P_1}{P_2}\right) \tag{3.34}$$

$$10^{\left(\frac{0.002}{0.059\,16}\right)} = \frac{P_1}{P_2} = 1.08 \tag{3.35}$$

Therefore, the 2 mV error in the voltage corresponds to a relative error of 8% in the calculated pressure for CO_2.

Of course, other volatile acids or bases can also cross the polymer membrane and change the pH at the electrode. Nitrous acid (HNO_2), acetic acid, sulfurous acid (H_2SO_3), and formic acid are common acids that are volatile at pH values between 4 and 5 and can interfere with the total carbonate determination.

Ammonia is another analyte that is commonly determined in a similar fashion. In the case of ammonia, the sample pH is raised above 11 using a buffer that sets both pH and ionic strength to force the equilibrium to favor the neutral form of ammonia that can cross the polymer film.

$$NH_{4(aq)}^+ + OH_{(aq)}^- \rightleftharpoons NH_{3(aq)} \rightleftharpoons NH_{3(g)} \tag{3.36}$$

The sensor must be in contact with the sample solution to reach the same concentration of NH_3 behind the membrane as in the sample solution. The solution between the gas permeable membrane and the glass electrode contains 0.1 M NH_4Cl. This level is high enough to fix the ammonium concentration. So the mixture formed from the sample NH_3 and the NH_4^+ inside sets the pH.

$$NH_{4(aq)}^+ \rightleftharpoons NH_{3(aq)} + H_{(aq)}^+ \tag{3.37}$$

Because the ammonium activity inside is constant, the ammonia activity is inversely proportional to the hydrogen ion activity:

$$K_a = \frac{a_{NH_3} a_{H^+}}{a_{NH_4^+}} \tag{3.38}$$

$$a_{NH_3} = \frac{K_a a_{NH_4^+}}{a_{H^+}} = \frac{C}{a_{H^+}} \tag{3.39}$$

$$\log(a_{NH_3}) = C' + pH \tag{3.40}$$

The response of these gas-sensing devices is a bit slower than for a direct measurement of pH because the gas must reach an equilibrium on both sides of the polymer membrane. A minute is usually adequate for measurements. As with the measurement of CO_2, waiting longer raises the risk of NH_3 loss to the atmosphere. Also, because the solubility of gases is very sensitive to temperature, precautions must be taken to maintain constant thermal conditions. Detection limits are typically around 10^{-5} M NH_3 [30].

The pH electrode is the most successful glass-membrane ISE. Glass electrodes with selectivity for Na^+ and Ag^+ also perform well and are commercially available. Relatively few ways have been demonstrated for manipulating ion exchange sites in glass or other ceramics in order to make those materials selectively adsorb other ions. New ISEs based on glass membranes are not on the horizon. Consequently, current research into producing new ISEs is mainly focused on liquid membrane devices rather than glass.

3.4. CRYSTALLINE MEMBRANE ELECTRODES

Some important, widely used ISEs are based on a third category of sensor membranes, namely, membranes made from insoluble inorganic salts. These crystal structures have a regular array of anions interspersed with cations that balance the charge. Under most conditions of interest, there are empty sites for analyte ions to occupy. These sites offer a cavity of optimum size for attracting another analyte ion to join the lattice. The selectivity of a crystalline matrix for the analyte ion depends on the unique fit of that ion into the crystal structure. Competition between an analyte ion and an interfering ion in solution for a surface site favors the ion that releases the most energy in binding. A closer contact with more neighboring ions of the opposite charge favors binding, while a large solvation energy opposes adsorption.

An important example of a crystal membrane device is the fluoride ISE. The membrane is based on a lanthanum trifluoride salt. The lattice site for fluoride is a small cavity

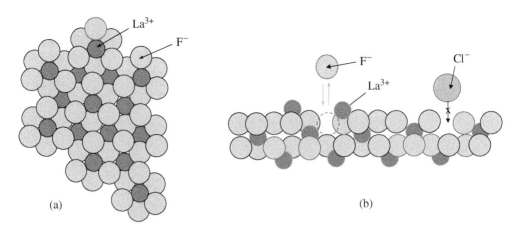

FIGURE 3.16 Conceptual views of lanthanum trifluoride lattice. (a) Top view. Each La^{3+} has six F^- ions as nearest neighbors; each F^- ion has two La^{3+} ions among its nearest neighbors. (b) Cross-sectional view of top layers showing open sites for anions. Selectivity of the crystal for binding fluoride ions from solution depends on the strength of the electrostatic interactions and the degree to which the anion fits the empty adsorption site in the crystal lattice.

that does not accommodate larger anions well. The small fluoride ion does have a large de-solvation energy that opposes its tendency to bind with the lattice. Nevertheless, the higher charge density of the fluoride together with contacts to multiple lanthanide ions makes the fluoride binding to the crystal highly favored over other anions.

The determination of fluoride ion concentration is an important and frequent analysis in countries that add fluoride to drinking water for protection against tooth decay. Such practice is usually tightly regulated to levels around 1 ppm fluoride ($\sim 53\,\mu M$) in order to prevent possible side effects. A fluoride membrane ISE is made from a membrane containing lanthanide trifluoride salt that has a low solubility (Figure 3.16). (The K_{sp} for LaF_3 is on the order of 10^{-18}.) [31]

$$LaF_{3(s)} \rightleftharpoons La^{3+}_{(aq)} + 3F^-_{(aq)} \tag{3.41}$$

As with other membrane devices, there must be some electrical communication across the membrane in order for it to function as part of a measurement circuit. The conductivity through a single crystal of LaF_3 is moderately high, largely as the result of the mobility of fluoride ions through the lattice (although the crystal is usually doped with EuF_2 to improve the conductivity further) [32]. In commercial fluoride ISEs, the LaF_3 crystalline membrane is sealed at the bottom of a plastic tube isolating a reference solution of 0.1 M NaF and 0.1 M KCl along with a silver/silver chloride reference electrode. On both sides of the ISE membrane, fluoride ions in solution establish an equilibrium with the surface.

$$F^-_{(membrane)} \rightleftharpoons F^-_{(aq)} \tag{3.42}$$

This process carries charge across the boundary. As with the glass electrode, there are two phase boundaries where potential differences form. In each case, the potential across the interface can be written in terms of the ion activity in the membrane and solution:

$$E_{PB} = \phi_{(memb)} - \phi_{(aq)} = \frac{RT}{(-1)F} \ln(a_{F^-_{aq}}) - \frac{RT}{(-1)F} \ln(a_{F^-_{memb}}) \tag{3.43}$$

As with the pH electrode, the activity of the analyte within the membrane material is assumed to be constant so that the phase boundary potential, E_{PB}, depends only on the fluoride in the aqueous solution:

$$E_{PB} = E_1 - 0.059\ 16 \log(a_{F^-_{aq}}) \tag{3.44}$$

where E_1 is a constant.

Keeping a constant fluoride activity in the solution inside of the sensor fixes the boundary potential on the inside (or reference side) of the membrane. Consequently, the membrane potential depends only on the sample solution activity of fluoride.

$$E_{memb} = E_{PB,spl} - E_{PB,ref}$$

$$= E_{1,spl} - E_{1,ref} - 0.059\ 16 \log(a_{F^-(aq)spl}) + 0.059\ 16 \log(a_{F^-(aq)ref}) \tag{3.45}$$

$$E_{PB} = E_1' - 0.059\ 16 \log(a_{F^-(aq)spl}) \tag{3.46}$$

Relatively few anions interfere with the fluoride ISE. The closest competitor is the hydroxide ion that interferes only at pH values above 8. The influence of interfering ions can be characterized by their selectivity coefficients. Their effect on the measured potential can be predicted using a variation of the Nicolsky–Eisenman equation:

$$E_{memb} = const - 0.059\ 16 \log \left(a_{F^-,spl} + \sum_j K_{Fj}^{POT} \cdot a_j^{z_F/z_j} \right) \tag{3.47}$$

The selectivity coefficient for hydroxide, K_{FOH}^{POT}, is 0.1. Even though K_{FOH}^{POT} is fairly large, hydroxide interference is negligible below pH 8. At pH 8, $a_{OH^-} = 10^{-6}$ and $K_{FOH}^{POT} \cdot a_{OH}^{-1/-1} = 10^{-7}$ [33]. Consequently, a 10% relative error in the estimate of $a_{F^-, spl}$ would occur for an analyte concentration of 10^{-6} M fluoride at pH 8. This level of fluoride activity is near the detection limit of fluoride at much lower pH (lower levels of a_{OH^-}). Procedures for fluoride determination recommend buffering samples at pH 5.5 using acetate and citrate buffers. Using citrate in the buffer has another benefit in addition to adjusting the pH. It chelates any aluminum or iron cations present that might, otherwise, interfere by complexing with the fluoride. A different problem arises, if the pH is too low. At low pH, a significant fraction of the fluoride is protonated (forming HF). Because the electrode responds to the anion form and not the HF form, the sensitivity drops rapidly with pH.

Other electrodes that rely on crystal lattice membranes include sensors for chloride (based on AgCl), sulfide (based on Ag_2S), bromide (based on $AgBr/Ag_2S$), cyanide and iodide (based on AgI/Ag_2S), thiocyanate (based on $AgSCN/Ag_2S$), cupric ion (based on CuS/Ag_2S), lead(II) (based on PbS/Ag_2S), mercury(II) (based on HgS/Ag_2S), bismuth(III) (based on Bi_2S_3/Ag_2S), and cadmium(II) based on (CdS/Ag_2S) [34]. These devices are polycrystalline, sparingly soluble salts that are pressed into thin pellets. Some membranes also include a polymer binder. Silver sulfide is added to many of the membranes described earlier in order to increase the conductivity of the membrane. The opportunity to create new ISEs based on crystalline membranes is limited by the relatively small number of suitable ionic matrices with a combination of low solubility, adequate conductivity, and selective adsorption sites. Consequently, the development of new ISEs is unlikely to be based on crystalline membranes.

3.5. CALIBRATION CURVES AND DETECTION LIMITS

The preparation of a standard calibration curve for a specific analyte ion consists of plotting the electrode voltage versus the common logarithm of the activity of the analyte. As indicated in Eq. (3.3), the slope ideally is $(2.303)RT/(zF)$ or a value of $\pm 59.16\,\text{mV}$ (at $T = 298\,\text{K}$) or 29 mV for a divalent ion ($z_i = 2$). It is often not convenient to calculate activity coefficients, so working in activity units is troublesome. A practical way of circumventing this problem is to set the ionic strength of all sample and standard calibration solutions to some constant level so that the activity coefficient is also constant. That strategy allows one to plot calibration data in concentration units. The slope remains the same as it would be for activity units. The only difference is a shift in the y-intercept. Holding the activity coefficient constant allows one to factor it out of the logarithm term and combine it with the other constants that contribute to the constant potential term.

$$\begin{aligned}
E_{\text{memb}} &= E_{\text{const}} - 0.059\,16\,\log(a_{X^-,\text{spl}}) = E_{\text{const}} - 0.059\,16\,\log(\gamma_{X^-,\text{spl}}[X^-]) \\
&= E_{\text{const}} - 0.059\,16\,\log(\gamma_{X^-,\text{spl}}) - 0.059\,16\,\log([X^-]) = E'_{\text{const}} - 0.059\,16\,\log([X^-])
\end{aligned}$$
$$(3.48)$$

However, whether one uses activity or concentration units, the linear range is limited (see Figure 3.17). The calibration curve bends at low analyte levels and eventually becomes horizontal. Of course, the slope of the calibration curve represents the sensitivity of the measurement method. There must be some mechanism that is acting to degrade the sensitivity of the sensor under these conditions. For sensors based on a crystalline membrane, the linear range is constrained by the solubility of the sparingly soluble salt used to make the membrane. Figure 3.17 shows data recorded with a silver chloride membrane in the absence of interfering ions. The membrane potential is dependent on Ag^+ ion activity and Cl^- ion activity. In Figure 3.17a, data were recorded for chloride standard solutions in the absence of silver ions. Figure 3.17b shows data from the sensor response to silver ions in the absence of chloride. In principle, a crystalline membrane device can be used as an ISE for either the anion or the cation that comprise the crystal. Although the graph is a linear function of either of those ions at higher activities, in both cases, the line bends to approach a constant potential. The explanation for this behavior is the fact that the solid AgCl in the membrane has a finite solubility. If the surrounding solution is very low in chloride (and silver ion) activity, then the amount of chloride ion coming out of the membrane from dissolution of the AgCl solid is the major source of the chloride activity in

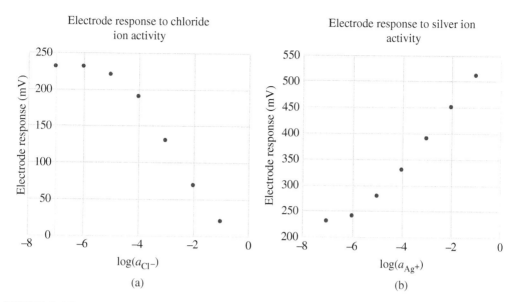

FIGURE 3.17   The response of an ISE made from a silver chloride membrane to silver and chloride activity. (a) Calibration curve for the ISE as function of chloride activity in the absence of silver ion. (b) Calibration curve for the same ISE as function of silver activity in the absence of chloride ion. Source: Morf et al. 1974 [35]. Reproduced with permission of American Chemical Society.

solution near the membrane.

$$AgCl \rightleftharpoons Ag^+ + Cl^- \tag{3.49}$$

Because it takes a finite amount of time for these ions to diffuse away from the membrane surface into the bulk solution, a steady-state level of chloride activity is established at the interface. Even in sample solutions with lower (bulk) chloride concentration, the crystal continues to dissolve and maintains the same chloride activity in the stagnant layer of solution at the sensor surface. As a result, the membrane potential reaches a constant value for solutions with low chloride activity.

When no other sources of silver or chloride ions are present, the activity of the chloride ion from the dissolution process can be estimated. At that point, the silver and chloride activities are equal, $a_{Ag^+} = a_{Cl^-}$.

$$K_{sp} = (a_{Ag^+})(a_{Cl^-}) = 1.8 \times 10^{-10} = (a_{Cl^-})^2 \tag{3.50}$$

$$a_{Cl^-} = \sqrt{1.8 \times 10^{-10}} = 1.3 \times 10^{-5} \, M \tag{3.51}$$

This estimate of chloride activity agrees reasonably well with the chloride activity near the bend in the calibration plot. It appears that the activity of chloride ion at the OHP will be buffered around 1×10^{-5} M. For this reason, this sensor loses sensitivity for determining chloride ion activity below 10^{-5} M [35].

Of course, the best practice is to use an experiment to establish the detection limit for an individual electrode. To establish the detection limit, one prepares a calibration curve of

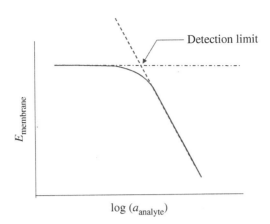

FIGURE 3.18 IUPAC method for estimating the detection limit for an anion ISE calibration plot. The x-coordinate corresponding to the intersection of the horizontal line in the region of very low analyte activity with the sloping line gives the logarithm of the minimum detectable analyte activity for this sensor.

the sensor voltage versus the logarithm of the analyte activity (or analyte concentration for solutions at a fixed ionic strength) using standard solutions that span a range to include the curving portion of the plot. The data corresponding to high levels of analyte activity should lie on a line with a slope close to the ideal of 59 mV/z per log unit (see Figure 3.18). At low analyte activity, the data approach a constant voltage. A horizontal line at that voltage is drawn to intersect the line with the Nernstian slope. The x-coordinate for this intersection gives the $\log(a_{analyte})$ from which one calculates the corresponding analyte activity. The International Union of Pure and Applied Chemists (IUPAC) recommends this method be used to define the minimum activity that can be reliably determined using this sensor [2]. This minimum quantifiable activity is referred to as the detection limit.

The other common constraint to making reliable determinations at low analyte activity is the presence of interfering ions. For example, Figure 3.19 shows a calibration plot for an iodide ISE in the presence of 0.095 M bromide – a potentially interfering ion. All solutions were adjusted to a constant ionic strength in order to keep the activity coefficients constant. In doing so, concentrations could be used rather than activities to yield a linear fit of the data.

In the absence of bromide ion, the Nernstian response of the electrode continues to activities below 10^{-6} M iodide [36]. The Nicolsky–Eisenman equation is generally applicable to all ISEs and can be used here to predict the behavior of the iodide electrode in the presence of bromide ions.

$$E_{memb} = E_1 + \frac{RT}{z_{I^-}F} \ln\left(a_{I^-,spl} + K_{IBr}^{POT} \cdot a_{Br}^{z_I/z_{Br}}\right) \qquad (3.52)$$

At high concentrations of iodide, $a_{I^-} \gg K_{IBr}^{POT} \cdot a_{Br}^{-1/-1}$, so the equation becomes:

$$E_{memb} = E_2 - 0.059\,16\,\log([I^-]) \qquad (3.53)$$

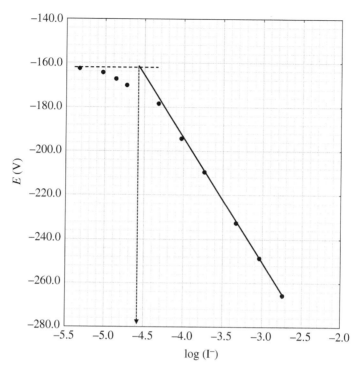

FIGURE 3.19 Calibration plot of response for an iodide selective electrode potential versus log[I⁻] in the presence of 0.095 M bromide. Concentrations were plotted instead of activities because the ionic strength was held constant. The linear portion of the curve extends to lower iodide concentrations in the absence of bromide. Source: Graph based on data from Papastathopoulos and Karayannis 1980 [36]. Reproduced with permission of American Chemical Society.

where E_2 includes the activity coefficient for the iodide ion. At the other extreme, the iodide activity becomes much smaller than the $K_{IBr}^{POT} \cdot a_{Br}$ term in the argument of the logarithm and the equation becomes independent of the analyte activity:

$$E_{memb} = E_3 - 0.059\,16\,\log([Br^-]) \tag{3.54}$$

where E_3 includes the activity coefficient for the bromide ion. The selectivity coefficient, K_{IBr}^{POT}, can be estimated as shown in Figure 3.19 in a process similar to finding the detection limit [2]. The x-coordinate corresponding to the intersection of the horizontal line with the extension of the Nernstian part of the calibration curve represents the conditions where $a_{I^-} = K_{IBr}^{POT} \cdot a_{Br}$ for the mixture of iodide and bromide. From these data (and only under these conditions), one can calculate the coefficient, K_{IBr}^{POT}, using

$$K_{IBr}^{POT} = \frac{a_{I^-}}{a_{Br^-}} = \frac{\gamma_{I^-} \cdot [I^-]}{\gamma_{Br^-} \cdot [Br^-]} \tag{3.55}$$

where γ_{I^-} and γ_{Br^-} are activity coefficients for iodide and bromide in this solution.

EXAMPLE 3.3

Refer to the calibration curve in Figure 3.19. If all of the standard solutions contained a concentration of 0.095 M KBr and 0.10 M $NaNO_3$ to control the ionic strength, estimate the value of the selectivity coefficient for bromide over iodide, K_{IBr}^{POT}, for the electrode.

The vertical line drawn from the intersection of the limiting voltage line and the extension of the linear part of the calibration curve in Figure 3.19 appears to cross the x-axis at $\log[I^-] \approx -4.60$.

Then the corresponding iodide level at this point is $[I^-] = 10^{-4.60} = 2.5_1 \times 10^{-5}$ M.

Compared to 0.095 M KBr and 0.10 M $NaNO_3$, the contribution to the ionic strength, μ, from the iodide (and its counter ion) is negligible. Then the μ is given by

$$\mu = 0.5 \sum c_i z_i^2 = 0.5\{[0.10](1)^2 + [0.10](-1)^2\}_{NaNO_3} + 0.5\{[0.095](1)^2 + [0.095](-1)^2\}_{KBr}$$

$$\mu = 0.19_5 \qquad (3.56)$$

The modified Davies equation is appropriate for estimating the activity coefficients at an ionic strength of 0.195 M.

$$\log(\gamma_i) = \frac{-0.511 z_i^2 \sqrt{\mu}}{1 + 1.5\sqrt{\mu}} + 0.2\mu z_i^2 \qquad (3.57)$$

However, both iodide and bromide have the same charge, z_i, and the ionic strength is the same for each, so their activity coefficients are equal. That is, $\gamma_{Br^-} = \gamma_{I^-}$. Consequently, the selectivity coefficient, K_{IBr}^{POT}, can be calculated using Eq. (3.55) directly from the ratio of the ion concentrations.

$$K_{IBr}^{POT} = \frac{a_{I^-}}{a_{Br^-}} = \frac{\gamma_{I^-} \cdot [I^-]}{\gamma_{Br^-} \cdot [Br^-]} = \frac{[I^-]}{[Br^-]} = \frac{2.5_1 \times 10^{-5}}{0.095} = 2.6_4 \times 10^{-4} \qquad (3.58)$$

3.6. A REVOLUTIONARY IMPROVEMENT IN DETECTION LIMITS

For ISEs based on a liquid membrane, detection limits are governed by both the presence of interfering ions and the leaching of the analyte from the internal reference chamber across the membrane into the sample solution. Migration of analyte ion through the membrane is possible because the ionophore and its complex are freely mobile. Normally, the internal chamber is filled with a moderately high level of analyte ion. Therefore, at low concentrations for the analyte ion on the sample side, there is a concentration gradient driving the analyte (from high to low) toward the sample side of the membrane (see Figure 3.20). This migration produces a finite build-up of analyte ions in the boundary layer at the sample solution side of the outer membrane interface in a manner similar to the dissolution process for crystalline membranes [3]. Even though these ions eventually diffuse into the bulk solution surrounding the membrane and are diluted, this process takes time and a

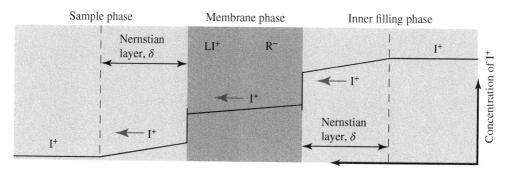

FIGURE 3.20   A concentration gradient of analyte ion, I^+ (not to be confused with iodide ion) across the liquid membrane can cause diffusion of analyte from the internal filling solution toward the sample solution. This leakage contributes to the local concentration of analyte in the sample solution at the membrane surface. This phenomenon prevents the sensor from responding to sample concentrations that are lower than the value established by this leakage. Source: Bühlmann and Chen 2012 [3]. Reproduced with permission of John Wiley & Sons.

steady-state concentration is established at the sensor surface. This level of analyte at the interface remains the same for solutions with even lower analyte concentration. That is, this "leakage" from the inside chamber becomes the dominant contribution to the analyte activity in the OHP on the sample side of the membrane. It reaches a constant value even as the analyte concentration in the bulk solution decreases further. Consequently, the membrane potential ceases to respond to further decreases in the analyte level in the bulk sample solution.

This mechanism was first described by Pretsch and coworkers in 1997 [11]. Once researchers recognized the problem, strategies for circumventing it were devised leading to lower detection limits. Sensors that respond to submicromolar analyte activities are now appearing. For example, the detection limits for a Pb^{2+} ISE were dramatically lowered by fixing the Pb^{2+} ion activity in the inner filling solution at a very low level by providing excess levels EDTA to complex the Pb^{2+} there.

The response of the electrode using a conventional reference solution of 1×10^{-3} M $PbCl_2$ and 0.10 M $MgCl_2$ leveled out at around 10^{-6} M Pb^{2+} as shown in Figure 3.21. However, in a similar electrode, the Pb^{2+} ion activity in the internal solution was buffered at 10^{-12} M by complexing the metal ion using a solution of $Pb(NO_3)_2$ with an excess of Na_2EDTA.

$$Pb^{2+} + EDTA^{2-} \rightleftharpoons Pb(EDTA) \qquad (3.59)$$

Just as a combination of weak acid and its conjugate base buffers the hydrogen ion activity in a solution, the presence of both $Pb(EDTA)$ and $EDTA^{2-}$ resists changes in the free Pb^{2+} activity. The conditions chosen (1 mM $Pb(NO_3)_2$, 50 mM Na_2EDTA and $K_{f,PbEDTA} = 9.1 \times 10^{-19}$ and pH 4.34) controlled the Pb^{2+} ion activity on the reference side at 10^{-12} M. One would expect that for sample solutions with Pb^{2+} activity above 10^{-12} M, the concentration gradient would favor the migration of lead ions through the membrane from the sample side inward; that is, any diffusion of lead ions would move away from the sample solution. Because this arrangement keeps the direction of diffusion

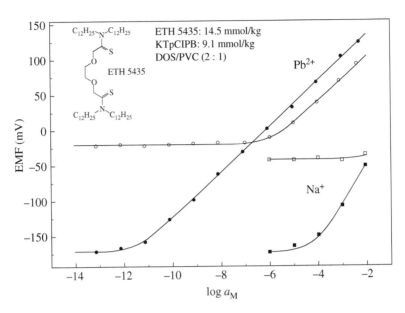

FIGURE 3.21 Improved detection limits for a lead ion selective electrode. The open circles indicate the response of the ISE for a "conventional" internal reference solution of a 1 : 1 mixture of 1×10^{-3} M $PbCl_2$ and 0.10 M $MgCl_2$. Filled black circles are data from a similar electrode in all respects except that the internal reference solution contained 1 ml of 0.1 M $Pb(NO_3)_2$ in 100 ml of 0.05 M Na_2EDTA at a measured pH of 4.34. The uncomplexed Pb^{2+} ion activity for this internal solution was calculated to be 10^{-12} M. The data plotted with square markers indicates the ISE response to Na^+ ions (open markers for sensor with the conventional filling solution and solid markers for sensor with the low-Pb^{2+} electrode filling solution). Source: Reproduced with permission from Sokalski et al. [11]. Copyright 1997, American Chemical Society.

of lead ions from the sample solution into the membrane, it prevents the membrane from contributing to the Pb^{2+} activity on the sample solution side of the membrane and the premature leveling of the calibration curve is avoided. Nernstian response was observed down to about 10^{-11} M.

This ionophore exhibits a huge preference for Pb^{2+}, but at a Pb^{2+} activity of 10^{-12} M, at high concentrations of sodium ion (at ~0.1 M) in the reference solution, sodium ion out-competes the Pb^{2+} ion for binding with the ionophore. Consequently, the inner solution/membrane interface potential is actually determined by the sodium activity rather than the activity of lead ions. Likewise, in the absence of Pb^{2+} in the sample solution, the membrane's outer interface potential responds to Na^+ ion activity. This behavior is shown in the lower right side of Figure 3.21. In the absence of Na^+ ions, the overall improvement in performance is an impressive lowering of the detection limit for Pb^{2+} ions by 6 orders of magnitude!

3.7. MORE RECENT ION SELECTIVE ELECTRODE INNOVATIONS

In all of the membrane electrodes described so far, an internal reference solution and an internal reference electrode have been used to complete the circuit. The internal conditions were always kept constant so that the internal membrane/solution interface potential would also be constant. However, there are limitations imposed by using a liquid filling solution. For example, reference solution volume changes caused by temperature variations or solution evaporation can lead to a drift in the signal. It is also difficult to miniaturize electrodes with internal volumes below one milliliter. Large differences in ionic strength lead to osmotic pressure that can damage the sensor membrane [37]. Many people have attempted to eliminate the internal solution by bonding the inside of the membrane directly to an electrical conductor that is connected to the outside measurement circuit. Even though many of these devices use liquid membranes, they have been referred to as "solid contact ISEs" or "all solid-state ISEs."

How do all solid-state ISEs work? To answer that question, one needs to consider what role that the inner solution of the conventional electrode plays in the operation of the sensor.

3.7.1. The Function of the Inner Reference Electrode

The inner reference solution is a key ingredient to the proper function of a reference electrode. The reference electrode serves as a transducer for charge carriers. That is, it is a device that enables current to move between a wire where electrons carry the charge and a solution where ions carry the charge. Why is that necessary? In order to measure electrical characteristics of a device, the device must be included in an electrical circuit, a closed loop through which current can pass. The act of measuring the potential of an electrochemical cell draws a finite current through all parts of the circuit. Fortunately, the amount of current required can be very small using modern electronics (on the order of 10^{-12} A or less with very good equipment). Keeping the measuring current small is important so that the charge separation across the membrane is not disturbed and, therefore, the voltage is not distorted by the measurement process. However, some charge must move throughout the circuit in order to communicate the value of the potential to the measurement equipment. Electrons carry the current in the voltmeter, but ions carry the current in the electrochemical cell. Consequently, there must be a mechanism that translates between these two types of charge carriers. Something must transform the ion current in the cell into electronic current in the wire leading to the meter. In a conventional ISE experiment, a reference electrode plays this role on both sides of the membrane. For example, when negative charge is moving (as electrons) from the meter into the cell, electrons move through the silver wire to the silver chloride coating on the electrode surface where they reduce some of the silver ions in the coating to form silver metal atoms and release an equal number of chloride ions into the solution (as in Eq. (3.60)):

$$AgCl + e \rightleftharpoons Ag + Cl^- \tag{3.60}$$

Negative charge is pumped into the solution in the process. At the reference electrode on the other side of the membrane, the reverse process occurs so that negative charge moves out of the solution (in the form of chloride ions) and into the silver wire (as electrons). Silver atoms donate the electrons to form silver ions that bind to chloride ions from the solution (Eq. (3.61)).

$$Ag + Cl^- \rightleftharpoons AgCl + e^- \tag{3.61}$$

Of course, only the analyte ions move into and out of the sensor membrane in order to carry the current there.

Because the reference electrode has a relatively large reserve of silver atoms and silver chloride solid, it can afford to transfer electrons with the outside circuit for the small amount of charge involved without the system deviating significantly from its rest potential (described by Eq. (3.62)). The original activity of the chloride must also be high enough that the small number of moles of chloride ion released or precipitated makes a negligible change in activity:

$$E_{Ag/AgCl} = E^o_{Ag/AgCl} - 0.059\ 16\ \log(a_{Cl^-}) \tag{3.62}$$

Under these conditions, the reference electrode is said to be "well poised" or to have a high redox capacitance. This mechanism allows charge to continue across the boundary between the solution where the current is carried by ions to the metal where the current is carried by electrons. In this sense, the reference electrode plays the role of an "ion-to-electron transducer." Providing this sort of transducer in the absence of a liquid electrolyte is the challenging part of eliminating the internal solution [38].

3.7.2. All Solid-State Reference Electrodes

The purpose of the filling solution is to support the redox reaction that translates between electronic current and ionic current. One way of eliminating the aqueous solution would be to find a redox reaction that produces ions that are soluble in the sensor membrane along with the ionophore. One such system was made by dissolving Co(II)/Co(III) tris(4,4'-dinonyl-2,2'-bipyridyl) complexes as salts of the hydrophobic anion, tetrakis(pentafluorophenyl)borate (TPFPB$^-$), along with the ionophore and polyvinyl chloride (the polymer matrix) in a volatile solvent [38]:

$$Co(III)L_3^{3+} + e^- \rightleftharpoons Co(II)L_3^{2+} \tag{3.63}$$

where L = and the counter ion, TPFPB$^-$ =

The membrane was formed by applying the mixture directly to the top of a carbon electrode and evaporating the solvent. In this manner, the membrane brings the redox couple and ionophore into direct contact with the inert electrode. This device was successful for determining K$^+$ ions with very good reproducibility from electrode to electrode [38]. Another approach is to use a redox polymer to replace the aqueous electrolyte and reference electrode. For example, poly(3-octylthiophene) is a highly conjugated linear polymer that can be oxidized or reduced [39].

where R = .

3.7.3. Eliminating the Inner Reference Electrode

A third strategy for eliminating the internal reference electrode is to exploit the electrical capacitance at an inert electrode/solution interface. Consider, for a moment substituting a platinum wire for the internal reference electrode of a conventional potassium ISE. How would the platinum respond to the tiny current inherent to the measurement process as described above? With no oxidizable or reducible ("redox active") species near the platinum electrode to use for transferring charge to or from the external circuit, no charge would cross the boundary there. However, as charge moves into the solution from the membrane, the excess ions would move to the platinum surface. The interface would behave as a capacitor where excess charge would build up as a layer of ions on the OHP. Electrons inside the platinum would move toward or away from the interface as needed to balance the charge on the solution side. The electrical potential energy difference across this double layer, E_{dl}, would be proportional to the charge that accumulates on either layer:

$$E_{dl} = \frac{Q}{C_{dl}} \tag{3.64}$$

where C_{dl} is the capacitance of the double layer.

The cell potential is the sum of all of the electrical potential energy transitions, including this double layer potential.

$$E_{meas} = E_{cell} = E_{dl} - E_{ext.Ref} + E_{jnc} + E_{memb} \tag{3.65}$$

$$= E_{dl} - E_{ext.Ref} + E_{jnc} + 0.059\,16\,\log(a_{K+spl}) \tag{3.66}$$

where $E_{ext.Ref}$ and E_{jnc} are constants. The measured potential is Nernstian (that is, the potential changes in proportion to $\log(a_{K+spl})$ with a slope near 59 mV) if E_{dl} is constant. The problem with this system is that the double layer voltage, E_{dl}, is not necessarily constant. Any net movement of charge within the cell, such as analyte ion leakage across the membrane, leads to a change in the charge at the double layer and, therefore, a change in the double layer voltage at the platinum/solution interface. That change in voltage creates an error. Fortunately, there is a remedy. If the capacitance in Eq. (3.64), C_{dl}, is very large, then for a fixed charge, Q, E_{dl} is small and changes in E_{dl} are even smaller. The strategy, then, is to make C_{dl} very large. One way of doing that is to use electrodes with very large surface area because the capacitance is proportional to the electrode surface area:

$$C = \frac{\varepsilon A}{4\pi d} \tag{3.67}$$

where ε is the dielectric constant for the medium, A is surface area of the electrode, and d is the distance between the charged layers.

Tiny particles have large surface area per gram of material. Carbon nanotubes and other porous carbon materials, such as three-dimensionally ordered macroporous carbon (3DOM) [13], can serve as high capacitance electrical contacts. ISEs using these electrodes in place of an internal reference electrode (and without an internal filling solution) have been demonstrated with very low (<1 mV) error in signal voltage. A diagram of a solid-state ISE based on a high capacitance 3DOM contact is shown in Figure 3.22.

The goal with all-solid-state devices is to make these sensors more robust and reproducible so that they have negligible error, require minimal calibration, and need little or no maintenance [13]. As such, they hold great promise for remote sensing for environmental analysis and, possibly, for sensors in wearable personal health monitors. One would also expect that the ISEs with solid contacts would have very low detection limits, since they

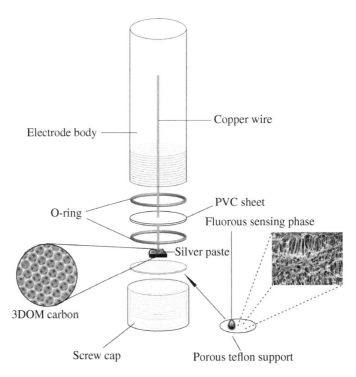

FIGURE 3.22 Assembly for an "all-solid-state ISE" using three dimensionally ordered macroporous (3DOM) carbon contact between the copper lead to the measurement circuit and the sensor membrane. This material provides a huge surface area and, therefore, a high electronic capacitance for transducing the signal from ion charge carriers to electrons. Source: Reproduced with permission from Lai et al. [13]. Copyright 2010, American Chemical Society.

have no internal electrolyte with analyte that can leach out into the sample solution. This type of work is still in its early stages as of this writing. However, progress is encouraging and may lead to an expansion of applications for ISE sensors.

3.7.4. Super-Hydrophobic Membranes

Another improvement in performance can be made by substituting a long-chain fluorinated solvent for the usual hydrophobic medium in the membrane. Not only are fluorinated solvents hydrophobic, but they are also immiscible with hydrocarbon solvents. They are the least polar molecules known. They prevent moisture and organic molecules from penetrating the membrane. Proteins, lipids, and other biomolecules in clinical or environmental samples are prone to adsorption on the surface of conventional, hydrocarbon-filled sensors. The accumulation of these nontarget molecules blocks the surface and may hold charged organic groups in the electrical double layer causing drift in the voltage or a loss of response to the analyte. Fluorinated media do not have this problem. Of course, the ionophores that are used in these membranes must also be fluorinated in order to dissolve there. Figure 3.23 shows several components suitable for constructing a silver ISE [13].

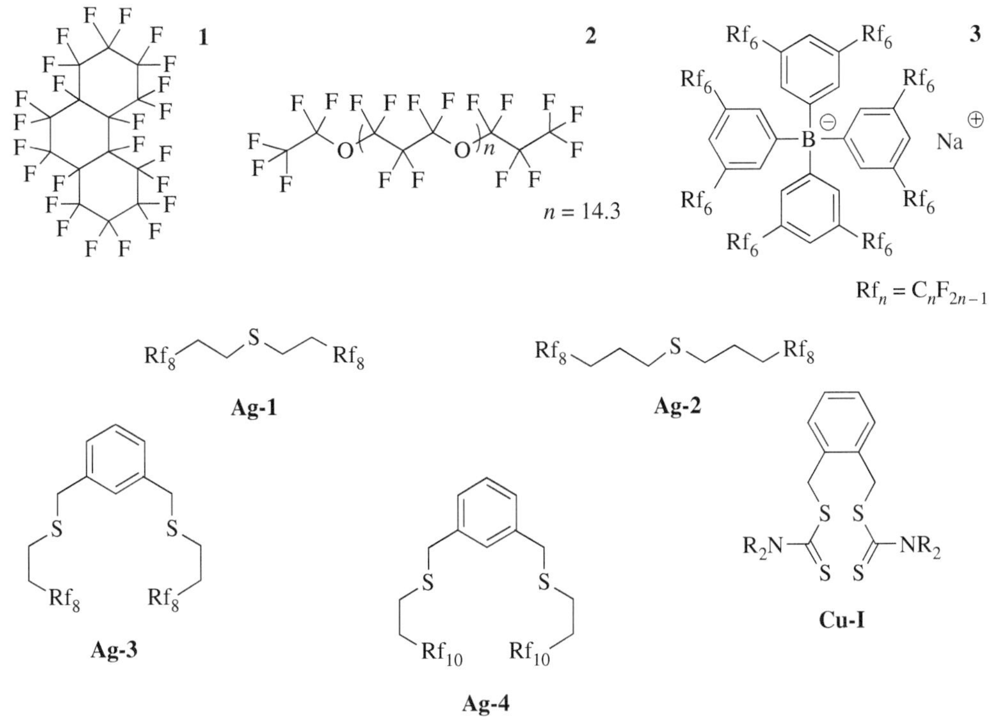

FIGURE 3.23 Components for fluorinated membrane ISEs. Structures 1 and 2 have been used as solvent matrices and the fluorinated derivative of tetraphenyl borate (structure 3) provides the "ionic sites" that compensate for charge when the analyte cation binds to the ionophore. Structures Ag-1, Ag-2, Ag-3, and Ag-4 are examples of fluorinated ionophores for silver. Cu–I is a selective agent for copper ion. Source: Reproduced with permission from Lai et al. [13]. Copyright 2010, American Chemical Society.

The fluorous side chains were attached to the ligands to make them soluble in the fluorinated matrix. However, hydrocarbon linkers between the fluorous chains and sulfur atoms were needed, since fluorine atoms closer than a distance of two carbon atoms from the sulfur atoms were too strongly electron withdrawing to permit strong coordination of the ionophore with the silver ion. The membranes were constructed from Teflon™ (a perfluoronated polymer) soaked in either compounds 1 or 2, that acted as solvents, plus one of four silver-binding ionophores (Ag-1, Ag-2, Ag-3, or Ag-4). In this particular paper [13], the authors dispensed with an inner filling solution and used a high capacitance contact. The selectivities for Ag^+ over many alkaline and heavy metal ions were much better than most other Ag^+ ISEs reported in the literature (e.g. using ionophore Ag-4 the selectivities were $K_{AgK}^{POT} = 2.5 \times 10^{-12}$ for potassium versus silver, $K_{AgPb}^{POT} = 6.3 \times 10^{-11}$ for lead versus silver, $K_{AgCu}^{POT} = 1 \times 10^{-13}$ for copper versus silver, and $K_{AgCd}^{POT} = 6.3 \times 10^{-14}$ for cadmium versus silver) [13].

3.8. ION SELECTIVE FIELD EFFECT TRANSISTORS (ISFETS)

Another type of all solid-state ISE that has had some commercial success is based on a field effect transistor (FET). This type of device has been available since the 1970s. A FET is a common semiconductor component that is used for electronic amplifiers. Transistors are made from doping pure silicon (or germanium) with small amounts of elements from the boron (group 13 or 3A) or nitrogen (group 15 or 5A) families. Figure 3.24 shows partial Lewis dot structures for different types of doped material. Pure silicon is a network solid in which all valence electrons are occupied in covalent bonds. Consequently, its electrons are not very mobile, and the material is a poor conductor of electricity. Arsenic is a common dopant that has five valence electrons. Adding a small mole fraction of arsenic atoms to the silicon produces sites where an extra valence electron is available, because it is not associated with a bond. The conductivity of the resulting material is higher than that of

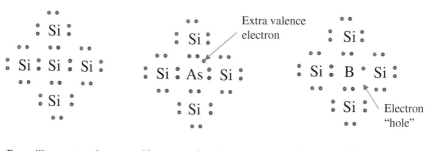

FIGURE 3.24 Partial Lewis dot structures for silicon and doped silicon semiconductors. Elemental silicon is a network solid that conducts poorly in its pure form because all of its valence electrons are occupied in covalent bonds. If the silicon is doped with a small fraction of arsenic atoms, the "extra" valence electrons from arsenic atoms are available to carry current. Doping with an element with only three valence electrons, such as boron, leaves one bond that is electron-deficient. Physicists refer to this as a "hole." Electrons from nearby atoms can "hop" into these holes leaving behind positive holes.

pure silicon. This type of material is known as an N-type semiconductor because the major charge carriers are negative (electrons). In a similar manner, doping with boron, which contributes only three valence electrons per boron atom, produces sites where one B—Si bond is electron-deficient. Physicists refer to the partially filled orbital as a "hole." The energy barrier preventing electrons in neighboring atoms from hopping into a hole is relatively small so that at room temperature, a significant amount of movement is observed. In a sense, holes are moving around in this material. Because holes are positively charged, this type of material is referred to as a P-type semiconductor.

Transistors are made from combinations of N-type and P-type material that form three distinct regions. In the transistor shown in Figure 3.25, the substrate is made of lightly doped P-type material. Two other regions are formed from N-type material. These two regions are called the source and the drain and are coated with aluminum metal in order to make electrical contact with outside circuitry. The region between the source and drain

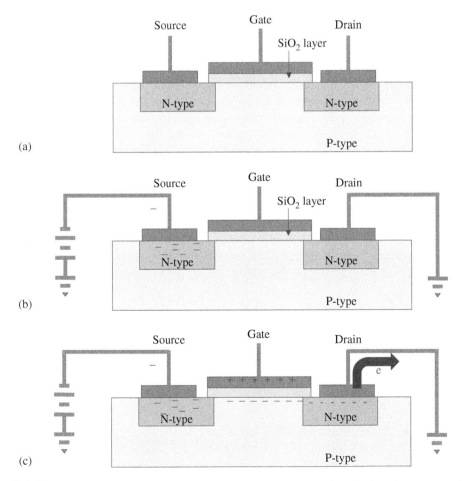

FIGURE 3.25 (a) Diagram of a conventional field effect transistor (FET) designed to operate in the enhancement mode. (b) No gate voltage; electrons are trapped in the source. (c) A positive voltage applied to the gate attracts electrons to carry current between the source and drain.

is called the gate. For conventional FETs, the gate is covered with a thin layer of silicon dioxide or silicon nitride onto which an aluminum coating is vapor-deposited for making contact to other electronic components. It is through this contact that a signal voltage is applied to the gate.

In a conventional FET, a voltage signal applied to the gate either increases or decreases the number of charge carriers in a channel between the source and the drain. If the device is designed to increase the number of charge carriers when a signal is applied, then it is called an "enhancement mode" device. (Although it is built on a P-type substrate, the transistor pictured here is an N-channel device. The name comes from the fact that negative charges [electrons] must be recruited to form a channel for current to move from the source to the drain.) Because the P-type substrate is lightly doped, the region between the source in the drain has few charge carriers and exhibits a high resistance normally. If the source is connected to the negative side of a battery or power supply (as in Figure 3.25b), then electrons move into the source but are trapped there. The high resistance between the source and the drain impedes any current between them. Now imagine that a positive voltage is applied to the gate as in Figure 3.25c. Positive charges build up on the metal contact, but go no further due to the nonconducting silicon dioxide layer. However, as in a capacitor, the positive charges attract electrons (from the source) to the other side of the oxide layer. Once inside the P-type zone of the gate below the silicon dioxide, these electrons experience the field between the drain and the source and migrate toward the drain. The net effect is a channel of current between the source and the drain. In that manner, a small signal voltage controls a large quantity of current between the source and the drain, but uses a negligible amount of current from the device that introduces the signal to the gate.

An ion-sensing field effect transistor (ISFET) is designed to use the charge from the selective adsorption of ions from a sample solution onto the gate to create a signal voltage that subsequently controls the source to drain current. Instead of a metal contact over the gate, the silicon dioxide layer is left open to contact a sample solution (see Figure 3.26a). Ion adsorption at this surface will control the charge at the gate and, therefore, control the current between the source and the drain. Proper design of the silicon dioxide layer produces a pH-sensitive coating. Modern semiconductor manufacturing methods can make very tiny ISFET pH sensors. Figure 3.26b shows a picture of a pH ISFET sensor embedded

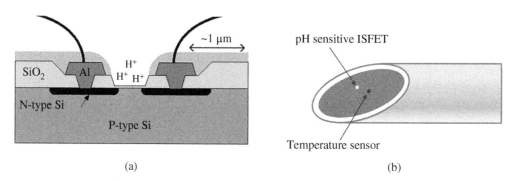

(a)   (b)

FIGURE 3.26 Ion-sensing field effect transistor (ISFET). (a) Diagram of the transistor showing the exposed silicon dioxide layer acting as a pH sensitive glass that controls the voltage at the gate. (b) An ISFET embedded in epoxy inside a hypodermic needle. Source: Lisensky [40].

in the tip of a hypodermic needle [40]. As with other all-solid-state ISEs, the challenge to making ISFETs for other analytes has been in creating a stable contact between the sensing membrane and the gate.

3.9. PRACTICAL CONSIDERATIONS

3.9.1. Ionic Strength Buffers

Because ISEs respond to the activity of an analyte, and not just its concentration, it is important to control (or account for) the ionic strength of the sample and standard solutions. When one has the luxury of working with discrete samples, adding a high level of electrolyte is an attractive way of fixing the ionic strength and, subsequently, the activity coefficient of the analyte. Adding a pH buffer at the same time also helps to control for complexation, precipitation (or protonation, if the analyte is the anion of a weak acid). In special cases, complexing agents are purposely included in the buffer in order to bind specific metal ions that might interfere with the analysis. For example, iron(III) and aluminum(III) bind fluoride ions in solution leading to an erroneously low signal for F^- when present. Citrate is often added to samples to prevent that problem because citrate competes with fluoride for binding to these metal ions. These additives are often combined into a "total ionic strength adjustment buffer" (TISAB). These ingredients can be mixed in a solution in advance and then added to all samples and standards in a constant volume ratio immediately before measuring the signal. This approach assumes that the ionic strength of the buffer is so much greater than that of the sample that the ionic strength and pH of the mixture is set by the TISAB. A further bonus of this approach is that the calibration curve can be plotted in terms of the concentration of the analyte rather than the activity. There is no need to evaluate the activity coefficient of the analyte. Table 3.2 gives example recipes for ionic strength adjustment buffers that were originally formulated for use with

TABLE 3.2 Ionic strength adjustment buffers

TISAB I	TISAB II	TISAB III
For use when magnesium, calcium, chloride, nitrate, sulfate, or phosphate are present at high levels	For use when iron(III) or silicate are present at high levels	For use when aluminum is present at high levels
500 ml deionized water	500 ml deionized water	500 ml deionized water
57 ml glacial acetic acid	170 g sodium nitrate	17.65 g cyclohexanedi-aminetetraacetic acid (CDTA)
58 g sodium chloride	68 g sodium acetate trihydrate	Add 40% NaOH, drop by drop until the salt is dissolved
0.3 g sodium citrate dihydrate	92.4 g sodium citrate dihydrate	300 g sodium citrate dihydrate 60 g sodium chloride
Dilute to 1 l	Dilute to 1 l	Dilute to 1 l

Source: Reproduced with permission from Buck and Cosofret [33]. Copyright 1993, IUPAC.

fluoride ISEs. An addition of 10 ml TISAB I per 100 ml of sample gives an ionic strength of about 0.1 M, a good target value.

3.9.2. Potential Drift

In principle, ISE membranes need only a second or less to reach an equilibrium with a new sample, if the system is well stirred. In practice, it is a common experience to have a system take many seconds or even minutes to reach a steady potential. Changes in potential in the same direction that occur over minutes despite constant analyte conditions are usually called "potential drift." This same term is also applied when repeated measurements with the same standard or sample solution are not reproducible, but trend in one direction. The source of the drift is often attributed to problems with a reference salt bridge [24].

Trustworthy results (as evident from highly linear calibration curves) can usually be obtained by waiting to record a potential until the signal has reached some predetermined rate of change such as 0.2–0.4 mV/min. Alternatively, good analytical data can also be obtained by recording potentials after a specific time interval from the time at which the sensor was first immersed in the sample solution [41].

PROBLEMS

3.1 A calcium ion ISE gives a response of +170 mV in a 100.00 ml sample solution. The same solution is spiked with 5.00 ml of 0.125 M calcium standard solution. After thorough mixing the same electrode combination gave a reading of +193 mV. Assuming a Nernstian response and no interferences, calculate the concentration of Ca^{2+} in the original sample solution.

3.2 A technician puts a new membrane on a nitrate ISE and records the following data for a series of calibration standards in an appropriate ionic strength buffer.

$[NO_3^-]$	E (mV)
5.50×10^{-7}	−133
6.00×10^{-7}	−133
1.19×10^{-6}	−134
2.50×10^{-6}	−141
6.00×10^{-6}	−150
2.38×10^{-5}	−183
1.18×10^{-4}	−224
$1.18E \times 10^{-3}$	−282

(a) Plot the appropriate form of the data for a calibration curve.
(b) Calculate the slope for the electrode response in the linear range.
(c) What is the apparent detection limit for this electrode?

3.3 A nitrate ISE with a new membrane gave the following response to nitrate standard solutions prepared in an appropriate ionic strength buffer. Each solution also contained 2.00×10^{-2} M ClO_4^-. Graph the data and calculate the apparent selectivity coefficient for the interfering ion.

$[NO_3^-]$	E (mV)
5.50×10^{-7}	-140
6.00×10^{-7}	-140
1.19×10^{-6}	-143
2.50×10^{-6}	-146
6.00×10^{-6}	-160
2.38×10^{-5}	-183
1.18×10^{-4}	-224
1.18×10^{-3}	-282

3.4 A colleague brings you her Pb^{2+} ISE based on a liquid membrane that uses the ionophore, ETH 5435. She said that it exhibited a slow, continuous voltage drift ($\sim 5\,mV/min$) when she was measuring trace lead levels on pond sediment samples in the field. You begin to speculate on possible causes and quickly list the following ideas as they come to mind. Describe the type of behavior that each of these issues would cause. Which, if any, might give rise to the behavior that your colleague described.

(a) The membrane may have been punctured, perhaps by something in the sediment.

(b) The silver ions from the reference electrode may have reacted with something in solution to precipitate in the salt bridge.

(c) Your colleague may have forgotten to soak the electrode a standard solution of Pb^{2+} after putting a new membrane on the device.

(d) The temperature may have been changing during the measurement.

3.5 How many Ca^{2+} ions cross the membrane/solution boundary corresponding to a 0.050 V increase in potential for a sensor with a $0.090\,cm^2$ area assuming a capacitance of $3 \times 10^{-6}\,F/cm^2$?

3.6 The following data were obtained by preparing each standard solution by diluting the appropriate volume of a 1.00×10^{-2} M Cl^- stock solution into 250 ml volumetric flasks and diluting to the mark with deionized water. Before measuring, 100.0 ml of each standard was spiked with 1.00 ml of 4 M $NaNO_3$ to adjust the ionic strength.

[Chloride]	E (mV)
1.00×10^{-3}	-113.0
1.00×10^{-4}	-54.0
1.00×10^{-5}	5.0
5.00×10^{-6}	22.8
1.00×10^{-6}	64.0
5.00×10^{-7}	81.8

(a) What concentration of chloride ion would be indicated for a sample that was also treated with ionic strength buffer and gave a signal of 15.5 mV?

(b) If the sample had not been treated with $NaNO_3$ to adjust the ionic strength before measuring, what would the potential reading have been (neglecting the influence on the junction potential for the reference bridge)?

3.7 A cell using a potassium ISE based on a liquid membrane using valinomycin as an ionophore and a silver/silver chloride electrode gives a voltage of -0.273 V in a solution of 7.50×10^{-6} M KCl standard solution (using an ionic strength buffer). Assuming that the electrode responds ideally, calculate the cell potential for the same conditions plus the presence of 2.00×10^{-2} M HCl as an interfering electrolyte.

3.8 What is the maximum activity of sodium ions that can be tolerated in order to measure K^+ ions at activities as low as 1×10^{-4} M with at most a 10% error using a valinomycin liquid membrane?

3.9 If the liquid junction potential for a combination pH electrode drifts by 5 mV, what is the relative error in the apparent H^+ concentration?

3.10 Consider using an ISE for determining bromide ion based on $AgBr/Ag_2S$ crystalline membrane. The K_{sp} for AgBr is 7.7×10^{-13}. What is the approximate detection limit for bromide ion in the absence of interfering ions? What would the bromide detection limit be in the presence of 0.01 M KCl, given that $K_{BrCl}^{POT} = 3 \times 10^{-3}$?

3.11 Using the Henderson equation and table of diffusion coefficients in Appendix C, calculate the junction potential for a salt bridge with 3 M NaCl solution contacting a sample solution with 0.2 M HNO_3.

REFERENCES

1. Arnold O. Beckman. Science History Institute. https://www.chemheritage.org/historical-profile/arnold-o-beckman (accessed 25 October 2017).

2. Lindner, E. and Pendley, B.D. (2013). *Anal. Chim. Acta* 762: 1–13.

3. Bühlmann, P. and Chen, L.D. (2012). Ion-selective electrodes with ionophore-doped sensing membranes. In: *Supramolecular Chemistry: From Molecules to Nanomaterials*, vol. 5 (eds. A.W. Steed and P.A. Gale), 2539. New York, NY: Wiley.

4. Palmer, M., Chan, A., Dieckmann, T., and Honek, J. *Biochemical Pharmacology*. University of Waterloo Available online: http://watcut.uwaterloo.ca/webnotes/Pharmacology/downloads/pharma-book.pdf.

5. Bakker, E. (2014). *Fundamentals of Electroanalysis*. Apple iBook Store (ebook).

6. Schaller, U., Bakker, E., Spichiger, U.E., and Pretsch, E. (1994). *Anal. Chem.* 66 (3): 391–398.

7. Bard, A.J. and Faulkner, L.R. (2001). *Electrochemical Methods*, 2e. New York, NY: Wiley.

8. Qin, W., Zwickl, T., and Pretsch, E. (2000). *Anal. Chem.* 72 (14): 3236–3240.

9. Lee, M.H., Yoo, C.L., Lee, J.S. et al. (2002). *Anal. Chem.* 74 (11): 2603–2607.

10. Späth, A. and König, B. (2010). *Beilstein J. Org. Chem.* 6 (32) https://doi.org/10.3762/bjoc.6.32.

11. Sokalski, T., Ceresa, A., Zwickl, T., and Pretsch, E. (1997). *J. Am. Chem. Soc.* 119: 11347.

12. Bühlmann, P. and Chen, L.D. (2012). Ion-selective electrodes with ionophore-doped sensing membranes. In: *Supramolecular Chemistry: From Molecules to Nanomaterials* (eds. P.A. Gale and J.W. Steed). Wiley. ISBN: 978-0-470-74640-0.

13. Lai, C.-Z., Fierke, M.A., da Costa, R.C. et al. (2010). *Anal. Chem.* 82 (18): 7634–7640.

14. Chen, L.D., Xu, U.Z., and Bühlmann, P. (2012). *Anal. Chem.* 84: 9192–9198.

15. Isard, J.O. (1967). The dependence of glass-electrode properties on composition. In: *Glass Electrodes for Hydrogen and Other Cations* (ed. G. Eisenman), 51–83. New York, NY: Marcel Dekker, Inc.

16. Cremer, M. (1906). *Z. Biol.* 47: 562.

17. Haber, F. and Klemensiewicz, Z. (1909). *Z. Phys. Chem.* 67: 385–431.

18. Perley, G.A. (1949). *Anal. Chem.* 21 (3): 394–401.

19. Eisenman, G. (1967). *Glass Electrodes for Hydrogen and Other Cations* (ed. G. Eisenman), 223–232. New York, NY: Marcel Dekker, Inc.

20. Haugaard, G. (1941). *J. Phys. Chem.* 45: 148.

21. Vesely, J., Weiss, D., and Stulik, K. (1978). *Analysis with Ion-Selective Electrodes*. Chichester: Ellis Horwood Limited.

22. Nickolsky, B.P., Shultz, M.M., Belijustin, A.A., and Lev, A.A. (1967). Recent developments in the ion-exchange theory of the glass electrode and its application in the chemistry of glass. In: *Glass Electrodes for Hydrogen and Other Cations* (ed. G. Eisenman). New York, NY: Marcel Dekker, Inc.

23. Ruzicka, J. (1997). *J. Chem. Educ.* 74 (2): 167–170.

24. Erich Springer (2014), pH Measurement Guide, Hamilton Co. http://separations.co.za/wp-content/uploads/2014/09/pH-Measurement-guide.pdf (accessed 10 August 2019).

25. Severinghaus, J.W. (2006). *CO$_2$ Electrodes in Encyclopedia of Medical Devices and Instrumentation*. John Wiley & Sons http://onlinelibrary.wiley.com/doi/10.1002/0471732877.emd326/abstract.

26. Young, C.C. (1997). *J. Chem. Educ.* 74 (2): 177–182.

27. Severinghaus, J.W. and Bradley, A.F. (1958). *J. Appl. Physiol.* 13 (3): 515–520.

28. Ricardo Hipólito-Nájera, A., Rosario Moya-Hernández, M., Gómez-Balderas, R. et al. (2017). *J. Chem. Educ.* 94 (9): 1303–1308.

29. Thermo Fisher (2008) User guide "Carbon dioxide ion selective electrode". https://www.instrumart.com/assets/ISEcarbondioxide_manual.pdf (accessed 10 August 2019).

30. Hanna Instruments (2008). Instruction manual for H4101 ammonia ion selective electrode. https://hannainst.com/downloads/dl/file/id/1137/man4101_r5.pdf (accessed 10 August 2019).

31. Lingane, J.J. (1968). *Anal. Chem.* 40: 935–939.

32. Frant, M.S. and Ross, J.W. Jr. (1966). *Science* 154 (3756): 1553–1555.

33. Buck, R.P. and Cosofret, V.V. (1993). *Pure Appl. Chem.* 65, 8, 1849-1858, 1993. (© 1993 IUPAC).

34. Strobel, H.A. and Heineman, W.R. (1989). *Chemical Instrumentation: A Systematic Approach*, 3e. New York, NY: Wiley.

35. Morf, W.E., Kahr, G., and Simmon, W. (1974). *Anal. Chem.* 46: 1538–1543.

36. Papastathopoulos, D.S. and Karayannis, M.I. (1980). *J. Chem. Educ.* 57 (12): 904–906.

37. Hu, J., Stein, A., and Buhlmann, P. (2016). *TrAC, Trends Anal. Chem.* 76: 102–114.

38. Zou, X.U., Zhen, X.V., Cheong, J.H., and Bühlmann, P. (2014). *Anal. Chem.* 86: 8687–8692.

39. Bobacka, J., Ivaska, A., and Lewenstam, A. (2008). *Chem. Rev.* 108: 329–351.

40. George Lisensky, ISFET (2004). http://archive.education.mrsec.wisc.edu/SlideShow/slides/pn_junction/ISFET2.html (accessed 10 August 2019).

41. Mueller, A.V. and Hemond, H.F. (2011). *Anal. Chim. Acta* 690: 71–78.

APPLICATIONS OF ION SELECTIVE ELECTRODES

4

4.1. OVERVIEW

Determinations by ion selective electrodes (ISEs) are operationally simple. However, there are several choices that the analyst must make before gathering data in order to insure accuracy and good precision. This chapter examines several analytical situations and the operating variables that can influence the measurement results. Some knowledge of the sample is helpful in deciding whether the ionic strength should be adjusted or whether any sort of pretreatment is necessary in order to screen out interferences. Preparing to do an analysis by ISE involves answering a series of interrelated questions. Is the expected sample concentration above the limit of quantification? What interfering ions are likely present? Can an ionic strength buffer be used? What pH range is appropriate? Are other sample pretreatment steps necessary to exclude sample loss to complexation with components in the matrix or to prevent changes in analyte concentration due to microbial activity? Of course, any additions to the sample must also be made to each of the calibration standard solutions as well.

In addition to considering sample treatment, one must also select the type of reference electrode and the type of electrolyte to use in the reference bridge. A number of factors affect the reliability of the reference. Here are some questions that can help guide one's choice. Are there any concerns with chloride leaching from the reference salt bridge into the sample solution? Are there components from the sample matrix that might compromise the reference electrode chemistry or clog the salt bridge? If so, what intermediate electrolyte can be used for a double junction salt bridge? Should a high flow salt bridge be used in order to prevent clogging or to maintain a more stable junction potential? Can the system be stirred to help stabilize the junction potential? What sort of temperature control will be possible? Does the $\pm 1\,^{\circ}$C variation that one might expect in the lab (which corresponds to ~2% relative uncertainty in analyte concentration) provide tight enough precision for the desired analysis? What procedure will be used to validate the method – analysis of certified samples or a percent recovery experiment? Can a standard calibration curve be used to determine sample concentration or must a method of standard additions be applied?

In order to bring out a wide range of issues, case studies in three different contexts are presented here: an industrial application, a clinical analysis setting, and an environmental monitoring project. Section 4.2 discusses the use of a fluoride ISE for determining fluoride

Electroanalytical Chemistry: Principles, Best Practices, and Case Studies, First Edition. Gary A. Mabbott.
© 2020 John Wiley & Sons, Inc. Published 2020 by John Wiley & Sons, Inc.

ion in an industrial setting and provides an adequate introduction for the student new to the subject. More advanced discussions are presented in the subsequent two case studies. Section 4.3 discusses testing a research-grade sensor for determining creatinine in blood and urine as a means of assessing kidney health. Section 4.4 contrasts an Environmental Protection Agency (EPA) procedure for the determination of nitrate ion in environmental water samples determined by ISE under lab conditions with an automated, remote platform system for monitoring several ions in the field for many days at a time.

Finally, because of their pervasive application, good lab practices for using pH electrodes are discussed at the end of this chapter. All parts of Section 4.5 are recommended reading for university coursework in instrumental analysis. These pages cover the maintenance, calibration, and application of pH electrodes, as well as the preparation and proper use of standard pH buffers.

4.2.   CASE I. AN INDUSTRIAL APPLICATION

Consider the application of a fluoride ISE in monitoring the level of fluoride in toothpaste for quality control purposes. In a recent study, an analytical team in Serbia developed and tested an ISE method using toothpastes with a certified value of 1450 mg fluoride/kg toothpaste [1].

What are the issues for an analytical chemist to consider in order to maximize the accuracy, precision, and efficiency of this analysis? A good starting point for this discussion is to think about the concentration level that one can reasonably expect in the sample. In the case of quality control of a manufactured item, the content should be easy to anticipate. In fact, the manufacturer has a target to meet with respect to active ingredients. In this situation, the analysis was part of the program created to help guarantee that the product meets that target.

4.2.1.   Will the Sample Concentrations Be Measurable?

One needs to determine in advance whether the expected value will be within the quantitative range for the sensor. In other words, is the typical concentration above the limit of quantitation (LOQ) for the method? Furthermore, it is often necessary to use a buffer to adjust the pH and the total ionic strength of each sample. Would it be possible to dilute the sample by 10-fold in such a buffer and keep the analyte concentration well above the LOQ? The expected value from a toothpaste sample following a dilution of a factor of 10 would be 145 ppm F^- or 7.6 mM F^-. The limit of detection (LOD) for a fluoride ISE is typically around 10^{-6} M. The LOQ is approximately 3.3 times the LOD or ~3 μM [2]. Therefore, diluting the sample by a factor of 10 keeps the fluoride sample concentration well above the LOQ. In fact, the Serbian team used a 200-fold dilution (producing a solution of ~380 μM F^-) that made it easier to prepare a homogeneously dispersed solution from the toothpaste.

4.2.2.   Ionic Strength Adjustment Buffer

Beyond dispersing the sample in a uniform aqueous solution, the buffer establishes other important conditions. The first of these conditions is intended to minimize interferences.

Looking at the manufacturer's literature on the selectivity for a fluoride electrode, one finds that the sensor also responds to hydroxide ion, but no other common ions appear to be a problem. Referring to the Nicolsky–Eisenman equation (3.4)

$$E_{memb} = E^{\circ} + \frac{RT}{(-1)F} \ln \left(a_{F^-_{spl}} + K_{FOH}^{POT} \cdot a_{OH^-}^{z_{OH}/z_F} \right) \qquad (4.1)$$

where $z_{OH} = z_F = -1$. The selectivity coefficient [3], K_{FOH}^{POT}, is 0.1 and the lowest fluoride activity that one might want to measure would be the activity associated with the LOD, namely, $a_{F^-_{spl}} = 10^{-6}$. The second term in the argument of the natural logarithm will be negligible as long as the first term is at least 10 times greater than the second term.

$$a_{F^-_{spl}} = 1 \times 10^{-6} > 10 \cdot K_{FOH}^{POT} \cdot a_{OH^-} = 10 \cdot (0.1) \cdot a_{OH^-} = a_{OH^-} \qquad (4.2)$$

Consequently, one should keep the hydroxide concentration at or below a value of 10^{-6} M which is the same as keeping the pH at or below 8. Many electrode manufacturers recommend a slightly acidic acetic acid/sodium acetate buffer of pH 5.5. These ionic strength adjustment buffer solutions (ISAB or "total ionic strength adjustment buffer," TISAB) contain a high level of NaCl that dominates the solution ionic strength. Also, frequently included is a strong chelating agent, such as cyclohexane diamine tetraacetic acid, CDTA, in order to bind metal ions [4]. Iron, magnesium, and aluminum ions can complex fluoride ions reducing the activity of free fluoride ion in solution thereby changing the electrode response. CDTA chelates these cations blocking this effect.

Of course, any other reactivity of the analyte must be considered. For example, it is well known that fluoride will react with silicate in glass [5]. Consequently, all containers used for samples and standard solutions in this study were made from plastic. Silica or alumina abrasives that are found in some toothpaste also scavenge fluoride ions. The adsorption sites on the surface of abrasive particles are more effectively blocked by binding them with organic ligands, such as tartrate, at higher pH [5]. In such situations, a buffer with sodium tartrate and tris(hydroxymethyl)aminomethane at pH 8.2 has been used for fluoride determinations.

4.2.3. Sample Pretreatment

Another consideration is whether the analyte in the sample is in the appropriate form for detection by the sensor. A fluoride ISE responds to the free fluoride ion species, F^-. Toothpaste may contain fluoride in the form of sodium fluoride, sodium monofluorophosphate, Na_2FPO_3, or stannous fluoride, SnF_2. Preliminary heating of the sample in acid (0.2–2 M HCl) is necessary to release the fluoride from Na_2FPO_3 before adjusting the pH. The final pH is adjusted well above 3.13, the pK_a of HF, in order to maximize the fraction of fluoride in the anion form [6].

The third condition that is set by the addition of a buffer is the ionic strength. Any ions present above millimolar levels in the sample will influence the activity coefficient of the analyte ion and, therefore, the response of the sensor. Consequently, the buffer usually contains a noninterfering salt, such as NaCl or KCl, in such a high concentration that it dominates the ionic strength when mixed with the sample. This buffer is prepared separately so that it can be added to each of the sample and calibration solutions in the same proportion adjusting the activity coefficient for the analyte to the same value in all cases.

Although electrode manufacturers offer TISAB solutions for sale, many workers prepare their own. In the work reported by the Serbian team, the TISAB was prepared by dissolving 58 g sodium chloride, 0.3 g sodium citrate, and 57 ml glacial acetic acid in 500 ml of triply distilled water [1]. The pH was adjusted to 5 with 5 M sodium hydroxide and the solution was brought to the mark in a 1000 ml volumetric flask and mixed. The procedure required weighing a 1-g portion of toothpaste into a 100 ml volumetric flask adding triply distilled water and 4 ml of 6 M HCl to the flask before heating it for 10 minutes in a water bath. The solution was cooled to room temperature, was brought to the mark with triply distilled water and mixed (10 drops of propanol were added to reduce foaming). Treated sample solutions were diluted 1 : 1 with TISAB and mixed before measuring. Consequently, the samples were diluted by 200-fold before measuring and the supporting electrolyte concentration was mainly set by the sodium chloride in the TISAB (~0.5 M in the solution measured) [1].

4.2.4. Salt Bridges

The composition of the supporting electrolyte in the sample also needs to be compatible with the reference electrode. If the sensor is a separate ISE, rather than a part of a combination device, one needs to choose an external reference electrode and an appropriate salt bridge between the reference solution and the test solution. The first concern is that any solution that leaks across the salt bridge from reference electrode does not interfere with the measurement of the analyte. Most commercial reference electrodes are based on either a silver/silver chloride or a calomel half reaction. In both of these cases, the reference filling solution (and, therefore, the solution in the salt bridge) usually contains KCl (or sometimes NaCl) solution at a fixed concentration (from 0.1 to 3.8 M for a saturated KCl solution). Fortunately, the fluoride electrode does not respond to chloride activity, so the presence of chloride ion is not a problem. The second question is whether ions crossing the salt bridge from the sample solution could interfere with the reference electrode half-reaction. For example, some anions can complex the reference metal ion or form insoluble salts with Ag^+ or Hg_2^{2+} ions thereby disrupting the equilibrium of the reference half reaction. However, this is also not a problem in this case for either Ag/AgCl or a calomel reference electrodes. Both AgF and Hg_2F_2 are not very stable, so fluoride will not interfere with the equilibrium between either chloride and silver or chloride and mercury for either reference electrode surface. Other sample components, including constituents that can bind silver or mercury (I) ions should also be prevented from reaching the reference compartment. A double junction salt bridge is used in cases where the mixing of small amounts of sample components with the reference electrode are a problem. Special salt bridges are discussed later in this chapter.

Frequently, electrode manufacturers provide sensors that are a combination of the ISE and the external reference electrode in one probe similar in construction to a combination pH electrode. The external reference half-cell used in the combination electrode in this case study included a saturated calomel electrode with a single junction salt bridge [7].

TABLE 4.1 Calculated junction potentials based on the Henderson equation for selected electrolytes in the sample and bridge electrolyte, at 25 °C

Bridge	Sample	E_j (mV)
3 M KCl	1 M CaCl$_2$	−0.28
3 M KCl	1 mM CaCl$_2$	−1.35
3 M KCl	1 M HCl	−16.3
3 M KCl	1 mM NaCl	−1.42
3 M KCl	H$_2$O (pH 7)	−2.80
3 M KCl	0.1 M NaCl	−0.13
3 M KCl	0.1 M Li$_2$SO$_4$	−0.44
0.1 M KCl	0.1 M NaCl	4.39
1 M LiOAc	0.1 M NaCl	−1.82
1 M KNO$_3$	0.1 M NaCl	3.74
1 M LiOAc	1 M HCl	−40.7
1 M KNO$_3$	1 M HCl	−27.6
1 M NaCl	H$_2$O (pH 7)	−2.80

Source: Reprinted with permission from Bakker [8].

It appears that the use of either calomel or silver/silver chloride reference electrode would work for this application.

Choices that the authors of this study made did lead to conditions that created ion gradients in the reference salt bridge [1]. How much of an effect on the junction potential did the TISAB have? If the concentrations of various ions are different on the two sides of the salt bridge, then diffusion will drive ions from the more concentrated region to the less concentrated region. Differences in diffusion coefficients between the cations and anions lead to a charge separation and, therefore, a potential difference across the salt bridge. In this case study, the electrolyte on the reference side was a saturated KCl solution (~3.8 M) and on the sample side, the electrolyte was 0.5 M NaCl. The diffusion coefficients of K$^+$ and Cl$^-$ are very similar, but those of Na$^+$ and Cl$^-$ differ by about 33%. An estimate of the junction potential calculated from the Henderson equation (see Appendix C) is about 0.26 mV in this case [8]. The junction potentials, E_j, are calculated for a few other sample/reference solution combinations and are displayed in Table 4.1.

A disadvantage of using saturated KCl in salt bridges is the frequent problem of KCl crystals forming inside the bridge whenever the temperature drops or evaporation occurs (a problem in long-term storage). These crystals impede ion movement and change the junction potential. Also, chloro-complexes of silver are more soluble in high KCl electrolyte. The silver in such complexes tends to precipitate out (as AgCl) on the far side of the salt bridge where lower chloride concentrations are encountered. This can block a salt bridge. This problem is the probable cause for the majority of failures in cell potential measurements [9].

The clogging of a salt bridge is not always recognizable in early stages. However, as channels in the bridge clog, the potential can drift. Sometimes AgCl solid collected in the bridge can be re-dissolved by a brief soaking of the salt bridge in ammonia. This precipitation problem is less for salt bridges with 0.1–1.0 M KCl in the reference solution.

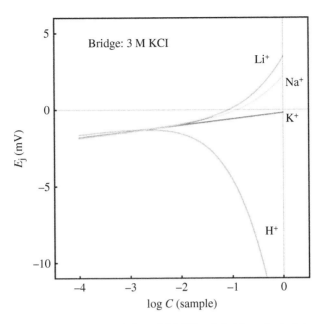

FIGURE 4.1 Calculated junction potential for a 3 M KCl salt bridge electrolyte as a function of the concentration of the sample electrolyte for various types of solutions of chloride salts in the sample. The junction potential is dominated by the reference bridge electrolyte except in the presence of moderate hydrogen ion concentration. Source: Reprinted with permission from Bakker [8].

A popular compromise is the use of 3 M KCl for the reference and bridging electrolyte. As can be seen in Figure 4.1, the junction potential for such a salt bridge is modest and is dominated by the high KCl concentration in the bridge accept at H^+ ion concentrations (above 0.01 M). Lithium acetate (LiOAc) is frequently used as an alternative whenever chloride is a contamination problem for the sample solution.

Choosing ionic components for the TISAB with matching diffusion coefficients helps minimize the junction potential. However, with so many components in the buffer for the toothpaste analysis, eliminating the junction potential was not realistic. Instead, the strategy in this case was to keep the junction potential constant [1]. If the junction potential is the same for all standard and sample solutions, then it does not create an error. The relatively large dilution factor for the sample and high electrolyte concentration of the TISAB made this objective relatively easy to meet.

4.2.5. Calibration

A calibration curve was prepared. The measured cell potential followed the form:

$$E_{cell} = E_1 - 0.0556 \log(a_{F^-,spl}) = E_2 - 0.0556 \log[F^-] \tag{4.3}$$

for an operating temperature of 293 K. In Eq. (4.1), E_2 is a constant that combines the external reference electrode potential, the junction potential, the internal phase boundary potential, internal reference potential, and the activity coefficient for fluoride. In this work, the slope of the calibration curve was −0.0556 instead of −0.0581 that would be expected at 20 °C. Although the slope appears to be low by about 5%, their data were quite repeatable and their calibration curve was linear with little scatter [1].

4.2.6. Temperature Control

Consistency in the measurement conditions is always good, but how tightly regulated do the conditions need to be? How immune the measurement response is to changes in an experimental condition is sometimes characterized as the robustness of a method with regard to that parameter. Temperature is a good example. Should sample and standard solutions be kept in a thermostatted water bath before and during measurements? Temperature affects the coefficient in front of the logarithm term in Eq. (4.1). A change of 1 K at 298.15 K represents only a 0.3% change in slope, but it corresponds to a variation in the calculated concentration of analyte of 1.3%; a change of 5 K corresponds to a change of 6.6% in the calculated concentration.

Likewise, the reference potential of both a saturated calomel electrode and a silver/silver electrode vary by approximately 3–5 mV for every 5 °C change around 25 °C. Furthermore, junction potentials and activity coefficients are affected by temperature. It is difficult to measure the effect on the reference potential independent of the effect on the junction potential. Table 4.2 shows the temperature effect on the measurement of the combination of reference and junction potentials. The data are mainly valuable as a guide to the user for the type of shift that one might see in the reference and junction potentials due solely to a change in temperature. For example, an error of 3 mV in the combined reference and junction potential would correspond to about a 5% error in the calculated fluoride concentration [9]. In the ideal situation, the external and internal reference electrodes are identical in composition so that the changes in voltage due to temperature fluctuations at one reference electrode are exactly offset by changes at the other reference electrode. However, one can expect a change in the junction potential with a change in temperature that can only be estimated and then, only when one has

TABLE 4.2   Reference half-cell potentials plus junction potentials for selected reference electrodes as a function of temperature

Electrode[a]	Reference + junction potential (V) at T (°C)							dE/dT (mV/°C)
	10	15	20	25	30	35	40	
Ag/AgCl @ 3.5 M KCl	0.215	0.212	0.208	0.205	0.201	0.197	0.193	−0.73
Ag/AgCl sat'd KCl	0.214	0.209	0.204	0.199	0.194	0.189	0.184	−1.01
Hg_2Cl_2/Hg @ 0.1 M KCl	0.336	0.336	0.336	0.336	0.335	0.334	0.334	−0.08
Hg_2Cl_2/Hg @ 1.0 M KCl	0.287	—	0.284	0.283	0.282	—	0.278	−0.29
Hg_2Cl_2/Hg @ 3.5 M KCl	0.256	0.254	0.252	0.25	0.248	0.246	0.244	−0.39
Hg_2Cl_2/Hg sat'd KCl	0.254	0.251	0.248	0.244	0.241	0.238	0.234	−0.67

[a]These reference potentials were measured against a standard hydrogen electrode at the same temperature. Reference electrolyte compositions correspond to the KCl concentrations at 25 °C.
Source: Sawyer et al. 1995 [10]. Reproduced with permission of John Wiley & Sons.

an estimate for each of the major ions in the bridging and sample electrolyte solutions. Table 4.2 provides reasonable estimates on the bias introduced by temperature changes.

The test solution and the electrodes should also be at the same temperature. A cold solution and a warm sensor will not give a steady cell potential until the two have reached equilibrium. The level of uncertainty associated with a 1° variation was considered acceptable for this case study; no temperature regulation was employed other than allowing the solutions to stand in the same lab for at least an hour. Good practice also dictates that in order to avoid problems with temperature variation, the standard solutions should be measured within a few minutes of recording measurements for the sample solutions. In the case of the analysis of many samples, the standards should be checked at frequent intervals between testing a group of sample solutions [11]. Stirring the solution during the measurement helps mix the sample and TISAB solutions, minimizes variations in temperature, and also helps maintain a steady concentration gradient and junction potential at the salt bridge. Since heat generated by electric motors in stirring plates can inadvertently heat the solution, many analysts use a piece of corrugated cardboard or foam sheet between the test beaker and the stirrer to isolate it thermally [12].

4.2.7. Signal Drift

Most ISE probes exhibit some drift in potential in the same direction after inserting the electrodes into a new solution even when temperature is not a problem. This slow approach to a steady value may continue for several minutes. The drift is often more serious in solutions of low analyte activity. Nevertheless, good precision and accuracy (and linear calibration curves) are usually obtainable without waiting for the complete stabilization of the signal. The key lies in consistently recording the signal at a fixed time interval from the moment at which the sensor is placed into the test solution. The authors actually studied the influence of varying the waiting time between immersing the electrode and measuring the signal and concluded that two minutes was optimum [1]. The source of this sort of potential drift is often attributed to changes in the junction potential. Obstruction of ion pathways in the salt bridge or changes in concentration gradients of charge carriers may be involved [13].

4.2.8. Validating the Method

Good practice in analytical chemistry involves validating any new method. Validation means demonstrating that the method measures accurately what it claims to measure. One approach to validation is to perform an analysis on a well-characterized sample provided by a reputable outside group. The US National Institute of Standards and Technology (NIST) is one source for a wide range of materials with certified content of selected analytes. However, certified samples were not used in this study. In this case, the validity of this sample treatment and calibration procedure was tested by performing "recovery assays" [1]. These experiments checked for matrix effects, transfer losses, or any procedural step that was not 100% efficient in extracting the fluoride in the toothpaste into the test solution. These tests were performed by taking two equal portions of sample and spiking one of them with a known amount of standard Na_2FPO_3 salt before this mixture was treated. This salt dissociates upon heating to yield a predictable increase in fluoride concentration. Usually, the amount of analyte added in this type of test is chosen so that it approximately doubles the original quantity of analyte in the sample. The concentrations of the analyte in both the spiked and the normal sample solutions were calculated from the equation fitting the cell potential data to the calibration curve. In this case study, the authors used the calibration curve to calculate the apparent concentration of the sample solution before and after spiking it with standard fluoride. The ratio of the apparent increase in fluoride concentration to the actual increment in concentration based on the amount of standard that was added was used to calculate a percent recovery.

$$\text{Percent recovery} = \left(\frac{C_{\text{spiked spl}} - C_{\text{spl}}}{C_{\text{added}}} \right) \cdot (100\%) \qquad (4.4)$$

where $C_{\text{spiked spl}}$ represents the estimated concentration of fluoride read from the calibration curve after spiking the sample with standard, C_{spl} was the concentration corresponding to the cell potential of the original sample solution according to the calibration curve, and C_{added} was the actual increase in mol/l of fluoride in the solution calculated from the known moles of standard added to the spiked sample.

The authors demonstrated that multiple, spiked samples yielded an average percent recovery of 97% [1]. The relative standard deviation for the percent recovery was similar to the relative standard deviation for repeated measurements of the sample solution. This outcome indicated that the process measures what it was intended to measure and that the scale being applied was accurate. Had the average percent recovery been significantly different from 100%, a flaw in the procedure or a problem with the sensor would be indicated. In some circumstances, a percent recovery below 95% is considered unacceptable.

The precision in the data was also characterized in terms of the measurement repeatability – the relative standard deviation of cell potentials for the same standard solution measured throughout the same day – and the reproducibility – the relative standard deviation of cell potentials for the same standard solution measured between days. Table 4.3 shows the repeatability and reproducibility for three widely different levels of fluoride concentration. Note that the relative errors were smaller for higher fluoride levels.

TABLE 4.3   Precision for fluoride measurements in case study

Fluoride (mg/l)	Repeatability (intra-day precision) RSD (%)	Reproducibility (inter-day precision) RSD (%)
1	0.46	0.75
5	0.02	0.04
50	0.0001	0.0002

Source: Reprinted with permission from Švarc-Gajića et al. [1]. Copyright 2013, Elsevier.

EXAMPLE 4.1

Consider the calibration curve in Figure 4.2 for some hypothetical data of standard solutions made with Na_2FPO_3 prepared in TISAB. The graph indicates that the response function is

$$E_{cell} = -0.274 - 0.0580 \log[F^-] \tag{4.5}$$

where $[F^-]$ is the fluoride concentration in mol/l.

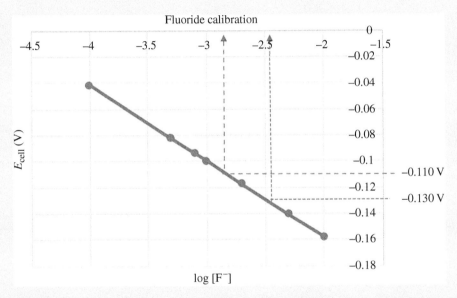

FIGURE 4.2 Calibration plot for the cell potential using a fluoride ISE and standard solutions prepared from Na_2FPO_3 and TISAB. The fluoride concentration data were in units of mol/l.

A percent recovery experiment was used to assess the accuracy of the procedure for a type of toothpaste sample. Two equal portions of toothpaste were treated by the protocol described. The cell potential for the sample (no spike) was −0.110 V, which corresponds to a concentration of 1.76×10^{-3} M fluoride on the standard curve.

$$-0.110\,V = -0.274 - 0.0580 \log[F^-] \tag{4.6}$$

or

$$[F^-] = 10^{\left\{ \frac{0.110-0.274}{0.0580} \right\}} = 1.76 \times 10^{-3} \tag{4.7}$$

A sample of toothpaste was spiked with standard Na_2FPO_3 and treated in the same manner. The increase in concentration based on the mass of Na_2FPO_3 added was calculated to be 1.80×10^{-3} M (that is, $C_{added} = 1.80 \times 10^{-3}$ M). The spiked solution was measured with the same ISE and gave a cell potential of −0.130 V. From the calibration curve, the corresponding concentration for the spiked solution was 3.87×10^{-3} M fluoride.

$$[F^-] = 10^{\left\{\frac{0.130-0.274}{0.0580}\right\}} = 3.87 \times 10^{-3} \tag{4.8}$$

What was the value for the percent recovery based on this experiment? Using the definition in Eq. (4.2),

$$\text{Percent recovery} = \left(\frac{C_{\text{spiked spl}} - C_{\text{spl}}}{C_{\text{added}}}\right) \cdot (100\%) = \left(\frac{3.87 \times 10^{-3} - 1.76 \times 10^{-3}}{1.80 \times 10^{-3}}\right)(100\%) = 117.2\% \tag{4.9}$$

The result for the percent recovery calculated in the exercise above implies a relatively large error. What is a reasonable course of action? Before drawing conclusions, the recovery experiment should be repeated a few more times to obtain an average percent recovery and a standard deviation for that number so that one can apply a t-test in order to decide whether a significant bias is indicated or whether the discrepancy is the result of badly scattered data. Imagine that four replicate spiked samples gave percent recovery values of $117._2\%$, $120._3\%$, $109._4\%$, and $116._1\%$. The average percent recovery and standard deviation are $115._{75}\%$ and 4.59%, respectively. Applying a t-test comparing the average to an accepted value of 100% for $n = 4$ gives an experimental t-value of 6.86 [14].

$$t_{\text{exp}} = \frac{|\text{average} - \text{accepted value}|}{\text{standard deviation}}\sqrt{n} = \frac{|115.75 - 100|}{4.59}\sqrt{4} = 6.86 \tag{4.10}$$

Because there was no expectation of a bias in a particular direction, a two-tailed t-test at the 95% probability level should be applied. The table value for the t-value at 3 degrees of freedom is 3.182. Thus, $t_{\text{exp}} > t_{\text{table}}$ and a significant difference is indicated [14].

The poor recovery results illustrated in Example 4.1 (a relatively large error of 17% in this case) could have been caused by a drift in the junction potential, for example, of a mere 4 mV. The important lesson is this: good analysis depends on maintaining the maximum reading precision in recording data (avoid rounding measurements) and in carefully matching the pH, ionic strength, temperature, and other matrix conditions as much as possible among the standard and sample solutions.

There are situations where matching the conditions between the standard and sample solutions is impractical. Fortunately, in those cases, the method of calibrating the electrode response based on spiking the sample is a valid way of determining the concentration of an analyte in a sample. This procedure is known as the method of standard additions. The method of standard additions is commonly used in spectrophotometry and many other instrumental analyses, but it is more challenging for potentiometric analysis since the signal is not a linear function of the analyte concentration [15].

4.2.9. Standard Additions for Potentiometric Analysis

Gunnar Gran suggested a way of expressing the relationship between the measured potential and the analyte concentration in terms of a function that could be treated linearly. Gran's technique applies to a standard additions experiment as well to any titration where a reactant or product is monitored potentiometrically. Consider here the case of a standard additions experiment in which an ISE for the analyte is being used to monitor cell potentials. (A pH titration is considered later in this chapter.) The cell potential is shown in Eq. (4.11).

$$E_{cell} = E_o + \frac{2.303RT}{z_x F} \log(a_x) + E_j = E_o + E_j + \frac{2.303RT}{z_x F} \log(\gamma_x) + \frac{2.303RT}{z_x F} \log(C_x) \quad (4.11)$$

where E_j is the junction potential, a_x the activity of the analyte, x, C_x is the concentration of the analyte, and E_o is the cell potential for $a_x = 1$ and no junction potential. γ_x and z_x are the activity coefficient, and the charge, respectively, for the same ion. Collecting the constant terms together, Eq. (4.11) becomes

$$E_{cell} = E_{cnst} + \frac{2.303RT}{z_x F} \log(C_x) \quad (4.12)$$

For an unknown concentration of analyte, C_o, in an initial volume, V_o, to which a standard solution of concentration, C_s, in a volume, V, is added the new concentration of analyte becomes

$$C_x = \frac{C_o V_o + C_s V}{V_o + V} \quad (4.13)$$

Substituting into Eq. (4.12) gives

$$E_{cell} = E_{cnst} + \frac{2.303RT}{z_x F} \log\left(\frac{C_o V_o + C_s V}{V_o + V}\right) = E_{cnst} + \frac{1}{m} \cdot \log\left(\frac{C_o V_o + C_s V}{V_o + V}\right) \quad (4.14)$$

where, ideally, $m = \frac{z_x F}{2.303RT}$. Equation (4.14) can be rearranged to give the following equation:

$$(V_o + V) \cdot 10^{mE_{cell}} = (C_o V_o + C_s V) \cdot 10^{mE_{cnst}} \quad (4.15)$$

or

$$(V_o + V) \cdot 10^{mE_{cell}} = \left(\frac{C_o}{C_s} V_o + V\right) \cdot 10^{mE_{cnst}} \quad (4.16)$$

A plot of the left-hand side, $(V_o + V) \cdot 10^{mE}$, versus V yields a straight line that intersects the x-axis at a negative value of V for which $|C_s V| = C_o V_o$. To make this more concrete, Figure 4.3 shows a Gran plot for a standard additions experiment for fluoride ions using a fluoride ISE [16]. The volume at the x-intercept is marked as V_e. Hence, the original analyte concentration can be calculated from

$$C_o = \left|\frac{C_s V_e}{V_o}\right| \quad (4.17)$$

Gran's method has two advantages over comparing the cell potential for a sample directly to a calibration curve. First of all, the sample matrix may be too complicated to mimic accurately with the standard solutions. The difficulty of matching the matrix of the standard solutions and the sample can lead to discrepancies in the junction potential and the activity coefficients. Both of these influence the constant term, E_{cnst}. This term may

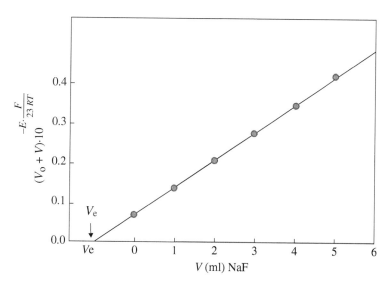

FIGURE 4.3 Gran's method for a standard additions experiment. Fluoride standard was added in 1 ml increments to an unknown sample and the function of $(V_o + V) \cdot 10^{mE}$ was plotted versus the added volume, V, of standard NaF, where $m = \frac{Z_f F}{2.303RT} = \frac{-F}{2.303RT}$. The magnitude of the x-intercept, V_e, represents the volume of standard that would contain the same number of moles of fluoride as in the original sample volume, V_o. Source: Adapted with permission from Liberti and Mascini [16]. Copyright 1969, American Chemical Society.

not be the same for the samples and the standard solutions. The matrices of the spiked and sample solutions are likely to be more similar and E_{cnst} is more likely to be constant for the two. The other advantage is that several measurements go into determining the concentration of the analyte with Gran's method. Consequently, the uncertainty in the calculated concentration is smaller for this determination.

One of the drawbacks of the approach used in Figure 4.3 is the assumption that one knows the value of m in the coefficient of the log-term in Eq. (4.14). The coefficient commonly differs somewhat from the ideal value of $(2.303RT)/(z_x F)$. Of course, this term can be evaluated from the slope of a calibration curve, but can one assume that the slope remains the same in the sample matrix? A more rigorous approach is to use the standard additions data to solve for the coefficient, $1/m$, as well as finding the original sample concentration, C_o [15]. Equation (4.16) was proposed by Gran, because it was a linear relationship that could easily be fit graphically or by using a least squares regression method. Fortunately, personal computers can handle nonlinear least squares where both m and the original sample concentration are determined as fitted parameters. A common approach is to re-express Eq. (4.14) as follows:

$$E_N = E_{cnst} + \frac{1}{m} \cdot \log \left(\frac{C_o V_o + C_s V_N}{V_o + V_N} \right) \tag{4.18}$$

where E_N and V_N are the cell potential and added volume of standard solution after the Nth spike of the sample. Note that the cell potential for the original (unspiked) sample is given by Eq. (4.19).

$$E_o = E_{cnst} + \frac{1}{m} \cdot \log(C_o) \tag{4.19}$$

Subtracting Eq. (4.19) from Eq. (4.18) yields:

$$(E_N - E_o) = \frac{1}{m} \log \left(\frac{C_o V_o + C_s V_N}{V_o + V_N} \right) - \frac{1}{m} \log(C_o) = \frac{1}{m} \log \left\{ \frac{C_o V_o + C_s V_N}{(V_o + V_N)C_o} \right\}$$

$$= \frac{1}{m} \log \left(\frac{V_o + K V_N}{V_o + V_N} \right) \tag{4.20}$$

where $K = C_s/C_o$. For every sample and spiked solution all of the terms in Eq. (4.20) are known except for m and K. These two terms and the uncertainties associated with them are readily obtained by fitting data using a nonlinear regression program. A full discussion of the fundamentals and implementation of nonlinear regression can be found in Refs. [17–19].

4.3. CASE II. A CLINICAL APPLICATION

In the previous case study, the sample matrix was a commercial product with a composition that is unlikely to vary much from sample to sample. Clinical samples represent a different type of challenge. Blood and urine samples are complex sample matrices and can vary in important ways among patients. This second case study examines the development of a sensor for creatinine. Creatinine is the end product of creatine metabolism in the muscles (see Figure 4.4). It is normally excreted at a relatively steady rate unless there is a problem with the kidneys [20].

   Measuring the creatinine levels in the blood or urine is part the process of assessing the health of one's kidneys. The complete assessment accounts for an individual's age, gender, and weight in the calculation of a diagnostic figure called one's glomerular filtration rate [21]. Roughly speaking high levels of creatinine in the blood or low levels in urine correlate with kidney disease or damage. The most widely used method for determining

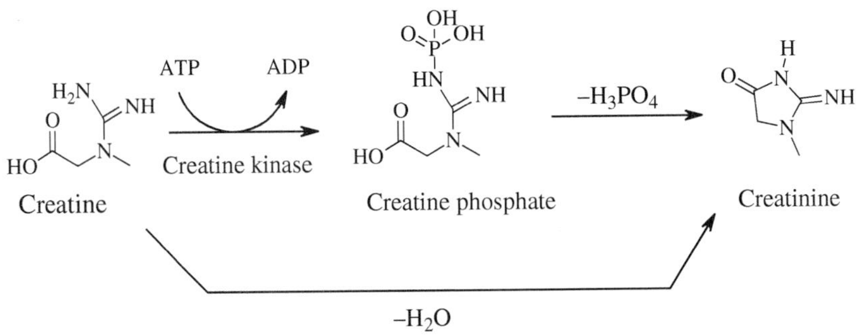

FIGURE 4.4 Two pathways for conversion of creatine to creatinine, the enzymatic route via the formation of creatinine phosphate and the nonenzymatic dehydration of creatine as it occurs in the muscles.

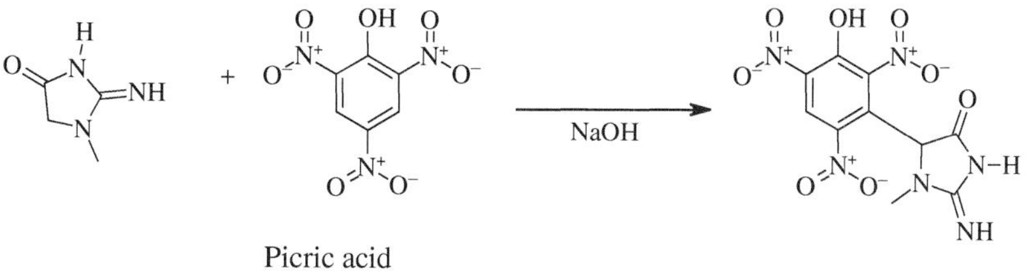

Picric acid

FIGURE 4.5 Jaffé reaction traditionally used to determine creatinine levels in blood and urine. The product absorbance is measured at 505 nm. Timing is important in the spectroscopic assay because the product is not stable [21].

creatinine concentration is a colorimetric technique based on the Jaffé reaction [20] as shown in Figure 4.5.

Among the drawbacks to this method are the fact that picric acid is hazardous. (It is a contact explosive when dry.) In addition, the colored product that is monitored is not stable. Consequently, the time interval between the moment of mixing the reagents with the sample and the measurement of the colored product must be tightly controlled. The timing is not a challenge for an automated (or flow injection) apparatus, but that implies more expensive equipment.

In 2017, a research group in Spain demonstrated an ISE method that successfully determined clinically relevant concentrations of creatinine in urine [22]. The sensor was based on a liquid membrane containing a phosphonate-bridged calix[4]pyrrole ionophore (see Figure 4.6). An X-ray crystal structure of the creatinine/ionophore complex shows the carbonyl oxygen on the creatinine in close proximity to the four hydrogen atoms on the pyrrole nitrogen atoms of the ionophore suggesting hydrogen bonding between the creatinine and the ionophore. In addition, the hydrogen atoms of the methylene group on the creatinine ring appear to be interacting with the π electrons of the two neighboring phenyl groups on the ionophore. Although this X-ray structure was recorded for the complex of the neutral creatinine molecule with the ionophore, the structure of the protonated creatinine (the creatininium cation) and the ionophore complex is thought to be similar. The formation constant for the creatininium/ionophore complex has been estimated to be 4.0×10^6 [22].

Of course, the sensor membrane develops a voltage only for the creatininium ion and not the neutral form. Consequently, controlling the pH of the sample solution is crucial to the formation of a signal. Figure 4.7 shows the relative sensitivity of the sensor (as measured by the potential response to a 1 mM creatinine solution) as a function of pH. The pK_a for the creatininium ion is 4.8, so 90% of the creatinine should be in the protonated form at pH 3.8, where the maximum sensitivity is indicated. In a buffer of pH 3.8, the sensor has a LOD of about 6×10^{-7} M creatininium ion [22].

One of the challenges to working with urine samples is the high level of electrolytes that might interfere. Some possible interfering cations are sodium, potassium, ammonium, and calcium ions. These are typically present at 0.14 M, 4 mM, 3–65 mM, and 5 mM, respectively, in urine samples. Of course, the selectivity coefficient combined with the concentration of the cation determine the level of interference. Table 4.4 summarizes the selectivity

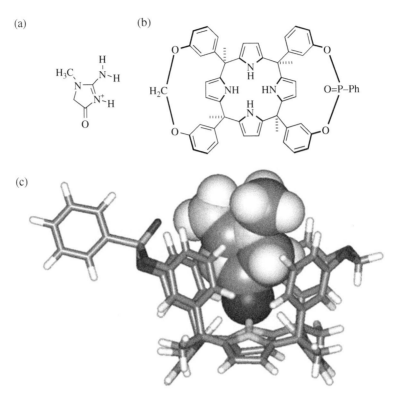

FIGURE 4.6 (a) Structure of creatininium ion, (b) Structure of calix[4]pyrrole ionophore, and (c) X-ray crystal structure of the creatinine/ionophore complex. The complex containing the charged form of the creatinine is thought to be similar. Source: Reprinted with permission from Guinovart et al. [22]. Copyright 2017, Elsevier.

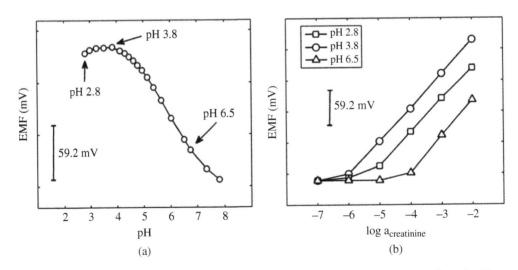

FIGURE 4.7 (a) Plot of E_{sensor} versus solution pH. The data show the sensor voltage for a 1 mM analytical concentration of creatinine versus pH. (b) Calibration plots for standard creatinine solutions at three different pH values. Source: Adapted with permission from Guinovart et al. [22]. Copyright 2017, Elsevier.

TABLE 4.4 Selectivity coefficients for cations interfering with the creatinine ISE and threshold creatinine levels where typical activity of these cations in urine samples leads to a 10% error

Interferant	Selectivity coefficient	Typical urine concentration (mol/l)	Creatinine $(a_{C, min})$ (mol/l)
Sodium	2.00×10^{-4}	0.16	3.19×10^{-4}
Potassium	3.16×10^{-3}	0.06	1.90×10^{-3}
Ammonium	5.01×10^{-3}	0.035	1.75×10^{-3}
Calcium	1.58×10^{-5}	0.005	7.92×10^{-7}

Source: Guinovart et al. 2017 [22]. Reproduced with permission of Elsevier.

constants that the Spanish group determined for these cations [22]. The table also indicates a minimum creatininium activity that the sensor can measure in the presence of the typical urine level of the interfering ion.

These threshold values were defined as levels where the measurements would be subject to a 10% error. Fortunately, normal creatinine levels in urine are much higher than the concentrations where the interference is a concern. Although blood levels are only 74.3–107 μmol/l (https://www.mayoclinic.org/tests-procedures/creatinine-test/about/pac-20384646), creatinine in urine is normally present in the range of 8.8–17.6 mM for men and 7.0–15.8 mM for women [23].

Renal tests are often based on the total amount of analyte excreted over a 24-hour period. The typical concentrations for these interferants were calculated from the average accumulation of that ion for a 24-hour period divided by an average volume of urine excreted in a 24-hour period.

Now consider the choice of an external reference electrode. In many cases, either a calomel or a silver/silver chloride electrode can be used as long as the salt bridge electrolyte and sample solution are compatible. In this application, the authors used a Ag/AgCl reference electrode with a 3 M KCl electrolyte. One issue to consider in choosing the type of reference electrode is the possible crossover of solution components between the sample and reference solutions. Many biochemically important compounds, such as proteins and DNA and the common buffer component, tris-hydroxymethyl aminomethane (known commercially as "Tris" or "THAM") bind silver and mercury ions. (Although Hg_2^{2+} ions do not form stable complexes with many organic ligands, this ion will disproportionate to form Hg^{2+} and Hg^0 because of the strong affinity of the Hg^{2+} for sulfhydryl and amino groups found in proteins and other biochemical species.) Sulfide ion can also be a problem in some environmental samples [24]. Some crossover does occur. Sample solution components that reach the reference solution might shift the potential of the reference electrode by changing the activity of the Ag^+ or Hg_2^{2+} or by adsorbing to the electrode surface. Of course, ions from the reference chamber might have detrimental effects on the sample solution as well. Another common complication is a reaction between reference and sample components leading to the formation of a precipitate inside the pores of the salt bridge itself. This can cause a change in diffusion through the salt bridge shifting the liquid junction potential.

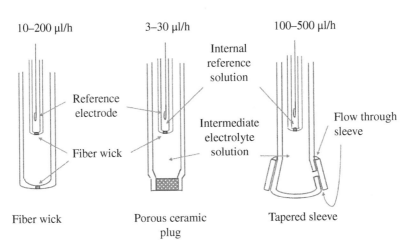

10–200 µl/h   3–30 µl/h   100–500 µl/h

Internal reference solution

Reference electrode

Intermediate electrolyte solution

Flow through sleeve

Fiber wick

Fiber wick   Porous ceramic plug   Tapered sleeve

FIGURE 4.8 Various types of double junction salt bridges. The additional chamber helps isolate the reference electrolyte from the sample. Typical electrolyte flow rates are indicated above each device. The higher volume flow rate helps to keep the junction open, but it also leads to more electrolyte leaching into the sample [25].

A urine sample is a complicated matrix with many species that might bind Ag^+. The best approach for avoiding incompatibilities between the sample and reference solutions is to use a double junction salt bridge. A double junction inserts an additional electrolyte chamber between the sample and the reference electrode as shown in Figure 4.8. The electrolyte chosen for the intermediate compartment is often potassium chloride, potassium nitrate, or lithium acetate, because the diffusion coefficients of the cations and anions in these salts are very similar to each other and that helps to minimize the junction potential. In this case study, the authors chose to use a 1 M lithium acetate solution for the intermediate electrolyte.

The salt bridges in Figure 4.8 have different flow rates for the volume of electrolyte that leaks from the inside chamber through the salt bridge to the sample [25]. Higher flow rates, such as that associated with a tapered sleeve salt bridge, are usually recommended for working with samples containing high levels of particulates or oily suspensions such as food products or soil slurries. Particulates can clog the salt bridge, but the problem lessens with higher rate of flow out of the reference chamber. The tapered sleeve double-junction salt bridge was used in this case study.

Preliminary research was necessary in order to develop the experimental protocol. A set of many authentic urine samples with a range of creatinine levels were first determined by the Jaffé method. Then, portions of each urine sample were diluted by different ratios with the pH 3.8 buffer (1 : 2, 1 : 10, and 1 : 100), and the creatinine levels in the diluted samples were measured with the sensor [22]. The signal from the diluted sample was compared to the standard calibration curve for the ISE. The results from the samples that were diluted 100-fold agreed with the results from the Jaffé method (after accounting for dilution). In the urine samples that were diluted with less buffer, the cell potential appeared to drift excessively indicating that something in the matrix was affecting the sensor membrane or the liquid junction potential (salt bridge). In contrast, the sensor had relatively little drift in samples diluted by 100-fold. Consequently, the 100-fold dilution factor was adopted for all

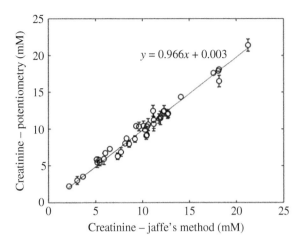

FIGURE 4.9 ISE method validation. Comparison of results for creatinine levels in 50 urine samples measured by the ISE and measured by the Jaffé reaction. The small degree of scatter and the closeness of the slope to unity indicate that the sensor response correlates well with the creatinine concentration established by the accepted method. Source: Guinovart et al. 2017 [22]. Reprinted with permission of Elsevier.

urine samples. The buffer used for this purpose contained 50 mM acetic acid and 50 mM magnesium acetate with the pH adjusted to 3.8. After dilution, a normal 24-hour average sample falls in the range of 70–180 μM creatinine. Of course, the potassium, ammonium, and other possible interferants in a urine sample also drop in concentration by 100-fold and are not a problem [22].

In order to validate their method, the workers in this study compared results from the ISE data with concentrations determined by the Jaffé method. Fifty samples were collected from patients that exhibited creatinine levels across the entire normal range. Each sample was split so that each sample could be determined by the Jaffé method and by the sensor. The results for the concentration as measured by the sensor were plotted against the value determined by the Jaffé method. Perfect agreement would yield a slope of 1.0. The observed slope was 0.966. Figure 4.9 shows that the agreement between the two methods was good over the entire concentration range.

One of the attractive aspects of ISEs is the fact that they are amenable to automated methods of analysis for situations where high sample throughput is important, such as in clinical laboratories. The most common clinical analytes that are regularly determined in blood and urine samples by ISE are K^+, Na^+, CO_2, Cl^-, Mg^{2+}, and Li^+. Of course, pH is also a major characteristic. Because of its importance to many fields, the calibration and maintenance of pH electrodes is discussed in a separate section later in this chapter.

4.4. CASE III. ENVIRONMENTAL APPLICATIONS

This section begins with a method recommended by the US EPA for determining nitrate ion in natural water samples by an ISE and then discusses how those principles were adapted in order to perform field measurements and continuous monitoring.

There are some moderate health concerns about nitrate and nitrite levels in drinking water [26]. Infants under six months of age appear to be particularly sensitive to nitrate-contaminated drinking water. The danger appears to be related to changes in the oxygen-carrying capacity of hemoglobin, although bacterial infection may have also played a role in increased sensitivity of infants in reported incidents where children were exposed to nitrate-contaminated drinking water. The International Agency for Research on Cancer (IARC) indicates that there is little evidence that nitrate itself is a carcinogen, but that under conditions where nitrate can be reduced to nitrite in the stomach which, in turn, can react with amines to form N-nitroso compounds, the risk of cancer is elevated [26].

Low levels of nitrates appear naturally in fruits, vegetables, cereals, and dairy products. Nitrate is also added to some types of food, such as cured meat, as a preservative. Unusually high ingestion of nitrate can cause a decrease in blood pressure, reduced ability to get oxygen to tissues, headache, cramps, vomiting, and even death. Consequently, the US Food and Drug Administration (FDA) set an upper limit for nitrate in bottled water at $10\,mg/l$ ($1.6 \times 10^{-4}\,M$). The EPA uses the same concentration for the maximum contaminant level for nitrate in well water [26].

Nitrate measurements are probably most valuable as an indicator of human-caused contamination. Background nitrate concentrations in natural water systems average around $1\,mg/l$ [27]. Elevated levels are very likely caused by agricultural or residential run-off of fertilizer or industrial pollution [27, 28].

4.4.1. US EPA Method for Nitrate Determination by ISE

There are different strategies for measuring nitrate using an ISE, and it is a valuable exercise to think about the special challenges with each approach. For example, one might be concerned with testing well water samples for nitrate ion. In this case, the most common procedure involves taking a small number of "grab" samples from a well and transporting them back to the lab for analysis. The US EPA publishes documents that address all of the appropriate steps in this sort of process including recommended methodology for the determination of nitrate ion by ISE [29].

There are situations where it might be appropriate to take samples at various depths in the water column in order to obtain a representative average for the system or in order to map variations with depth. In contrast, for sampling water that is already plumbed into a house, it is common practice to run the water for five minutes in order to flush out the pipes before filling a sample bottle. Sample containers must be clean and prepared in a manner that ensures no change in the analyte concentration will occur during storage or transport of the sample to the lab. Generally, fluorocarbon polymer bottles are less likely to adsorb analyte ions or desorb contaminants into the solution than glass bottles. The EPA method does not specify the type of container to be used, because adsorption of nitrate ions to container walls is less of a problem than for other analytes such as trace concentrations of metal ions.

Nitrate ions are not easily converted to another species by protonation or by forming complexes or insoluble salts. However, nitrite ions can be oxidized during sample storage to nitrate ions. Nitrite is usually eliminated from the sample by adding 1 ml of 0.1 M sulfamic acid to 100 ml of the solution at the time of sampling [29].

$$NaNO_2 + HSO_3NH_2 \rightarrow N_2 + NaHSO_4 + H_2O \tag{4.21}$$

The most likely pathway for nitrate loss involves bacteria that use the nitrate to metabolize organic matter. In those cases, nitrate is converted to either nitrite or molecular nitrogen. A precaution against nitrate loss by microbial activity involves either filtering samples through membranes with pores $\leq 0.2\,\mu m$ (to remove bacteria) or adding boric acid as a preservative. Filtering of samples is sometimes done to remove fine particulates known as colloids. Some particulates act as ion exchangers removing analyte ions in a nonspecific manner. Protein colloids that can clog salt bridges or adhere to the sensor surface are a concern in food or biological samples. Organic colloids are not a likely problem in a well water sample. However, deep well water might be depleted of O_2. In some cases, anoxic water contains modest levels of soluble Fe^{2+} species. These ferrous ions oxidize easily in the presence of dissolved oxygen once the water contacts air. The result forms hydroxides in a gelatinous precipitate that can carry other ions out of solution. Adding 1 ml of 1 M boric acid solution per 100 ml of sample at the time of sampling and storing samples at $\leq 6\,°C$ lowers the pH to about 4, keeping iron hydroxides from precipitating. The lower pH also protonates any cyanide, bisulfide, bicarbonate, carbonate, and phosphate that might be present removing them as possible interferants. Furthermore, boric acid acts as a preservative discouraging bacteria. However, even when using a preservative, the EPA method recommends that samples be analyzed within 48 hours of sampling [29].

Well water, generally, has a low ionic strength. Accordingly, an ionic strength adjustment (ISA) is recommended for maintaining a consistent activity coefficient. For that purpose, 1 ml of ISA solution consisting of 2 M $(NH_4)_2SO_4$ should be added per 50 ml of sample and mixed well at the time of measuring – for a net sulfate concentration of 0.04 M.

For decades, commercial nitrate ISEs have been based on liquid membranes incorporating very hydrophobic quaternary amines, such as tridodecylmethylammonium nitrate (TDMAN). These membranes use no ionophore [27]. The quaternary amine acts as an ion exchange site that favors those anions that are more easily extracted from water. That is, unlike an ionophore, the amine does not provide any specific interaction with the anion to make the membrane more selective to a specific anion. Consequently, any anion is a candidate for pairing with the quaternary amine cation in the membrane. The anion that will be favored will be the one that requires the least amount of energy to strip the water from its solvation sphere upon crossing into the hydrophobic phase. The relevance here is that the possibility for interference from other anions increases with the lipophilic nature of the interfering anion. The competitive nature of common inorganic ions follows a regular trend frequently called the Hofmeister series [27, 30]: $ClO_4^- > SCN^- > ClO_3^- > I^- > NO_3^- > Br^- > Cl^- > F^- > HCO_3^- > SO_4^{2-}$. This series comes from studies made by Franz Hofmeister over 130 years ago on the effect of various inorganic salts on the solubility of proteins. Anions appear to have a stronger effect than cations, but both influence the precipitation of proteins. Adding salts to an aqueous solution has been used as a method for removing protein from solution by a process known as "salting out." The strength of the precipitating agent follows the reverse order to that given above – namely, sulfate precipitates protein the most effectively. Sulfate ions

TABLE 4.5   Selectivity coefficients for nitrate liquid ISE based on TDMAN ion exchange and no ionophore

Interferant	$K_{NO_3J}^{POT}$	Interferant	$K_{NO_3J}^{POT}$	Interferant	$K_{NO_3J}^{POT}$
Perchlorate	10 000	Nitrite	1.4	Phosphate	0.02
Iodide	200	Bisulfide	1	Acetate	0.005
Chlorate	20	Bicarbonate	0.1	Fluoride	0.001 6
Cyanide	10	Carbonate	0.05	Sulfate	0.001
Bromide	1.4	Chloride	0.03		

Values calculated for the Mettler Toledo DX262-NO3 ISE from data in Ref. [31]. Selectivity values vary in the literature. For example, the EPA reports constants for some interferants for an Orion nitrate ISE that are smaller [29] $K_{NO_3J}^{POT} = 4.0$, $K_{NO_3Br}^{POT} = 6.3 \times 10^{-2}$, $K_{NO_3Cl}^{POT} = 1.3 \times 10^{-3}$.

also compete most strongly with proteins for solvent molecules. However, the salting out mechanism is more complicated than that. The influence of ions on the hydrogen bonding of water with carbonyl groups in the protein backbone also plays an important role in salting out [30]. The Hofmeister series indicates that nitrate ions require more energy to desolvate than perchlorate, chlorate, and thiocyanate. Consequently, the membrane responds more strongly to perchlorate, chlorate, and thiocyanate than to nitrate. The fact that these three ions are not commonly found above trace levels in natural water systems makes it possible to use this membrane for nitrate determination.

The selectivity coefficients for a variety of inorganic anions are shown in Table 4.5 [31]. The presence of moderate levels of halide ions often requires their removal before measurement with this sensor. Even though the membrane is more selective for nitrate than chloride, higher concentrations of chloride can interfere. The addition of silver sulfate, Ag_2SO_4, (usually at a ratio of 1 ml of 0.05 M Ag_2SO_4, per 50 ml of sample solution) is recommended for removal of most interferants (other than perchlorate and chlorate) by precipitation of their silver salts [29]. Notice that even though the sensor is only weakly sensitive to sulfate, using sulfate in the ionic strength adjusting buffer limits the detection of nitrate ion to about 4×10^{-5} M.

$$\left(a_{SO_4^{2-}} \cdot K_{NO_3SO_4}^{POT} = 0.04\, M \cdot 1 \times 10^{-3} = 4 \times 10^{-5}\, M \right)$$

Whenever significant levels of organic carboxylic acids are present, they can also be a problem. An ISA buffer containing $Al_2(SO_4)_3$, in addition to $(NH_4)_2SO_4$, can be used to complex the carboxylates with aluminum blocking their interference [31].

Because the ISE responds to chloride, a double-junction salt bridge with an outer electrolyte of 0.1–1 M lithium acetate is recommended to minimize contamination of the sample by leakage from the reference electrode [29]. The preparation of standards and all measurements use polyethylene containers. New ISE membranes are pretreated by storing the sensor for at least one hour in 100 mg/l ($\sim 1.6 \times 10^{-3}$ M) nitrate standard solution. Overnight storage is usually in the same solution.

Calibration of the ISE is performed using a series of standard solutions made fresh daily from reagent grade (dry) KNO_3 salt in deionized (18 MΩ) water. Before measurement, 0.5 ml preservative and 1 ml of ISA are mixed with 50 ml portions of each standard solution. In addition, a method blank and four separate control solutions are prepared [29]. If the samples are filtered, the associated blank and other control solutions must also be filtered. A method blank provides a check for nitrate contamination in the reagents. It is prepared by mixing 1 ml of boric acid preservative solution with 100 ml deionized (18 MΩ) water and 2 ml ISA solution. The blank is tested with the sensor and compared with the calibration curve. If the signal indicates a level of 1 mg NO_3^-/l (1.6×10^{-5} M) or more, the level of contamination is unacceptable. This blank should also be tested after every 10 samples, in order to flag problems with any carry over between solutions. Two "calibration verification standards" should also be prepared [29]. One of these solutions is made *from a different source* than the calibration standards and should have a nitrate level in the middle of the standard concentration range. This solution is called the "initial calibration verification standard" (ICV) and is tested immediately after the calibration standards are measured. The second of these controls is the "continuing calibration verification standard" (CCV). It is made from the same stock solution that the calibration standards are made from and should be analyzed after every 10 samples are measured. The corresponding nitrate concentration should be calculated at the time that these solutions are tested. The concentration calculated from the reading should be within 10% of its actual concentration. If not, then the source of the error should be corrected. A common issue is excessive signal drift. Remeasuring the concentration of nitrate in each of the standard solutions can indicate whether the sensor has drifted, in which case, all of the sample solutions tested since the last acceptable test with the CCV should be re-determined.

The last two controls are known as a "matrix spike" (MS) and a "matrix spike duplicate" (MSD) [29]. While in the field at the point where a sample is gathered, two extra samples are collected in the same manner. Both of these additional samples are first spiked with an equal amount of standard that would produce a concentration of nitrate in the low-to-middle range of the nitrate standard solutions, if there were no nitrate in the original sample solution. These solutions should then be treated with preservative and handled in the same manner in every other way as all of the other sample solutions. These two spiked samples play essentially the same role as the percent recovery test discussed for the fluoride case study in Section 4.2.8. One calculates the concentrations of the nitrate based on the calibration curve for the sample and for the spiked samples. One can also calculate the theoretical concentration of the spiked sample based on the known addition of moles of analyte to the original solution. If the "observed" concentration of the spiked solution and the "theoretical" concentration do not agree with the each other, then a matrix effect or other bias is indicated. Consequently, the sample concentrations cannot be accurately calculated from the standard calibration curve. If no criterion has been previously decided, then a relative discrepancy of 20% in the concentrations calculated by the two methods is usually considered the maximum tolerable disagreement. In addition, the apparent concentrations in the MS and the MSD should also agree to within 20% of each other [29].

4.4.2. Field Measurements

Performing the analysis as close to the time and point of sampling as possible decreases the chance of the sample changing before the measurement is taken. Adsorption of the analyte to container walls, changes in temperature and pressure leading to losses of volatile components and microbial activity transforming species over time are some of the concerns that make field testing desirable. ISEs provide appealing sensing devices for environmental analysis because they can be used outside the lab. They are simple to use, rugged and portable. In some cases, the ideal would be a continuous monitoring system to observe changes that occur over the course of a day, weeks, or even months. Implementing such a system introduces another level of difficulty because operating a sensor continuously in the field requires automation.

However, the measurement of chemical species in the field raises a different set of challenges for the analytical chemist. Two of the biggest challenges are the issues of temperature control during the analysis and the need for recalibration. Measurements based on manual sampling are also limited in the frequency and total number of samples that can be taken [32]. Ideally, the most attractive approach in monitoring a natural water system, such as an estuary, would be a robust system of sensors implanted in remote locations or incorporated in a submersible device that could be moved from site to site reporting measurements continuously for days at a time. A Swiss group has demonstrated an intermediate approach in which a submersible sampling system pumps samples from various depths to an automated flow-through analysis device on a platform [33].

This autonomous flow system with an array of eight different ISEs and a commercial reference electrode was tested for lake water monitoring [33]. Figure 4.10a depicts a single flow channel as it pushes sample solution across the surface of a single ISE. The sampled water branched out into eight separate channels, each with an ISE for a different analyte: nitrate, ammonium, carbonate, calcium. Two separate channels were used to determine the pH, one with a commercial glass electrode and another using a liquid membrane pH sensor. After flowing past the ISE, each stream was recombined into the same beaker where a single reference electrode with a double junction salt bridge was positioned. The voltage of each sensor was read multiple times for each solution with respect to the reference electrode. The signals for five selected channels are shown in Figure 4.10b and an expanded view of the carbonate ion electrode response is displayed in Figure 4.10c. Measurements were recorded for water taken at various depths in a predetermined cycle multiple times each day. After each sampling cycle, two calibration solutions were pumped through the sensor array to check for drift. A platinum resistance thermometer was placed in the reference beaker so that the reference potential and estimates of activity coefficients in the calibration solutions could be corrected for changes in temperature. A separate pump continuously replenished the reference beaker with 3 M KCl during the entire duration of the experiment. The performance of the automated system was compared to results of samples taken in the conventional manner and brought back to the lab for analysis of calcium by atomic absorption spectroscopy and for nitrate by ion chromatography [33].

In this work, the authors applied no sample treatment. The lake water had relatively low ionic strength, but no adjustment buffer was used. Usually, when analytes are present in trace levels ($\leq 1 \mu M$, especially in the case of metal analytes), lab measurements have been the default approach because of the need for sample treatment in order to remove interfering ions and/or adjust the pH and ionic strength. Nevertheless, the data obtained here by the automated system were consistent with the levels of calcium and nitrate found by the lab-based (non-ISE) methods [33]. The authors noted that no filters were used to

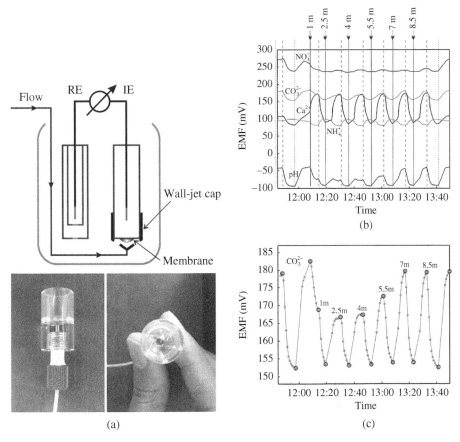

FIGURE 4.10 Autonomous multichannel analyzer. (a) Upper diagram: flow from one channel directed across the surface of single ISE and out into the beaker where the reference electrode was placed. Lower photos: view of single channel fitting designed to bring solution flow to the corresponding ISE. (b) Sensor response curves for one cycle of measurements at various depths. Vertical dashed lines indicate the time when calibration solution 1 was introduced to each sensor; dotted lines indicate transition to calibration solution 2 and the solid vertical lines indicate the introduction of sample solution from a fixed depth. (c) Expanded view of carbonate sensor response for the same cycle. Peak levels coincide with readings from sample solutions take at different depths. Lowest points indicate replicate readings of calibration solution 1. Source: Pankratova et al. 2015 [33]. Reproduced with permission of Royal Society of Chemistry.

remove cells or particulates, but that filtration might be necessary to avoid biofouling for longer studies or for lakes with high levels of biological activity.

An automated field analysis, such as this, can reveal interesting temporal and special variations. Rising ammonium levels and falling nitrate concentrations are often a useful indicator of oxygen depletion. In the water column, nitrate levels fell slightly near the sediment at a depth of 8.5 m. In addition, calcium and nitrate results showed interesting two-dimensional profile patterns [33]. Both nutrients appeared to be depleted in the upper layers of water where light levels and phytoplankton activity are higher presumably because bioactivity was removing nutrients from the water. Day and night cycles of calcium levels were also evident. Unfortunately, the background of 100 μM KCl in the lake limited the detection of ammonium ion to about 10^{-5} M. The chloride background also hampered nitrate monitoring in the sediments where nitrate levels were thought to be around 10^{-6} M.

4.5. GOOD LAB PRACTICE FOR pH ELECTRODE USE

Many of the ideas discussed in the case studies above regarding sample treatment and method validation also apply to pH electrode measurements. Because the application of pH measurements is so widespread, a discussion about good laboratory practice for pH electrode use is in order here.

4.5.1. Electrode Maintenance

Regular upkeep is vital to good performance. One should pour out and replace the reference solution in the outer reference chamber regularly (e.g. once per month) or whenever cleaning the salt bridge. (Newer solid polymer or gel salt bridge reference electrodes that are saturated with KCl are exceptions to this rule. The internal electrolyte does not need refilling in these devices [34].) A thin flexible tubing attached to a syringe is an effective device for draining the chamber and removing bubbles. Rinsing two or three times with electrolyte before refilling helps remove contamination that may have leaked across the bridge. Any bubbles near salt bridge should be removed. Bubbles block the flow of electrolyte and can cause unstable behavior. Whenever the electrode is in use, the filling port should be open to the atmosphere to maintain a pressure equal to the surroundings. The open port keeps the pressure inside the reference constant allowing the slow electrolyte flow through the salt bridge to be more uniform and discourages clogging [35–37]. Keeping the filling port closed during storage avoids evaporation. Electrodes should not be stored in deionized water or tap water, because leakage across the bridge dilutes the reference electrolyte.

The pH sensitive glass should always be kept wet. Drying can cause cracks in the gel layer and can clog the reference bridge. Either of these issues may lead to erratic behavior. For short-term storage, the glass bulb and salt bridge should be covered with KCl solution at a concentration that matches the electrolyte in the salt bridge (often 3 M) and pH 4 or pH 7 buffer. For storing >2 weeks, the opening to the container should be wrapped with parafilm to prevent evaporation [38]. Often, a combination electrode that has dried out can be rejuvenated by replacing the electrolyte in the reference chamber and immersing the glass bulb and salt bridge in electrolyte and allowing it to soak for 24 hours. For situations where electrolyte soaking is not adequate, some manufacturers sell dilute solutions of ammonium fluoride that regenerate the hydrated gel layer. This process actually removes some glass and should be undertaken only according to the manufacturer's directions [35].

A dirty salt bridge will slow the response of a combination electrode [38]. Unclogging fiber or ceramic salt bridges is often a matter of removing AgCl or other precipitates. A simple remedy is to soak the electrode in warm water (≤55 °C) saturated with KCl. Silver ions will form soluble chloro-complexes. Sometimes sulfide containing samples lead to a clogged salt bridge. Those precipitates are typically dark in color. Soaking in a thiourea/HCl solution can remove the precipitate [34]. Occasionally, the barrier to flow appears only at the outer surface of a fiber wick. In these cases, abrading the fiber wick (do not touch the pH glass bulb!) lightly with fine sand paper and rinsing will restore the electrolyte flow [31].

Adsorbed material on the glass sensor can make its behavior unstable. The pH-sensitive bulb should *not* be rubbed to clean it. Rubbing can damage the hydrated layer at the surface. Samples containing bacteria, food, protein, or oily substances can coat the pH-sensitive glass and degrade its response. Soaking electrodes exposed to organic material in a mild detergent periodically is recommended. For calcium carbonate deposits a short soak in 0.1 M HCl is effective. Oily grunge might require a rinse with alcohol [34]. It should be noted, however, that alcohol dehydrates the glass, so contact with alcohol solutions should be kept short. The probe should be soaked in aqueous electrolyte subsequently in order to rehydrate it [38]. Protein films can often be removed by soaking the electrode for a few hours in 1% pepsin solution containing 0.1 M HCl [34]. A soak in hydrogen peroxide or sodium hypochlorite may be necessary for very resistant dirt [34]. Electrode manufacturers also sell cleaning solutions for electrodes used in dirty solutions [34, 36]. After any of these cleaning procedures, the electrode should be thoroughly rinsed with deionized water and then soaked for a day in the reference filling solution [34].

4.5.2. Standard Buffers

NIST has defined a number of buffers solutions that have reliable pH values when prepared from pure, dry reagents [39]. Table 4.6 shows their pH values for a range of temperatures. The compositions of these solutions are listed below the table.

It is important to keep buffers and alkaline solutions covered in order to prevent absorption of CO_2 from the air. Carbon dioxide hydrolyzes to produce carbonic acid that can shift the pH of the buffer [35, 38]. One should also avoid contamination of standard buffers by throwing away any portion of buffer solution that was poured out of the storage bottle. Never return solution to the storage bottle. Some manufacturers sell "technical" buffers that have different recipes from those defined by NIST. The pH of these technical buffers usually are guaranteed to within only ±0.02 pH units [34].

Of course, the temperature also affects the equilibrium constant, K_w, for the dissociation of water. K_w is exponentially dependent on temperature so that while it is 1.008×10^{-14} at 25 °C, it has a value of 1.846×10^{-15} at 5 °C and 2.089×10^{-14} at 35 °C. Water is more acidic at higher temperatures [40].

4.5.3. Influence of Temperature on Cell Potentials

Temperature appears in the Nernst equation, but there are other aspects of the measurement of pH that also depend on temperature. Both the reference electrode potential and the liquid junction potential are temperature-dependent (see Table 4.2 and the discussion on temperature control in Section 4.2.6). One reason for using the same type of reference electrode and electrolyte conditions for both the inner reference electrode and the external reference electrode is the fact that changes in temperature effect both in the same way and the changes in potential at one are canceled by changes at the other. However,

TABLE 4.6 NIST standard buffer pHs

Buffer		pH at temperature (°C)					
		0	5	10	15	20	25
Oxalate	A		1.669	1.671	1.673	1.677	1.681
Phthalate	B	4.006	4.001	3.999	3.999	4.001	4.006
Tartrate	C						3.557
Phosphate	D	6.982	6.949	6.921	6.898	6.878	6.863
Phosphate	E						7.000
Phosphate	F	7.534	7.501	7.472	7.449	7.430	7.415
Borate	G	9.463	9.395	9.333	9.277	9.226	9.180
Carbonate	H	10.316	10.244	10.178	10.118	10.062	10.011

Buffer		pH at temperature (°C)					
		30	35	37	40	45	50
Oxalate	A	1.686	1.692		1.697	1.706	1.714
Phthalate	B	4.012	4.021	4.025	4.031	4.043	4.057
Tartrate	C	3.552	3.549	3.548	3.547	3.547	3.549
Phosphate	D	6.851	6.842	6.839	6.836	6.832	6.831
Phosphate	E						
Phosphate	F	7.403	7.394	7.392	7.388	7.385	7.384
Borate	G	9.139	9.102	9.081	9.070	9.042	9.018
Carbonate	H	9.965	9.922		9.884	9.850	9.819

A. Oxalate	0.05 m $KHC_2O_4 \cdot 2H_2C_2O_4 \cdot 2H_2O$
B. Phthalate	0.05 m $KHC_8H_4O_4$
C. Saturated solution of potassium hydrogen tartrate at 25 °C	
D. Phosphate	0.025/0.025 m KH_2PO_4/Na_2HPO_4
E. Phosphate	Approx. 0.020/0.0275 m KH_2PO_4/Na_2HPO_4 IUPAC soln.
F. Phosphate	0.008695 m KH_2PO_4 and 0.03043 m Na_2HPO_4
G. Borate	0.01 m $Na_2B_4O_7$
H. Carbonate	0.025/0.025 mol/kg $NaHCO_3/Na_2CO_3$

Source: Reprinted with permission from Wu et al. 1988 [39].

diffusion coefficients change with temperature and, subsequently, change the liquid junction potential.

The pH meter's automatic temperature correction feature only applies a correction to the slope in the calibration curve. It does not adjust for changes in junction potential or changes in pK_a that depend on temperature and change the true pH of solution [38]. The temperature of the sample solution should be the same as the calibration standards. A mismatch in the solution and electrode temperature causes a drift in the potential until they equilibrate [38]. Most pH electrodes work up to 80 °C, but high temperature work limits the lifetime of the probe. Only all glass body electrodes should be used above

80 °C. For high temperature work, Ag/AgCl reference electrodes with saturated KCl are recommended [31].

The general purpose, pH-sensitive glass electrode has a resistance of about $100\,M\Omega$ [37]. The pH-sensitive glass increases in resistance with decreasing temperature by a factor of ~ 2.5 for every 10 °C decrease. That is, a combination pH electrode that has a resistance of $100\,M\Omega$ at 25 °C has a resistance of $250\,M\Omega$ at 15 °C. Higher resistance slows the electrode response time [38]. As the membrane resistance increases (with age, dirt, or damage), it approaches the input impedance of the pH meter (usually $\sim 10^{12}\,\Omega$) [34, 38]. When the electrode impedance increases to within 1% of the input impedance of the meter, the meter loads the signal and the slope (mV/pH unit) decreases.

4.5.4. Calibration and Direct Sample Measurement

Rinsing electrodes with deionized water between measurements and wicking excess liquid away with a clean tissue before immersing it in the next solution helps prevent errors caused by carry over between solutions. One should never rub the pH sensitive glass bulb; that can leave a static charge on the membrane and cause erratic behavior or even damage the surface [38].

Even a well-functioning new combination electrode has a nonlinear response at extreme pHs. Many electrodes based on Na_2O glass are less sensitive to pH (and more sensitive to Na^+ ions) above pH 11. Sensors based on Li_2O glass exhibit less error at high pH. At a pH below 2, anions tend to adsorb to the glass at cationic sites causing a decrease in sensitivity – an error. (Sometimes the mechanism causing this acid error is described as an adsorption of neutral acid molecules.) In addition to these limitations, the calibration curve may not be linear over the intermediate range of pH (2–10). Multiple pH standards should be used to establish the slopes for regions of interest [35]. It is wise to bracket the pH range of samples with calibration buffers and use the slope of the calibration curve calculated from buffers with pH values closest to the sample [38].

4.5.5. Evaluating the Response of a pH Electrode

Different models of pH meters vary in the features that they provide. However, all commercial instruments offer a certain minimum combination of controls that can be used for careful calibration of the electrode and meter. It should be possible to check any meter and electrode combination with the following procedure.

1. *Selecting standard buffers*. Table 4.6 indicates the accepted pH value for several standard buffer solutions at various temperatures. For work in the acid range, one chooses certifiable buffers at pH 7.00 and an acid value, such as phthalate at pH 4.00. Alternatively, for work in the basic region the second buffer might be a borax buffer at 9.22. The calibration should be performed over the region of interest so that all of the test solutions will fall between the pH of the two calibration buffers. If both acidic and basic solutions are to be tested, separate calibrations should be performed for each region. The slope (mV/pH unit) is not necessarily the same for different regions.

2. *Maintaining temperature control.* The temperature of the buffer solutions is measured with a thermometer. After connecting the electrodes to the meter and turning on the power, the electrodes are immersed in the pH 7.00 buffer. The solution is stirred gently with a magnetic stirrer. A sheet of corrugated cardboard under the beaker helps insulate the solution from the heat of the stirring motor.

3. *Setting the zero or meter offset.* With the combination electrode in pH 7 buffer, the display is switched to the millivolts position. Normally, the signal levels off after about two minutes. At this point, the display should be adjusted to 0.0 mV using the offset or zero control. Ideally, the instrument's response is a linear function of pH, namely, $y = mx + b$ where are the output, y, is in millivolts, x is the pH of the solution, m is the response slope in mV/pH unit, and b is the y-intercept or offset voltage. One starts with pH 7.00, because the solution inside the bulb is also usually buffered at pH 7.00. Consequently, the potential across the entire membrane should be zero when the pH outside and inside match. The pH 7.00 buffer is used to set the intercept. If the glass membrane and reference electrodes are working properly (and the internal and external reference electrodes are the same type with filling solutions of the same composition), then the offset adjustment will require 30 mV or less. A larger discrepancy may indicate a mismatch in reference solutions and/or a partly clogged salt bridge. Procedures for cleaning salt bridges were discussed in Section 4.4.1. Since two experimental parameters, namely the slope and intercept, must be calibrated, this process requires at least two buffers. Some of the mechanisms for the nonzero mV reading at pH 0 can be easily reversed. A large change in the zero off-set caused by infiltration of sample solution across the salt bridge into the reference electrolyte can be overcome by replacing the reference solution. Similarly, cleaning a salt bridge may restore proper electrode behavior. An increase in offset voltage can also be caused by contact corrosion and can be removed by cleaning the metal with sandpaper [34]. On the other hand, degradation of the outer glass gel layer by exposure to aggressive solutions or a change in the reference electrolyte composition inside the pH sensitive glass bulb may lead to an electrode that will not calibrate properly. It should be noted that heating can cause a release of hydroxide from the glass surface on the inside chamber shifting its buffer pH and the inner membrane potential. Collectively, these problems associated with the inside glass surface or internal electrolyte are called the "asymmetry" potential. There is not much one can do about changes inside the pH-sensitive bulb.

4. *Measuring the signal for the second buffer.* Next, the electrodes should be removed from the first solution and rinsed with deionized water. The edge of a tissue should be used to touch off any clinging drops before putting the electrodes into the second buffer. Again, the readout should be recorded in millivolts following a settling time of two minutes.

5. *Calculating the observed slope.* The observed slope is calculated from:

$$m_{obs} = \left(\frac{V_2 - V_1}{pH_2 - pH_1} \right) \tag{4.22}$$

where V_2 is the display in millivolts for the second buffer, $V_1 = 0 =$ the displayed voltage for the pH 7.00 buffer. pH_2 and pH_1 are the pH-values of the two standard buffers at the measured temperature for the solutions as read from the table.

6. *Calculating the percent slope.* The theoretical slope, in mV/pH unit, for a glass electrode is given by:

$$m_{theo} = \left(\frac{RT}{zF} \right)(2.303) = \frac{(8.314 \, J/(mol \, K))(T)}{(1)(96\,485 \, C/mol)}(2.303)\left(\frac{1000 \, mV}{V} \right) = 0.1984 \cdot T$$

(4.23)

where T is the solution temperature in kelvins.

One then calculates a figure of merit called the percent slope. It is the percentage of the observed slope compared to the theoretical slope:

$$\text{Percent slope} = \frac{m_{obs}}{m_{theo}}(100\%)$$

(4.24)

This calculation is made for the sake of checking the condition of the electrode. In some tightly regulated businesses, such as the pharmaceutical industry, the percent slope must be in the range of 95–100% to be acceptable [35]. At best, the slope is only 99.8% of theoretical and then only when the electrode is new [34].

7. *Setting the slope.* If the result is inside the acceptable range, the instrument display is switched to the pH mode and the "slope" control is adjusted until the readout matches the table value for the pH of the buffer at the measure temperature for the experiment.

4.5.6. Calibrating a Combination Electrode and pH Meter

An abbreviated procedure can be used whenever the electrode is known to be functioning well.

1. *Setting the signal at pH 7.00.* Starting with a pH 7.00 buffer, the electrode is immersed in the solution while it is being stirred. After the reading is stable (usually ≤2 minutes), the pH is adjusted to read the value corresponding to the true pH at the solution temperature using the offset (or "zero") control.

2. *Setting the signal for a second pH.* The electrode is removed, rinsed with deionized water, and immersed in a second standard buffer. The choice of this standard should be made so that the two standard buffers will bracket the pH of the sample solutions. This solution should be stirred as the previous solution was. After the reading is stable, adjust the display with the slope control to match the pH of the standard at the temperature of the system. If some samples fall outside that range, then those should be set aside and measured again after checking the pH with a third standard buffer that will bracket the pH of these samples together with the one of the other standard solutions. If the meter reading for the third buffer agrees with its theoretical value, then the initial calibration is fine for testing the rest of the samples. (One should not readjust the zero or slope controls on the meter.) If the meter reading does not agree with the theoretical value for the third standard buffer, then the conditions fall outside the linear range of the electrode. In that case, it would be wise to set the meter to the mV mode and plot the readings for several standard buffers throughout the pH range of interest and compare the system's mV response for sample solutions to a calibration curve based on the mV response versus standard pH values.

3. *Verifying the calibration.* After the slope control has been set using the second standard, the validity of the calibration procedure is tested while stirring a third buffer with a known pH that falls within the range of the calibration buffers. The pH on the meter should match the accepted value within 0.05 pH units. A larger discrepancy indicates some nonlinearity that maybe the result of an aging or damaged glass membrane. In order to rely on the electrode under those conditions, one should use multiple standards in the region of interest and plot the apparent pH versus the accepted buffer pH in order to correct the electrode readings for unknown test solutions.

4. *Measuring sample pH.* Samples should be measured while stirring in a manner similar to the procedure for measuring the pH of the standards. As before, the sensor should be rinsed with deionized water, wicking any droplets away with the edge of a tissue (without rubbing the glass bulb) between solutions. Periodically (perhaps, after every 10 samples), the pH of a standard solution with a value in the range of the samples should be measured again in order to check for drift. The US Pharmacopeia (USP) rules require measurement precision of ±0.02 pH units [35]. If the reading for the standard buffer fails to match the original measurement to that degree, then one should recalibrate, and re-measure the samples since the last time the standard buffer yielded an acceptable reading.

4.5.7. Low Ionic Strength Samples

Samples with low electrolyte concentrations will lead to electrolyte ions leaching from the salt bridge into the sample. That will cause a liquid junction potential and probably signal drift (until a steady state gradient builds up). The use of a high flow rate reference bridge, such as a tapered sleeve double junction is recommended for those situations. One should stir the solution in order to keep the conditions near the bridge as constant as possible [35, 38]. One alternative is to add pure, dry KCl to set the ionic strength of the sample (near that of the calibration standards) [38]. Low ionic strength samples may also mean low buffer capacity, consequently, such samples may also be susceptible to pH changes caused by adsorption of CO_2 from the atmosphere or by contamination from glassware.

4.5.8. Samples Containing Soil, Food, Protein or Tris Buffer

The pH of food and soil samples can be measured by making a slurry. The food industry may have specific protocols for a given material that one needs to adhere to. However, for soil, a common approach is to mix 5 g soil and 25 g deionized water, stirring for 10 minutes in order for the water to equilibrate with the soil [38]. Before measurement the stirring is stopped and the particles are allowed to settle. The pH-sensitive glass bulb is immersed in the particles. If using a combination glass electrode, the salt bridge should be kept above the sediment. However, a separate reference electrode with a high flow, tapered sleeve salt bridge is recommended [38].

A separate reference electrode with a sleeve type double junction salt bridge is also recommended for working with food, especially oily material that easily clogs frits or fiber wick type bridges. The same is true for working with samples containing protein or Tris (or THAM) buffer. The Tris molecules complex silver ions resulting in potential drift or erratic behavior. Frequent cleaning of the salt bridge with detergent is appropriate in these situations [38].

4.5.9. pH Titrations

Whenever both high-precision and high-accuracy analyses are required, few analytical procedures can outperform a titration. Careful work can often achieve relative errors as low as 0.1%. With automated or manual titrations that use potentiometry as a means of endpoint detection, there are graphical procedures that can assist the user. Consider the titration of a weak acid, HA, with a strong base, such as NaOH.

The slope in the plot of pH versus titrant volume will reach a maximum at the equivalence point. It coincides with the inflection point of the curve. As such, the equivalence point can often be estimated from a graph of the data by visual inspection (see Figure 4.11a). One can also plot a derivative of the curve near the equivalence point (Figure 4.11b). The derivative strategy works well for automated titrations. Both inspection and the derivative approach give the best precision when lots of data points are taken near the equivalence point.

4.5.10. Gran Plots

Gathering lots of points in the narrow volume region near the equivalence point may be difficult to do in some cases. In other cases, the changes in pH near the equivalence point are small and the inflection point may be difficult to pick out. An alternative approach is a graphical method that does not rely on taking many points close to the equivalence point. Instead, it linearizes the data in the end-point region. The equivalence point is found by extrapolation. This method was first introduced in the 1950s by Gunnar Gran [41, 42] and is now known as a Gran Plot.

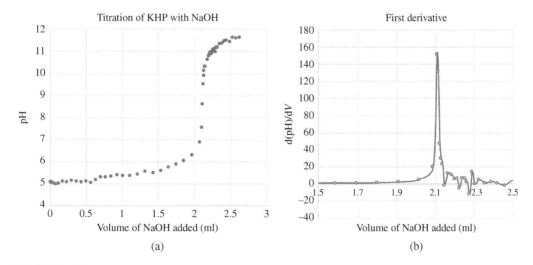

FIGURE 4.11 pH titration of potassium hydrogen phthalate (KHP) with NaOH. (a) Conventional titration data generated by an autotitrator. (b) Derivative of titration data near the steep portion of the curve. The endpoint volume is indicated by the volume associated with the position of the peak.

Here is the underlying principle of the Gran plot for an acid/base titration. Consider the titration of a weak acid, HA, by a strong base, such as NaOH. At any point in the titration, the acid form is in equilibrium with its conjugate base, A^-.

$$HA \rightleftarrows H^+ + A^- \tag{4.25}$$

$$K_a = \frac{(a_{H^+})(a_{A^-})}{a_{HA}} \tag{4.26}$$

Or in terms of the hydrogen activity:

$$a_{H^+} = K_a \left(\frac{\gamma_{HA}}{\gamma_{A^-}} \right) \left(\frac{[HA]}{[A^-]} \right) \tag{4.27}$$

It will be assumed here that for weak acids one can neglect the amount of HA lost (and the amount of conjugate base formed) due to self-ionization compared to the amount remaining in the HA form after reaction with NaOH. Then, the weak acid and base concentrations can be expressed at any point during the titration as follows:

$$[HA] = \frac{\text{original moles HA} - \text{moles NaOH added}}{\text{total volume}} = \frac{V_{HA}C_{HA} - v_{NaOH}C_{NaOH}}{V_{HA} + v_{NaOH}} \tag{4.28}$$

and

$$[A^-] = \frac{\text{moles NaOH added}}{\text{total volume}} = \frac{v_{NaOH}C_{NaOH}}{V_{HA} + v_{NaOH}} \tag{4.29}$$

where

V_{HA} = the original volume of the weak acid, HA,
C_{HA} = the original concentration of the weak acid, HA,
v_{NaOH} = the volume of titrant added to the mixture, and
C_{NaOH} = the concentration of the NaOH titrant in the buret.

Substituting the expressions for [HA] and $[A^-]$ into Eq. (4.27) gives

$$a_{H^+} = K_a \left(\frac{\gamma_{HA}}{\gamma_{A^-}} \right) \left\{ \frac{(V_{HA}C_{HA} - v_{NaOH}C_{NaOH})/(V_{HA} + v_{NaOH})}{(v_{NaOH}C_{NaOH})/(V_{HA} + v_{NaOH})} \right\}$$

$$= K_a \left(\frac{\gamma_{HA}}{\gamma_{A^-}} \right) \left\{ \frac{(V_{HA}C_{HA} - v_{NaOH}C_{NaOH})}{(v_{NaOH}C_{NaOH})} \right\} \tag{4.30}$$

Multiplying both sides of Eq. (4.30) by the titrant volume yields

$$v_{NaOH} \cdot a_{H^+} = K_a \left(\frac{\gamma_{HA}}{\gamma_{A^-}} \right) \left\{ \frac{(V_{HA}C_{HA} - v_{NaOH}C_{NaOH})}{(C_{NaOH})} \right\}$$

$$= K_a \left(\frac{\gamma_{HA}}{\gamma_{A^-}} \right) \left\{ \frac{(V_{HA}C_{HA})}{(C_{NaOH})} - v_{NaOH} \right\} \tag{4.31}$$

Note that the equivalence point volume, v_e, is equivalent to $V_{HA}C_{HA}/C_{NaOH}$. Also, recall that

$$pH = -\log(a_{H^+}) \quad \text{or} \quad a_{H^+} = 10^{-pH} \tag{4.32}$$

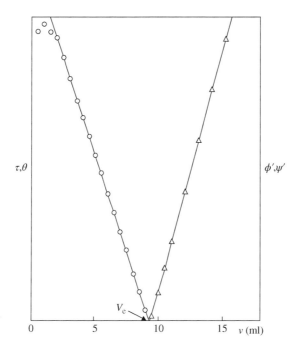

FIGURE 4.12 Gran plot for the titration of acetic acid by a standardized NaOH. The vertical axis before the equivalence point (on the left side) corresponds to $v_{NaOH} \cdot 10^{-pH}$ and after the equivalence point, the vertical axis is $(V_{HA} + v_{NaOH})10^{pH}$. The equivalence point volume, v_e, is indicated by the intercept of the linear segments with the volume axis. Source: Adapted with permission from Rossotti and Rossotti [43]. Copyright 1965, American Chemical Society.

Substituting into Eq. (4.31) gives

$$v_{NaOH}10^{-pH} = K_a \left(\frac{\gamma_{HA}}{\gamma_{A^-}} \right) \{v_e - v_{NaOH}\} \tag{4.33}$$

Equation (4.33) indicates that a plot of $v_{NaOH}10^{-pH}$ versus the titrant volume, v_{NaOH}, gives a straight line with an x-intercept at the equivalence point volume, v_e (see Figure 4.12).

Beyond the equivalence point, the values of $v_{NaOH}10^{-pH}$ approach a horizontal line. One can also plot a function that tracks the rise of the hydroxide concentration as $(v_{HA} + v_{NaOH})10^{pH}$ versus v_{NaOH}. This graph also forms a straight line that crosses the volume axis at $v_{NaOH} = v_e$; however, the magnitude of the y-values will be quite different from the graph of the data before the equivalence point. It is sufficient to plot only one or the other.

PROBLEMS

4.1 A solution of TISAB with 1 M NaCl is available to treat drinking water samples for fluoride determination using a fluoride ISE and a Ag/AgCl reference electrode in 3 M KCl. The expected fluoride concentration in these samples is 1 mg/l. How much TISAB should be used per 100 ml sample solution to reach a reasonable IS and minimize junction potentials?

4.2 After treating the sample solution 1 : 1 with TISAB containing 1 M NaCl, what would the approximate junction potential be for a salt bridge containing 3 M KCl?

4.3 Your colleague is preparing to perform some field measurements of chloride in stream water using an ISE. He is aware that he cannot use measurements of standards in the lab at 20 °C to compare the chloride measurements in the field at 10 °C, so he proposes to take a TISAB solution and one standard along to perform a single point calibration at the same time and place as his measurements on stream water (after treatment with TISAB). Explain to him why that will not give him reliable data.

4.4 A well-used nitrate ISE was being used to test some ground water samples for possible leaching from a hazardous waste site. A percent recovery procedure was performed to decide whether a standard calibration curve could be used for determining the concentration of nitrate in the ground water samples. All standards and samples were treated with 10 ml TISAB (1 M NaCl) per 100 ml of test solution and stirred thoroughly just before measurement. In addition to the ground water sample, a spiked sample and a spiked duplicate sample were used. In each of the spiked samples, 1.00 ml of a 1.00×10^{-2} M standard NaNO$_3$ solution was added to 100 ml of the sample. The calibration curve was linear and fit the equation:

$$E_{cell} = 0.265 - (0.0574) \log(a_{NO_3^-})$$

The sample, spiked sample, and spiked duplicate sample gave cell potentials of 0.473_8, 0.464_6, and 0.464_2, respectively. Calculate the percent recovery and the percent discrepancy (between the spiked and spiked duplicate samples).

4.5 In experiments with standard solutions, a Li$^+$ ISE gave a calibration curve of

$$E = 0.39 + (0.0577) \log[Li^+]$$

An analyst assumed that the sensitivity of the ISE (represented by the coefficient of the log term) is the same in urine samples as it was in the standard solutions, but that differences in the activity coefficient and junction potential were too great to use the calibration curve to determine the [Li$^+$] in urine samples. She decided to perform a standard additions experiment. Here are the data. Use a Gran plot of the data to determine the original [Li$^+$] in the urine sample. Increments of 0.0788 M standard Li were added to 50.0 mL of sample.

V_N (ml)	E_{cell} (V)
0.00	0.144
1.00	0.177
2.00	0.190
3.00	0.198
5.00	0.209

4.6 Refer to Table 4.4 of selectivity coefficients for cations that interfere with the creatinium ISE described in Section 4.3. At what concentration of creatinine would 0.3 M ammonium interfere to cause a 10% error in the measured creatinine concentration?

4.7 For a membrane without an ionophore, why is the membrane potential more sensitive to chlorate ion than sulfate ion?

4.8 In addition to preparation of the standard solutions and treatment of the sample solutions, a good environmental analysis requires various control solutions that check the reliability of the analysis. List five different types of control solutions that are tested along with samples in a good environmental analysis by ISE. Briefly, explain the purpose of each type of control sample.

4.9 Refer to Table 4.4 for the selectivity coefficients for ions that interfere with the determination of creatinine by the experimental ISE described in Section 4.3. Assume that Na^+, K^+, NH_4^+, and Ca^{2+} are present in a urine sample at 0.16, 0.06, 0.035, and 0.005 M, respectively. What is the lowest concentration of creatinium ion that can be measured in the presence of these combined interferences, if one is willing to accept only a 10% error in creatinine concentration?

4.10 Imagine that you were asked to set up a flow injection system in order to measure K^+ ion levels in blood samples in a hospital lab. List at least five different challenges for determining K^+ in 100 blood samples per day.

4.11 Why is the filling port for the outer reference chamber of a combination glass electrode kept open during measurements?

4.12 Here are data from a pH titration of potassium hydrogen phthalate with NaOH using an automated titrator. Only data for part of the titration has been provided. Prepare a Gran plot and find the equivalence point volume.

V_{NaOH}	pH
1.912	4.7
2.046	4.8
2.18	4.95
2.314	4.97
2.462	5.12
2.61	5.27
2.758	5.41
2.906	5.64
3.055	6.14
3.132	7.16
3.154	9.86
3.162	10.42

REFERENCES

1. Švarc-Gajića, J., Stojanovića, Z., Vasiljevićb, I., and Kecojević, I. (2013). *J. Food Drug Anal.* 21 (4): 384–389.

2. Hanna Instruments (2018). Fluoride ISE. https://hannainst.com/hi4110-fluoride-combination-ion-selective-electrode.html (accessed 08 August 2019).

3. Nico2000 Ltd., Technical specifications for the fluoride ion-selective electrode, ELIT 8221. http://www.nico2000.net/ise_specs/fluoride.pdf (accessed 08 August 2019).

4. ThermoFisher Scientific Thermo scientific orion fluoride ion selective electrode user guide, 254792-001, Revision B, September 2016. https://www.fondriest.com/pdf/thermo_fluoride_ise_manual.pdf (accessed 08 August 2019).

5. Pérez-Olmos, R., Soto, J.C., Zárate, N., and Díez, I. (2008). *J. Pharm. Biomed. Anal.* 47: 170–176.

6. Blaedel, W.J. and Meloche, V.W. (1963). *Elementary Quantitative Analysis*, 2e. New York, NY: Harper and Row.

7. Jenway and Bibby Scientific. A comparison of ion-selective electrode analysis methods. pH/ION METER Application note: A04-001A. jenwayhelp@bibby-scientific.com (accessed 08 August 2019).

8. Bakker, E. (2014) Fundamentals of electroanalysis: 1. Potentials and transport. ibook. https://itunes.apple.com/us/book/fundamentals-electroanalysis-1-potentials-transport/id933624613?mt=11 (accessed 08 August 2019).

9. Cammann, K. (1979). *Working with Ion-Selective Electrodes: Chemical Laboratory Practice*. Berlin, Heidelberg: Springer Science & Business Media.

10. Sawyer, D.T., Sobkowiak, A., and Roberts, J.L. Jr. (1995). *Electrochemistry for Chemists*, 2e, 192. New York, NY: Wiley, Chapter 5.

11. Vitha, M.F., Carr, P.W., and Mabbott, G.A. (2005). *J. Chem. Educ.* 82: 901.

12. Mettler Toledo (2017). A guide for ion selective electrodes, theory and practice of ISE applications. https://www.mt.com/us/en/home/library/guides/lab-analytical-instruments/Ion-selective-electrode-guide.html (accessed 08 August 2019).

13. Rundle, C.C. (2011) A beginners guide to ion-selective electrode measurements. http://www.nico2000.net/Book/Guide1.html (accessed 08 August 2019).

14. Harris, D.C. (2010). *Quantitative Chemical Analysis*, 8e. New York, NY: W. H. Freeman.

15. Bader, M. (1980). *J. Chem. Educ.* 57: 703–706.

16. Liberti, A. and Mascini, M. (1969). *Anal. Chem.* 41: 676–679.

17. Tellinghuisen, J. (2000). *J. Chem. Educ.* 77: 1233–1239.

18. de Levie, R. (2008). *Advanced Excel for Scientific Data Analysis*, 2e. New York, NY: Oxford University Press.

19. Billo, E.J. (2011). *Excel for Chemists*, 3e. Hoboken, NJ: Wiley.

20. Rossini, E.L., Milani, M.I., Carrilho, E. et al. (2018). *Anal. Chim. Acta* 997: 16–23.

21. Narayanan, S. and Appleton, H.D. (1980). *Clin. Chem.* 26: 1119–1126.

22. Guinovart, T., Hernández-Alonso, D., Adriaenssens, L. et al. (2017). *Biosens. Bioelectron.* 87: 587–592.

23. (a) The Royal College of Physicians and Surgeons of Canada Clinical laboratory tests – reference values. http://www.royalcollege.ca/rcsite/documents/credential-exams/clinical-lab-tests-reference-values-e.pdf (accessed 08 August 2019); (b) Lentne, C. (ed.) (1981). *Geigy Scientific Tables: Units of Measurement, Body Fluids, Composition of the Body, Nutrition*. Basel: Ciba-Geigy.

24. VWR (2012). SympHony™ ion selective electrodes ISE manual. https://us.vwr.com/assetsvc/asset/en_US/id/10947624/contents (accessed 08 August 2019).

25. Sawyer, D.T., Sobkowiak, A., and Roberts, J.L. Jr. (1995). *Electrochemistry for Chemists*, 2e. New York, NY: Wiley.

26. Agency for Toxic Substances and Disease Registry (2015). Toxic substances portal – nitrate and nitrite. https://www.atsdr.cdc.gov/phs/phs.asp?id=1448&tid=258 (accessed 14 April 2019).

27. Schazmann, B. and Diamond, D. (2007). *New J. Chem.* 31: 1–6.

28. Cuartero, M., Crespo, G.A., and Bakker, E. (2015). *Anal. Chem.* 87: 8084–8089.

29. US EPA 2007. Potentiometric determination of nitrate in aqueous samples with an ion-selective electrode. METHOD 9210A, revision 1. https://www.epa.gov/sites/production/files/2015-12/documents/9210a.pdf (accessed 08 August 2019).

30. Xie, W.J. and Gao, Y.Q. (2013). *J. Phys. Chem. Lett.* 4: 4247–4252.

31. Mettler Toledo (2015). Nitrate ion selective electrode application guide principle, practical knowledge and tips for applications with nitrate ISE. https://www.mt.com/us/en/home/library/guides/lab-analytical-instruments/nitrate-ion-selective-electrode-application-guide.html (accessed 08 August 2019).

32. Cuartero, M. and Bakker, E. (2017). *Curr. Opin. Electrochem.* 3: 97–105.

33. Pankratova, N., Crespo, G.A., Afshar, M.G. et al. (2015). *Environ. Sci. Processes Impacts* 17: 906–914.

34. Springer, E. (2014). *pH Measurement Guide.* Hamilton Co.. http://separations.co.za/wp-content/uploads/2014/09/pH-Measurement-Guide.pdf.

35. Mettler Toledo (2018). Perform your next pH measurement in compliance with USP. *White Paper on pH measurements.* www.mt.com (accessed 08 August 2019).

36. Thermo Scientific. pH measurement handbook (2014). https://assets.thermofisher.com/TFS-Assets/LSG/brochures/pH-Measurement-Handbook-S-PHREFBK-E.pdf (accessed 08 August 2019).

37. Hanna Instruments. Hanna Instruments Library of pH electrode guides. https://blog.hannainst.com/ph-guides (accessed 08 August 2019).

38. Bier, A. (2010). *Electrochemistry, Theory and Practice.* Loveland, CO: Hach Company. https://www.mt.com/us/en/home/library/guides/lab-analytical-instruments/pH-Theory-Guide.html.

39. Wu, Y.C., Koch, W.F., and Durst, R.A. (1988). *Standard Reference Materials: Standardization of pH Measurements*, 2e. Gaithersburg, MD: National Bureau of Standards, NBS Special Publication (SP) 260-53. https://doi.org/10.6028/NBS.SP.260-53e1988.

40. Weast, R.C. (ed.) (1967). General chemical. In: *Handbook of Chemistry and Physics*, 48e, D-92. CRC.

41. (a) Gran, G. (1950). *Acta Chem. Scand.* 4: 559–577. (b) Gran, G. (1952). *Analyst* 77: 661.

42. Gran, G. (1988). *Anal. Chim. Acta* 206: 111–123.

43. Rossotti, F.J.C. and Rossotti, H. (1965). *J. Chem. Educ.* 42: 375–378.

CONTROLLED POTENTIAL METHODS

5

5.1. OVERVIEW

The previous chapters on potentiometry describe methods of analysis based on measuring an electrical potential energy difference between a solution and an electrode or membrane surface as a result of an exchange of ions or electrons. Ideally, that voltage is measured without drawing current, so that the equilibrium establishing the charge separation is not disturbed. The voltage is sometimes called the "open circuit potential" to emphasize the zero-current condition. Another family of electrochemical methods of analysis is based on applying a voltage to an electrode surface to force an electron transfer reaction to take place. In these cases, the system is deliberately pushed away from its equilibrium or resting state, and the rate of the reaction that occurs in response is monitored in the form of a current. These techniques are generally known as controlled potential electrochemical methods.

The electrochemical cell for controlled potential measurements can be simply a working electrode made of an inert metal, such as gold or platinum or a carbon surface immersed in a sample solution including a supporting electrolyte and a reference electrode (and its reference salt bridge).

The role of the working electrode is to serve as a source or sink for exchanging electrons between electroactive species and the outside electrical circuit. In practice, there are usually special considerations for the design of the working electrode and other accessories that are needed to ensure precise and accurate control of the applied potential. The most common modification is the use of a third electrode (usually a Pt wire) to carry the current for the reference electrode. This precaution is suitable for experiments that draw current above a microamp. Higher current can partially polarize the reference electrode pulling it away from its reference potential. The third electrode is called the auxiliary or counter electrode. Figure 5.1 shows a three-electrode system. The circuit principles and other instrumental considerations are addressed in Chapter 7.

High levels of current can also lead to an error in the applied potential in the form of resistive heating of the solution. Energy in the form of heat is lost to the system in overcoming the solution resistance. This heat loss is known as Joule heating. The product of the cell current and the solution resistance is equal to the amount of energy, in volts, lost to Joule heating and represents an error in the amount of energy that was intended to be applied

Electroanalytical Chemistry: Principles, Best Practices, and Case Studies, First Edition. Gary A. Mabbott.
© 2020 John Wiley & Sons, Inc. Published 2020 by John Wiley & Sons, Inc.

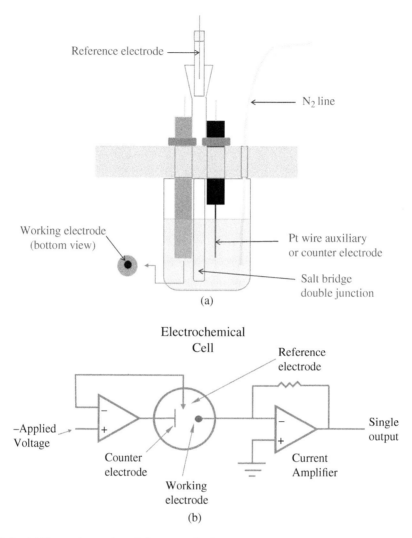

FIGURE 5.1 (a) Three-electrode cell for controlled potential experiments. The working electrode is a metal or carbon disk shrouded in an insulating material such as glass or a polymer. The reference is shown here with a double junction salt bridge. (b) The external control and measurement circuit is built around operational amplifiers. Details are discussed in Chapter 7.

to the working electrode. (The error caused by Joule heating is also called the "iR" loss or Ohmic loss). One normally adds readily soluble salts to the solvent to form a "supporting electrolyte" solution in order to keep the solution resistance down. An aqueous 0.1 M KCl solution, for example, typically has a resistivity of about 50–100 Ω/cm. So, a cell with a 0.1 M KCl solution, and electrodes spaced by 1–2 cm carrying current of 100 μA would be expected to have an iR-loss of $(10^{-4}$ A$)(100\,\Omega) = 10^{-2}$ V $= 10$ mV. Often a deviation of 10 mV in the applied potential can be tolerated, but other times not. Consequently, one should be able to estimate the upper levels of current that an experiment is likely to generate and the net resistance of the system in order to compensate, if necessary.

Here is a brief guide to some of the main ideas and take-home lessons about controlled potential methods. These concepts are developed more thoroughly in the sections that follow. Controlled potential methods include amperometric techniques where the applied voltage is fixed, and the current is measured. For example, an important application of amperometry is an electrochemical detector for high performance liquid chromatography (HPLC). The current signal is proportional to the concentration of the analyte, but signal strength is also dependent on the velocity of the solution flowing past the electrode and the diffusion coefficient of the analyte. The convection of the bulk solution plays an important role in the mass transport of reactant to the electrode surface. Interestingly, there is always a layer of solution at the electrode surface that remains stagnant, in spite of the motion of the solution. The reactant moves through this layer to the surface by the random walk of individual molecules – a mechanism called diffusion. In Chapter 1, the concept of diffusion was introduced. The diffusion coefficient is the proportionality constant linking the flux of a species and the concentration gradient for that material in Fick's first law. The faster the solution moves, the thinner this stagnant layer becomes and the steeper the concentration gradient is. Under the conditions where a reactant is transformed immediately upon reaching the electrode surface, the concentration gradient for that species is essentially the ratio of the bulk concentration of the reactant, C_o, to the stagnant layer thickness, δ. Consequently, for a moving solution the maximum current, i_l, is:

$$i_l = nFAD_oC_o/\delta$$

where n is the number of electrons transferred per reactant, F is Faraday's constant, A is the electrode area, and D_o is the diffusion coefficient for the reactant.

Another form of amperometry is performed in solution without stirring. It is called chronoamperometry. It monitors the current response as the voltage is stepped to a potential where electron transfer is initiated. Because material near the surface is continuously removed by electrolysis, a zone deplete in the reactant grows from the electrode surface and extends out into the solution. This depletion zone defines the diffusion layer thickness. As a result, the current decreases with the square root of time in chronoamperometry.

$$i = \frac{nFAC_{ox}D_{ox}^{1/2}}{\pi^{1/2}t^{1/2}}$$

This experiment has been applied to quantitative analysis in special cases, such as for the determination of blood glucose in personal health–monitoring devices. The other common application of chronoamperometry is for measuring diffusion coefficients or for determining the active area, A, for an electrode surface. Sometimes, the current signal is integrated to yield a less noisy measurement. That version of the experiment is known as chronocoulometry. However, in general, the currents for convective systems are much more sensitive than unstirred solutions for quantitative analysis.

If the applied voltage is scanned over a range of potentials and the current response is measured, then the technique is called voltammetry. Both stirred and unstirred voltammetry experiments are valuable for studying electrochemical reactions. Cyclic voltammetry (CV) involves scanning the voltage at a stationary electrode in a quiet solution. The current plotted against the applied voltage (a voltammogram) in an unstirred solution produces a peak for the oxidation or reduction of the reactant in solution. The rate of electron transfer is a dependent on the applied voltage and the formal potential for the reactant. The voltage is also scanned back in the reverse direction in order to observe the electrolysis of the product formed on the forward scan. The peak current in the forward scan is proportional to the bulk concentration of the reactant, C_o, and the square root of the voltage scan rate, $v^{1/2}$.

$$i_p \propto D_o^{1/2} v^{1/2} C_o$$

One of the most widely used applications of the CV experiment is measurement of the formal potential of redox couples. Whenever the reverse scan produces a peak, the average of the peak potential for the scan in the anodic (positive) direction, E_{pa}, with the potential for the peak from the cathodic scan, E_{pc}, is a good estimate of the formal potential of the redox pair.

$$\frac{(E_{pc} + E_{pa})}{2} \cong E^{o\prime}$$

Another extremely valuable application of CV is for the study of chemical reaction mechanisms in which the electron-transfer process is coupled to a chemical reaction. For example, an electron-transfer step may be followed by the protonation, deprotonation, or other reaction of the product with another species in solution. Examining the shapes of the voltammograms as a function of scan rate can lead to important inferences about the associated chemical step, and it can often provide a good measure of the rate constant for that process.

Quantitative analysis at sub-millimolar levels of analyte based on voltammetry requires techniques for discriminating the signal current (Faradaic current) from background current. The main sources of unwanted background current are associated with the charging of the electrical double layer and adsorption of solution species onto the working electrode or other changes in the electrode surface. Strategies that step or pulse the applied voltage during a scan are quite effective. The current is measured a few milliseconds following the pulse after the charging current has decayed. The differential pulse and square-wave techniques are the two methods that are most popular. Pulsing the potential can be a valuable strategy even for amperometric experiments. Some electroactive molecules form products that adhere to the working electrode, blocking further electrolysis and leading to a loss of signal. A method known as pulsed amperometric detection (PAD) uses a program that changes the potential in a cycle of three steps. First, the electrode is stepped to an extremely positive value that cleans off the electrode. Next, the voltage is returned to a slightly negative value to reduce any surface metal oxides that were formed, and finally, the voltage is stepped to a positive voltage that is suitable for oxidizing the analyte in solution at which time the current is measured.

Voltammetric methods have extended quantitative analysis into the nanomolar range (and below) by pre-concentrating the analyte into a thin film at an electrode surface. These methods are generally known as stripping analysis. Forms of stripping analysis have been applied to metals that can be reduced at mercury film electrodes and, subsequently, dissolve into the mercury, anions that form precipitates as mercury salts on mercury electrodes as well as organic molecules that adsorb spontaneously on electrode surfaces.

Sections 5.2 through 5.4 introduce the fundamental ideas that are essential to the understanding of amperometry and voltammetry. In addition, one should know about solid electrodes (covered in Section 5.5.2) and carbon electrodes, in particular (Section 5.5.2.1). Surface pretreatment can be crucial to the performance of carbon working electrodes. Section 5.5.2.3 describes the different cleaning and activation methods that one should know about. Today, several controlled potential methods are very commonly used in analytical laboratories and are worthy of inclusion in the reading assignment of university instrumental analysis courses. These include PAD (Section 5.6), stripping analysis (Section 5.7), and dissolved oxygen sensors (Section 5.8.2).

The sections not yet mentioned provide a window into some advanced topics. Some are special applications that are very widely employed in industry on a daily basis, such as the coulometric Karl Fisher method for moisture determination (Section 5.8.4). Other sections contain the basis for some of the most exciting, new developments in the field today, and only time constraints have kept them off of the required reading list. These topics include ultramicroelectrodes (Section 5.5.3) that are making precise spatial and quantitative analysis of neurotransmitters possible in the brains of living animals and the development of fast scanning voltammetry (Section 5.5.4) for the study of rapid electrochemical reaction mechanisms. Another active area is the application of enzymes for electrochemical sensors. The reliability and robustness of enzyme electrodes (Section 5.8.3) are progressing to the point where they are now suitable as implantable sensors in human patients.

Finally, a curious twist on the application of ion selective electrodes (ISEs) may circumvent problems with potential drift. In recent investigations, researchers have deliberately scanned the voltage on specialized all-solid-state ISEs to demonstrate a method parallel to CV (Section 5.9) and in some cases analogous to stripping voltammetry. These "ion transfer voltammograms" produce peaks in these scans in which the average potential of the forward and reverse peak is equivalent to the open-circuit potential for the electrode. This potential is related to the analyte concentration through the Nernst equation. The results are obtained without waiting for a drifting signal to stabilize – an issue in conventional ISE use. The bonus of this approach is the fact that by using two different ionophores at the same time in the same ISE, two different analytes can be determined in the same voltammogram. If one of these analytes is present at a known concentration, it can be used as an internal standard enabling one to correct for uncertainties in reference potential, junction potential, or ionic strength.

5.2. SIMILARITIES BETWEEN SPECTROSCOPY AND VOLTAMMETRY

The most common subset of controlled potential methods is called voltammetry. This technique involves scanning the voltage and monitoring the current. Voltammetry might be called the electrochemical equivalent to absorption spectroscopy [1]. In spectroscopy, one plots a spectrum using the wavelength or energy of the light that illuminates the sample on the horizontal axis, and a function of the intensity of the light that passes through the sample is plotted on the vertical axis (see Figure 5.2). The vertical axis is related to the rate at which the sample absorbs photons depending upon the type of spectroscopy. The signal on the vertical axis is proportional to the concentration of the analyte. This axis yields

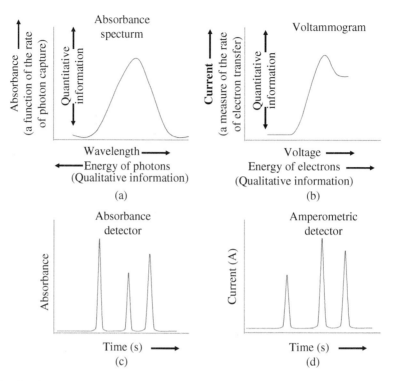

FIGURE 5.2 Graphical analogy between spectroscopy and voltammetry. (a) The height of a peak in an absorbance spectrum relates the amount of analyte to the absorbance, a quantity that is a function of the rate of capture of photons per second. (b) A peak (or a plateau) in the current appears in a voltammogram relating the concentration of analyte to the current, a quantity related to the number of electrons transferred per second. (c) A liquid chromatogram using an absorbance detector at a fixed wavelength. (d) A liquid chromatogram using an amperometric detector at a fixed voltage.

quantitative information while the position on the horizontal axis contains qualitative information about the system. The wavelength axis indicates where the sample absorbs or emits and that is related to the identity of the analyte.

In voltammetry, the graph is called a voltammogram. On the horizontal axis, one plots the energy applied to electrons at the electrode surface (as opposed to the energy of photons) and records the response of the system as the current (in amperes or coulombs per second) for electrons transferred across the interface. The current is a measure of the rate of the surface reaction.

There are also spectroscopic methods in which the wavelength of the instrument is fixed, and changes in absorbance or intensity are monitored. A familiar example is a UV absorption detector used to monitor analytes emerging from a liquid chromatograph (see Figure 5.2c,d). The electrochemical analog is a chromatographic detector in which a fixed voltage is applied to an electrode, and the current is monitored. As analytes emerge from the separation column and flow by the electrode, they are oxidized (or reduced), and the current peak produced is used for quantitative analysis. When the voltage on the sensor is fixed and the current is measured as a function of time, the method is called amperometry.

Sometimes, the current is integrated to get the charge that is transferred. This approach is attractive because charge is related to the number of moles of analyte that has been oxidized or reduced through Faraday's law. If one converts the entire sample from one redox form to another, the charge passed in the process can be used for calculating the amount of the original electroactive compound. Consider the reduction of a species in solution that is initially in the oxidized form (Ox). For example, Fe^{3+} can be reduced to Fe^{2+}.

$$Ox + ne \rightleftharpoons Red \tag{5.1}$$

$$Q = nFN = \int i\,dt \tag{5.2}$$

The charge, Q, in coulombs, associated with electron transfer is proportional to the number of moles, N, of the reactant converted where F is Faraday's constant (96,485 C/mol of electrons), and n is the number of moles of electrons transferred per mole of reactant. When the charge is monitored in an electrochemical experiment, the method is called coulometry.

5.3. CURRENT IS A MEASURE OF THE RATE OF THE OVERALL ELECTRODE PROCESS

Another analogy between spectroscopy and voltammetry is helpful here. In a spectrum, the vertical axis is a function of intensity or the number of photons per second being absorbed or emitted. In fluorescence spectroscopy, the detector output is directly proportional to the number of photons emitted from the sample per second. In absorption spectroscopy, the vertical axis is a more complicated function. It is actually the logarithm of a ratio of intensities, namely the number of photons per second illuminating the sample divided by the number of photons per second transmitted through the sample. This manner of defining absorbance yields a term that increases linearly with analyte concentration. It is also a quantity that grows as the fraction of photons captured per second increases. So, the vertical axis in spectroscopy is still a function of a rate for the process of light interacting with the analyte. In voltammetry, the vertical axis is also a measure of a rate. It is usually the current and, as such, is proportional to the number of electrons transferred per second at the electrode surface.

There are two important consequences of recognizing that the vertical axis (that is, the current) is a measure of the rate of electron transfer. First of all, the current will be proportional to the amount of electroactive material at the electrode surface. The more the reactant available, the faster the rate. The second insight is that the current will be dependent on the rate constant for the electron transfer process between the electrode and the electroactive material.

5.3.1. Rate of Electron Transfer

The rate constant in an electrode reaction depends on the electrode voltage. Consider a simple one-electron exchange where an oxidized form of a ruthenium complex, $Ru(NH_3)_6^{3+}$, is reduced to $Ru(NH_3)_6^{2+}$, at a platinum electrode.

$$Ru(NH_3)_6^{3+} + e^- \rightleftharpoons Ru(NH_3)_6^{2+} \tag{5.3}$$

Current is a measure of the rate of electrons moving across the electrode/solution interface. Therefore, the magnitude of the current is a function of a rate constant. This is worth exploring because it provides a link between current and voltage for an electrode reaction. At a given electrode voltage, there is a rate for the forward (i.e. reduction) reaction.

$$i_f = nFAk_f C_{Ox,s} \tag{5.4}$$

In Eq. (5.4), $C_{Ox,s}$ is the concentration of the oxidized form, $Ru(NH_3)_6^{3+}$, at the electrode surface. The number of moles of electrons, n, transferred per mole of reactant is, of course, unity for this simple case, and k_f is the rate constant for the forward process. The magnitude of the current for the back reaction, the oxidation process, is also controlled by a rate constant, k_b, and the concentration, $C_{Red,s}$ of the reduced form, $Ru(NH_3)_6^{2+}$, at the electrode surface.

$$i_b = -nFAk_b C_{Red,s} \tag{5.5}$$

The forward and back currents represent electrons moving in opposite directions, so the negative sign was introduced for the current in the back direction.

As with reactions between species in solution, the rate constant can be described as a function of the activation energy barrier separating the reactants and products. It follows the form described by the Arrhenius equation.

$$k_f = A \cdot \exp\{-\Delta G_f^{\ddagger}/RT\} \tag{5.6}$$

$$k_b = A \cdot \exp\{-\Delta G_b^{\ddagger}/RT\} \tag{5.7}$$

As with any kinetic process, the rate constant is an exponential function of the activation energy barrier, $\Delta G_f^{\ddagger}$, in the forward direction or $\Delta G_b^{\ddagger}$ in the back direction and the amount of thermal energy available (expressed as RT where T is the absolute temperature, R is the ideal gas constant). The ideal gas constant, R, is also equal to $k_B N_A$, the product of Boltzmann's constant and Avogadro's number. It has a value of $8.314 \, J/(mole \, K)$. The pre-exponential term is related to the frequency of reactant encounter. (In a simple homogeneous reaction, $A = k_B T/h$ where k_B is Boltzmann's constant and h is Plank's constant.) It is also useful to examine the potential energy well diagrams for the reactants and products as shown in Figure 5.3.

In the first diagram, the reactants and products are depicted as having equal free energy on average. This picture is merely one possibility because the energies of the reactants and products vary as the applied potential varies. The free energy of the reactants is tunable because the electrons reside in the working electrode, and their free energy is a function of the electrode voltage. Electronic energy can be added or removed by the outside circuit. The situation diagrammed in Figure 5.3a is a convenient point of reference. It represents an applied potential that is equal to the formal potential for the half reaction

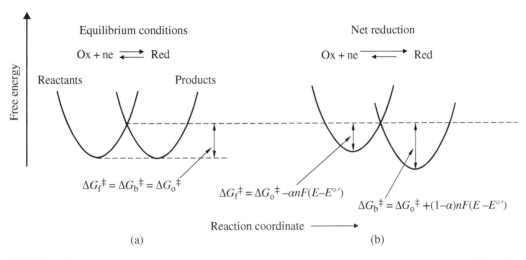

FIGURE 5.3 Plot of the potential energy well for the reactants and the potential energy well for the products separated by an activation barrier. The crossing point defines the activation energy for the forward and back reactions. (a) At equilibrium, the potential wells are equidistant from the activation state when the applied potential equals the formal potential for the redox couple. (b) The activation energy gets smaller for the reduction reaction when the electrode potential is more negative than the formal potential. The activation energy for the back reaction is now larger than when $E = E^{o'}$. Source: Strobel and Heineman 1988 [2]. Adapted with permission of John Wiley and Sons.

$E = E^{o'}$. At this point, the activation barriers for the forward and back reactions are equal, and the rate constants are equal.

$$k_f^o = A \cdot \exp\{-\Delta G_f^{\ddagger}/RT\} = A \cdot \exp\{-\Delta G_b^{\ddagger}/RT\} = k_b^o = A \cdot \exp\{-\Delta G_0^{\ddagger}/RT\} = k^o \quad (5.8)$$

Now consider applying a more negative voltage to the working electrode (where $E < E^{o'}$). Using an outside source of energy (a battery or power supply) to make the electrode more negative (by pushing more electrons into the electrode) raises the free energy of the electrons and, consequently, raises the free energy of the reactants as in Figure 5.3b. The energy well moves up vertically, and the activation barrier for the forward process gets smaller by an amount that is proportional to the shift in voltage away from the formal potential. That is, the activation energy barrier is proportional to $(E - E^{o'})$. The term $(E - E^{o'})$ represents the amount that potential well has moved (in the vertical direction) from the reference state in volts. This term is sometimes called the "overpotential" for the system.

$$\Delta G_f^{\ddagger} = \Delta G_0^{\ddagger} - \alpha nF(E - E^{o'}) \quad (5.9)$$

The voltages, E and $E^{o'}$, are expressed in volts (which is the same as J/C). Multiplying by the number of moles of electrons, n, per mole of reactant and Faraday's constant, $96\,485$ C/mol of electrons, converts the units to joules per mole of reactant (which are also the units for $\Delta G_0^{\ddagger}$ and the product, RT). There is also another coefficient, α. It is called the transfer coefficient. It accounts for the fact that only a fraction of the free energy supplied to the reactants goes into lowering the activation barrier. The diagram suggests that the position of the crossing point for the two potential energy wells changes as the reactant energy well moves vertically. If the well for the products was represented by a horizontal line, then the energy between the bottom of the well and the activation barrier would decrease in the forward direction by the same amount of energy as supplied to the reactants. If the well for the products was represented by a vertical line, then the top of the activation barrier in the forward direction would merely shift straight up by same amount of energy supplied to the reactants. That is, there would be no decrease in the energy barrier in the forward direction as the electrode became more negatively charged. Because the potential energy wells have oblique walls, the activation barrier as measured from the crossing point changes by only a fraction, α, of the applied potential for the forward reaction. The transfer coefficient is sometimes called a symmetry factor. Whenever $\alpha = 0.5$, the walls of the potential wells of the reactants and products have the same steepness. Common values for α range from 0.3 to 0.7 [3].

Accounting for the energy barrier shift, the rate constants can be expressed as follows:

$$k_f = A \cdot \exp \left\{ \frac{[-\Delta G_o^{\ddagger} - \alpha nF(E - E^{o\prime})]}{RT} \right\} \tag{5.10}$$

$$k_b = A \cdot \exp \left\{ \frac{[-\Delta G_o^{\ddagger} + (1 - \alpha)nF(E - E^{o\prime})]}{RT} \right\} \tag{5.11}$$

Notice that the term, $\Delta G_o^{\ddagger}$, for the free energy barrier at the standard state (namely, where $E = E^{o\prime}$) is a constant. It can be factored out and combined with the pre-exponential term, forming a new constant, k^o, known as the standard heterogeneous rate constant. The name is a reminder that the associated reaction takes place at an interface between two different phases and is influenced by the chemical nature and conditions of the two phases.

$$k_f = A \cdot \exp \left\{ \frac{[-\Delta G_o^{\ddagger} - anF(E - E^{o\prime})]}{RT} \right\} = A \cdot \exp \left\{ \frac{-\Delta G_o^{\ddagger}}{RT} \right\} \cdot \exp \left\{ \frac{-anF(E - E^{o\prime})}{RT} \right\}$$

$$= k^o \cdot \exp \left\{ \frac{-anF(E - E^{o\prime})}{RT} \right\} \tag{5.12}$$

On the other hand, driving the electrode to a more negative voltage, increases the activation barrier for the back reaction, or oxidation process by only a fraction, $(1 - \alpha)$, of the overpotential. Therefore, the back-reaction rate constant becomes the following:

$$k_b = A \cdot \exp \left\{ \frac{[-\Delta G_o^{\ddagger} + (1 - \alpha)nF(E - E^{o\prime})]}{RT} \right\} = k^o \cdot \exp \left\{ \frac{(1 - \alpha)nF(E - E^{o\prime})}{RT} \right\} \tag{5.13}$$

The current in the forward direction was associated with a reduction process in this discussion. For the sake of clarity in future discussions, the forward process will be referred to as the reduction or cathodic process, and the reduction current will be represented as i_c

or cathodic current. Eqs. (5.4) and (5.12) can be combined to give the cathodic current as a function of the applied voltage.

$$i_c = i_f = nFAk_fC_{Ox,s} = nFAC_{Ox,s} \cdot k^o \exp\left\{\frac{-\alpha nF(E - E^{o\prime})}{RT}\right\} \tag{5.14}$$

Likewise, the current for the back reaction (oxidation or anodic current) will be called the anodic current and notated as, i_a. Because the anodic current is going in the opposite direction compared to the cathodic current, we introduce a negative sign for the anodic current.

$$i_a = i_b = -nFAk_bC_{Red,s} = -nFAC_{Red,s} \cdot k^o \exp\left\{\frac{(1 - \alpha)nF(E - E^{o\prime})}{RT}\right\} \tag{5.15}$$

5.3.2. The Shape of the Current/Voltage Curve

The measured current, i, for this electrode reaction is actually the sum of the current for the two opposing processes.

$$i = i_c + i_a = nFAC_{Ox,s} \cdot k^o \exp\left\{\frac{-\alpha nF(E - E^{o\prime})}{RT}\right\}$$

$$-nFAC_{Red,s} \cdot k^o \exp\left\{\frac{(1 - \alpha)nF(E - E^{o\prime})}{RT}\right\} \tag{5.16}$$

Equation (5.16) is known as the Butler–Volmer equation [3]. Figure 5.4 shows the relationship between the anodic and cathodic contributions to the net current when both the

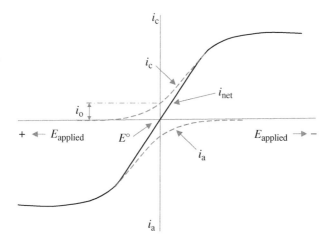

FIGURE 5.4 The net current is the sum of the individual anodic and cathodic current components for an applied electrode voltage in the neighborhood of the formal potential, $E^{o\prime}$. The conditions depicted here are for equal concentrations of the oxidized and reduced forms. The current in either direction at $E = E^{o\prime}$ is called the exchange current, i_o [4].

oxidized and reduced forms of the redox couple are present at equal concentrations. The net current goes to zero when the applied potential matches the formal potential, $E^{o\prime}$. Of course, under those conditions, both the anodic and cathodic currents, themselves, are not zero. They merely off-set each other at that point. The absolute value of either at $E = E^{o\prime}$ is known as the exchange current, i_o, and is dependent on the standard heterogeneous rate constant, k^o. The larger the value of k^o, the steeper the curve in Figure 5.4.

$$i_o = nFAk^o C_{Ox,s} = nFAk^o C_{Red,s} \tag{5.17}$$

5.3.3. Rate of Mass Transport

One concludes from the aforementioned discussion that the rate of the forward process (the reduction of the $Ru(NH_3)_6^{3+}$) is faster as the electrode potential becomes more negative. However, this is not the complete story. There is another step in the overall process leading to the generation of current that needs to be addressed. The movement of the reactant from the bulk solution to the electrode surface is also important. Because the rate of the electrode reaction is measured as the current, one would expect to see an exponential rise in the reduction current as the voltage applied to the electrode becomes more negative. However, that is not what one sees.

What does a complete voltammogram look like? It turns out that there are some differences depending on whether or not the solution is stirred during the voltage scan. It is a bit simpler to discuss a stirred solution, so consider an experiment in which a platinum electrode is fixed in the wall of a channel or tube as diagrammed in Figure 5.5b. Imagine that an aqueous solution of the ruthenium complex, $Ru(NH_4)_6^{3+}$, is pumped continuously past the electrode. This setup is known as a flow-through electrode system. It provides a reproducible system that can be modeled accurately by equations. In general, stirred experiments are called hydrodynamic or convective electrode systems. A reference electrode completes the electrical connection to the outside circuit. In the outer circuit, there is a sensitive current meter and an adjustable voltage source that allows one to scan the voltage or apply a fixed value to the cell. It is common to generate a current/voltage curve (voltammogram) by automatically scanning the voltage applied to the working electrode linearly with time and plotting the current response as function of the applied voltage. (The circuitry needed to perform this task is described in Chapter 7). Figure 5.5c represents a current/voltage curve for the reduction of the $Ru(NH_4)_6^{3+}$ complex. (Note that it is common to plot the voltammogram with the negative voltage region on the right when a reduction process is being considered.)

Imagine starting the scan on the left end of the graph at a relatively positive voltage, such as $+1$ V. Here, $E \gg E^{o\prime}$. Eq. (5.14) indicates that the forward rate constant is very small at this point, so the current is virtually zero. As the voltage is scanned in the negative direction, the rate constant grows, but the current is not perceptible until the voltage gets to within about $200\,mV$ of $E^{o\prime}$. Beyond that point, the current rises from the baseline in an exponential manner. However, at more negative voltages, the current begins to deviate from the exponential curve and eventually becomes horizontal at voltages much more negative than $E^{o\prime}$. What is going on?

The answer is that as the voltage becomes more negative, another step in the overall process becomes rate-determining. The picture up to this point has been over-simplified.

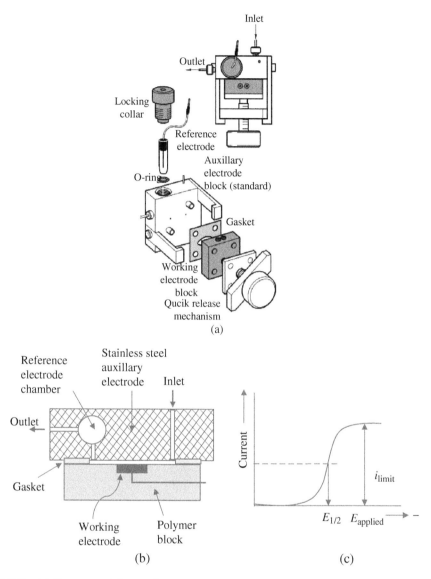

FIGURE 5.5 (a) Commercial flow cell assembly. Source: Adapted with permission of Bioanalytical Systems, Inc. [5]. Copyright 2006, Bioanalytical Systems, Inc. (b) Cross-sectional view of flow cell with reference electrode chamber and electrical connections. (c) Voltammogram for the reduction of a $Ru(NH_4)_6^{3+}$ complex at a platinum electrode in a flowing system. The potential where $i = (1/2)i_{limit}$ is known as the half-wave potential.

If $Ru(NH_4)_6^{3+}$ is reduced at the electrode surface, the current can be maintained only if more reactant is transported to the surface. Figure 5.6 shows that the overall process is a series of steps, namely, mass transport of reactant from the bulk solution to the outer Helmholtz plane (OHP), electron exchange and mass transport of the product away from the surface. In some electrochemical reactions, there are additional steps, such as adsorption (movement of the reactant from the outer to the inner Helmholtz plane [IHP]), proton

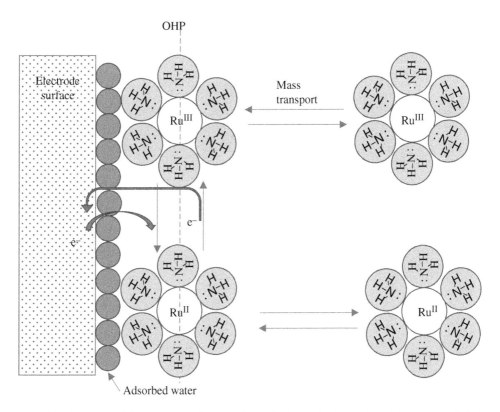

FIGURE 5.6   The current in a voltammetry experiment is dependent on, at least, two sequential steps: mass transport of reactant to the electrode surface followed by electron transfer. Adsorption, ligand exchange, and proton transfer are other possible steps that can also be involved [3].

transfer or ligand exchange. However, in even the simplest of reactions, more reactant must be brought to the surface in order to maintain the current.

As the rate of electron transfer becomes very fast (at $E \ll E^{o\prime}$ for a reduction), the mass transport step becomes rate limiting. The rate of the mass transport step depends on the intensity of the stirring. In the case of a flow-through electrode, the faster the volume flow rate, the greater the convection and the faster the mass transport. The hydrodynamics can be characterized by how closely the movement of the bulk solution stirs the layer next to the electrode surface. A thin layer of solvent adheres to the solid. In fact, the motion of the bulk solution has little effect on the solvent within a few microns of the surface. (This zone is much, much thicker than the single layer of solvent molecules in direct contact with the electrode. The presence of this stagnant film is sometimes referred to as the "no slip" condition by engineers.) The faster the flow rate in this system, the thinner the stagnant layer. The ruthenium(III) complex must diffuse across this stagnant layer in order to reach the electrode. Hence, electrochemists also refer to this zone as the diffusion layer [3, 4]. Figure 5.7 depicts the diffusion layer for two different flow rates under conditions where the electron transfer rate is so fast that the ruthenium(III) is immediately converted to ruthenium(II) as it arrives at the surface. That is, under these conditions, the overall process (as measured by the current) is limited by the mass transport step. The current is proportional to

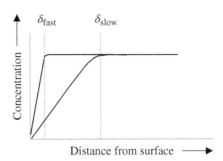

FIGURE 5.7 The concentration profile for a reducible species near the electrode surface at a voltage where the electron transfer step is very fast. The mass transport step at this point is rate limiting. The diffusion layer thickness, δ, depends on the intensity of the stirring for a hydrodynamic electrode system. Concentration profiles are shown at a flow-through electrode at fast and slow flow rates. The current is proportional to the concentration gradient at the surface. Therefore, the thinner the diffusion layer, δ, the higher the current.

the flux of material, that is the rate at which material moves by diffusion through a plane perpendicular to the direction of flow. The flux is described by Fick's first law:

$$J_{O,x} = -D_O \frac{\partial C_O}{\partial x} \tag{5.18}$$

where D_O and C_O are the diffusion coefficient and concentration of the oxidized species, respectively, and x is the direction of movement. From Figure 5.7, the concentration gradient near the electrode surface is approximately equal to the ratio of the bulk concentration of the analyte over the diffusion layer thickness.

$$J_{O,x} = -D_O \frac{\partial C_O}{\partial x} \approx -D_O \frac{C_O}{\delta} \tag{5.19}$$

$$i_L = -nFAJ_{O,x} = nFAD_O \frac{C_O}{\delta} = m_O nFAC_O \tag{5.20}$$

The faster the stirring, the thinner the diffusion layer, δ, the more rapid the mass transport and the higher the current [3]. When the current is mass transport limited, it becomes independent of the applied voltage. At this point, the current reaches a limiting plateau because the stirring is constant. The mass transport coefficient, m_O, is equal to the ratio of the diffusion coefficient for the oxidized species (in cm^2/s) and the diffusion layer thickness, δ, in centimeters. On the current limiting plateau, the concentration of the reactant at the surface is virtually zero because of its extremely rapid conversion at the electrode. Consequently, the ratio of C_O/δ is the slope of the concentration profile for the reactant near the electrode surface.

An important inference is that the limiting current is proportional to the bulk concentration and inversely proportional to the diffusion layer thickness. The diffusion layer thickness and the mass transport coefficient have been carefully described for several different hydrodynamic electrode systems. One of the more popular ones is a rotated

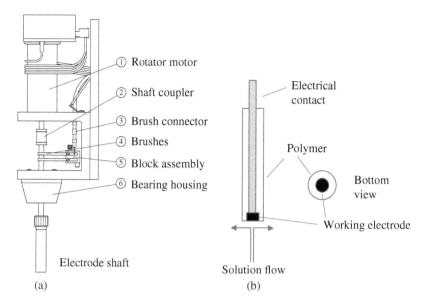

FIGURE 5.8 (a) Rotated disk electrode assembly. Source: Adapted with permission ofrom Bioanalytical Systems, Inc. [6]. Copyright 2000, Bioanalytical Systems, Inc. (b) Cross-section and bottom views of electrode shaft. The rotating electrode draws solution up to and across the rotating disk.

disk electrode (RDE) (see Figure 5.8). For a RDE, the mass transport coefficient is the following: [3]

$$m_{\mathrm{o}} = \frac{D_{\mathrm{O}}}{\delta_{\mathrm{RDE}}} = 0.617 D_{\mathrm{O}}^{2/3} \omega^{1/2} v^{-1/6} \tag{5.21}$$

where ω is the rotation rate in rad/s ($\omega = 2\pi f$, where f = the rotation rate in rev/s) and v is the kinematic viscosity (the ratio of the solution viscosity to its density) expressed in cm^2/s. Consequently, the limiting current for a RDE is as follows:

$$i_\ell = m_{\mathrm{O}} nFAC_{\mathrm{O}} = 0.62 D_{\mathrm{O}}^{2/3} \omega^{1/2} v^{-1/6} nFAC_{\mathrm{O}} \tag{5.22}$$

Equation (5.22) is known as the Levich equation. In order for the units to work out, the concentration of the analyte in the bulk solution, C_{O}, is in $\mathrm{mol/cm}^3$. An important conclusion is that the limiting current will increase with the square root of the rotation rate. For flow-through electrodes, m_{O} is proportional to the cube root of the flow rate [7].

For any hydrodynamic system, the voltammogram takes the shape of a sigmoidal curve or wave similar to that of a flow-through electrode as illustrated in Figure 5.5c. At the foot of the wave, the current is controlled by the rate of the electron transfer process. On the plateau, the current is controlled by mass transport. Both steps contribute to the overall rate in the intermediate region. It can be shown that for very rapid electron transfer, the ratio of the concentrations for the oxidized and reduced species *at the surface of the electrode* is equal to the ratio predicted by the Nernst equation for the electrode potential. In such cases, the current and potential in the rising portion of wave (where $0.1 i_1 < i < 0.9 i_1$) follow Eq. (5.23) [3].

$$E = E^{o\prime} - \frac{RT}{nF} \ln \left(\frac{D_O}{D_R} \right) - \frac{RT}{nF} \ln \left\{ \frac{i}{i_1 - i} \right\} = E_{1/2} - \frac{RT}{nF} \ln \left\{ \frac{i}{i_1 - i} \right\} \tag{5.23}$$

In Eq. (5.23), D_R and D_O are the diffusion coefficients for the reduced and oxidized species, respectively. The potential halfway up the steep portion of the curve (where the current is half the value of the limiting current) is often called the half-wave potential, $E_{1/2}$.

$$E_{1/2} = E^{o\prime} - \frac{RT}{nF} \ln \left(\frac{D_O}{D_R} \right) \approx E^{o\prime} \tag{5.24}$$

The half-wave potential is a good estimate of the formal potential because the diffusion coefficients are often similar.

In the discussion so far, the concentration of the product, ruthenium(II) complex, in the bulk solution was assumed to be zero. Of course, the same principles hold when the concentration of the reduced species is nonzero. The oxidation of $Ru(NH_4)_6^{2+}$ at the electrode produces a limiting current when the applied voltage is much more positive than $E^{o\prime}$. This anodic limiting current, $i_{l,a}$, is dependent on the transport coefficient and the bulk concentration of the reduced form.

$$i_{l,a} = m_R nFAC_R = nFAD_R \frac{C_R}{\delta} \tag{5.25}$$

In the presence of finite concentrations of both species in the bulk solution, one can write a more general expression for the current/voltage curve [2].

$$E = E^{o\prime} - \frac{RT}{nF} \ln \left(\frac{D_O}{D_R} \right) - \frac{RT}{nF} \ln \left\{ \frac{i - i_{l,a}}{i_{l,c} - i} \right\} = E_{1/2} - \frac{RT}{nF} \ln \left\{ \frac{i - i_{l,a}}{i_{l,c} - i} \right\} \tag{5.26}$$

where the subscripts a and c refer to the anodic (oxidation) process and cathodic (reduction) process, respectively.

Hydrodynamic electrode systems are frequently used for quantitative analysis. Such applications are especially appealing in cases where the voltage is fixed, and the current is monitored because high sensitivity can be achieved by intense stirring. The main value of Eqs. (5.23, 5.26) to the analyst is their guidance in keeping current measurements for quantitative analysis far enough beyond the $E_{1/2}$ value so that the current being measured is on the current-limiting plateau for the species of interest. Under those conditions, the current is directly proportional to the concentration of the analyte. One avoids working at a value close to $E_{1/2}$, where a small change in the applied voltage can lead to a large change in the current and, therefore, effect the sensitivity of the analysis.

5.3.4. Electrochemical Reversibility

One other important concern should be discussed at this point. One of the conditions invoked in describing the shape of the aforementioned current/voltage curve was that the electron transfer process was assumed to be so facile that the ratio of the concentrations for the oxidized and reduced species *at the surface of the electrode* is equal to the ratio predicted by the Nernst equation at the applied electrode potential. These are special conditions and deserve a special label. Electrochemists refer to these conditions as an electrochemically "reversible system." This term really refers to the kinetics of the electron transfer process. Reversibility means that the Nernst equation applies to the ratio of oxidized and reduced concentrations at the electrode surface. It can be experimentally demonstrated that a system is reversible in the electrochemical sense whenever the heterogeneous rate constant, k^o, is greater than 2×10^{-2} cm/s. But, why is that important?

Whenever the k^o value is slower, the rising portion of the curve is less steep as seen in Figure 5.9. Furthermore, the potential on the curve half-way up the wave will deviate significantly from the formal potential. When the wave is less steep, the applied voltage necessary to reach the current-limiting plateau (where quantitative analysis is the most sensitive and most reproducible) must be more extreme. The more extreme the applied potential, the greater the chances that some species other than the analyte will also react and contribute to the current. Even in the absence of other electroactive solutes, the solvent exchanges electrons with the electrode at extreme voltages, limiting the useful voltage range for one's analysis.

Some electron transfer reactions are sluggish. The reasons for that can be quite complicated, but even normally reversible electron transfer can be inhibited by a dirty electrode surface. (Methods for cleaning and treating solid electrode surfaces in order to improve their performance are discussed in Section 5.5.2.3.) The type of the material used for the working electrode can make a substantial difference as well. In extreme cases ($k^o < 2 \times 10^{-5}$ cm/s), the sloping portion of the curve extends over several hundred mV, and electrochemists say the reaction is "irreversible." [3] It is important to emphasize that this is a special context for the term. It does not necessarily mean that the reaction will not go in the other direction. Often one can observe the reverse reaction by applying a voltage far to the other side of the $E_{1/2}$. Of course, there are electrode reactions where electron transfer is accompanied by bond breaking and loss of substituents. Then, reversing the process may be unlikely in the absence of significant concentrations of those separate pieces. Systems with current/voltage curves of intermediate steepness are called "quasi-reversible" systems. Systems that appear quasi-reversible are actually the most commonly observed with solid working electrodes in aqueous solution. Although these terms are related to the numerical value of k^o, they are mainly used as a qualitative

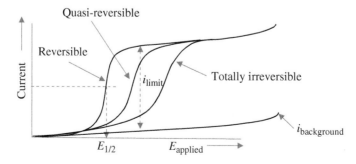

FIGURE 5.9 Hydrodynamic voltammograms for reversible, quasi-reversible, and irreversible electrode reactions. Source: Bard and Faulkner 1980 [8]. Adapted with permission of John Wiley & Sons.

descriptor for the shape of the voltammogram when working with hydrodynamic electrodes.

In real voltammograms, there is also a background current. At extreme applied voltages (beyond +1 and −1 V), the electrolysis of the solvent or supporting electrolyte contributes current. Trace levels of electroactive contaminants and changes in the electrode surface also can contribute to the background current. Even the very act of scanning the voltage produces a current associated with establishing the separation of charge at the interface that defines the potential energy difference. (The origins of this charging current and techniques for circumventing its influence on the signal are discussed in Section 5.4.) The lesson for the analyst is to always run a blank. That is, whenever using voltammetry for quantitative analysis, always record a current/voltage curve under the same conditions as used for scanning the sample, but in the absence of the analyte. The appropriate signal is measured as the difference between the limiting current plateau in the sample scan compared to the current at the same potential recorded in the blank (see Figure 5.9).

5.3.5. Voltammetry at Stationary Electrodes in Quiet Solutions

Stirring the solution provides a powerful way of enhancing the current signal and, therefore, the sensitivity of a voltammetric analysis. However, stirring the sample may not be practical in some situations, such as during *in vivo* analysis of neurotransmitter molecules in the brain. In a quiet solution, products of electrode reactions accumulate near the electrode surface. Consequently, voltammetry in quiet solutions provides an opportunity for probing chemical reactions of species that are formed from electron transfer processes and that may not be easily made or manipulated with standard methods of synthesis. This type of voltammetry has been useful for studying redox reactions important to biochemical processes over a wide range of experimental conditions. Because the metabolism of food and drug molecules is largely governed by oxidation processes, voltammetric studies can help identify key intermediates that might otherwise be missed during *in vivo* studies. For example, voltammetry has been used to investigate side reactions in the biochemical degradation of the neurotransmitter molecule dopamine as a possible source of toxins that might lead to neuronal damage following heavy ethanol consumption [9].

Voltammetry can be used to measure rates of reaction and the relative stability of various intermediates. Another area where the study of electrochemical reaction mechanisms is crucial is the field of catalysis. Redox catalysis is of interest in many areas such as organic synthesis, energy production (energy-efficient oxygen reduction for fuel cells), carbon dioxide reduction, electrochemical decomposition of organic molecules for waste water treatment, and water oxidation (O_2 generation). Improvements in efficiencies for these technologies may result from a fundamental understanding of the reaction pathways [10].

Curiously, electrode reactions in a quiet solution are more complicated to describe than in stirred solutions. In a quiet solution, the mass transport of electroactive material to and from the working electrode surface is a function of both the applied potential and the time spent at a voltage where electrolysis occurs. In order to build an intuitive sense of mass transport behavior in a quiet solution, it is helpful to consider first a simple experiment where the potential is stepped from a voltage where the current is zero (because the electron-transfer rate is essentially zero) to a voltage where the electron-transfer is very fast and the current is mass-transport-limited.

5.3.5.1.   *Potential Step Experiments: Chronoamperometry.*

The study of current–time curves following a potential step is generally known as chronoamperometry. Chronoamperometry was an early tool for measuring important physical parameters, such as diffusion coefficients and the electrode surface area. It serves as a basis for the theory of many modern voltammetric methods. Consider an experiment in a quiet solution in the presence of an electroactive species, Ox, in which the potential is stepped from a value where no electrolysis occurs to a potential well past where the electron transfer is so fast that the current is limited entirely by the rate of mass transport.

The current following the potential step decreases as the square root of time, t, according to the Cottrell equation for a planar electrode as shown here [3].

$$i = \frac{nFAC_{ox}D_{ox}^{1/2}}{\pi^{1/2}t^{1/2}} \tag{5.27}$$

Why should the current change with time? The initial voltage is chosen to be well before the formal potential, and the final potential is chosen to be well after the formal potential for the electroactive material. Those voltages would be well before the current wave in the corresponding hydrodynamic voltammogram and well out on the current limiting plateau, respectively, as shown in Figure 5.10b. The initial current following the step would be high (Figure 5.10d), but the electrode reaction would immediately consume

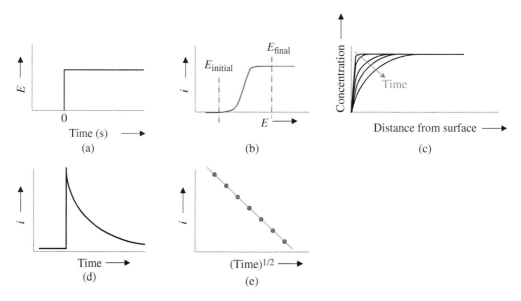

FIGURE 5.10   The chronoamperometry experiment. (a) Potential step. (b) The initial and final voltage levels for the potential step are chosen to be well before the formal potential and well after $E^{o\prime}$, respectively, as indicated on this hydrodynamic voltammogram. (c) Concentration profiles near the electrode surface as a function of time following the potential step. (d) Current as a function of time following the potential step. (e) Faradaic current versus $t^{1/2}$ following a potential step [4].

most of the reactant near the surface, and continued electrolysis would depend on the rate at which reactant is replenished. In this case, mass transport is solely dependent on the rate of diffusion. If it were possible to take snapshots of the concentration profile of the reactant species near the electrode surface, one might see that the concentration profile is quite steep at times very soon after the potential step (Figure 5.10c). As time increases, the zone of depleted reactant extends further into the solution, but the concentration profile becomes less steep. Fick's first law (Eq. (5.19)) indicates that the flux of material to the surface will depend on the steepness of this concentration profile, so the current decreases with time. Figure 5.10d represents the current response as described in the Cottrell equation. A plot of the current versus $t^{-1/2}$ (Figure 5.10d) is frequently used to extract any one of the following: the diffusion coefficient, D_{Ox}, the electrode surface area, A, or n, the number of electrons per reactant, when the bulk concentration, C_{Ox}, and the other parameters are known.

One can also integrate the current during the experiment and plot the charge versus time. That variation on the method is called chronocoulometry [3]. Here we include two additional sources of charge.

$$Q = \frac{2nFAD_O^{1/2}C_{Ox}t^{1/2}}{\pi^{1/2}} + Q_{dl} + nFA\Gamma_O \tag{5.28}$$

where Q_{dl} is the charge associated with the double layer. Electroactive material attached to the surface accounts for the last term. Γ_O is the concentration of any material attached to the surface that is reduced during the step, in mol/cm^2.

Because in the course of integrating, noise in the current signal averages out, the coulometric signal appears less noisy. That is, there is a better signal-to-noise ratio for the charge measurement than for the current sampled at a given time in a chronoamperometric experiment. Although the process of integrating also includes the double layer charging, Q_{dl}, this nonfaradaic component is relatively easy to evaluate in chronocoulometry. As Eq. (5.28) indicates, the value of the y-intercept in a plot of Q versus $t^{1/2}$ is equal to $Q_{dl} + nFA\Gamma_O$.

EXAMPLE 5.1

Chronoamperometry is a good method for determining the effective surface area of an electrode. This application is particularly attractive when the geometry of the electrode is a bit irregular. It is also the case that the surface roughness can make the effective area differ from the geometric area. The Cottrell equation indicates that a plot of the current versus $t^{1/2}$ will produce a straight line with a slope that is proportional to the electrode area, A. Here are current–time data for a step from +0.5 V to 0 V versus SHE (standard hydrogen electrode) for a solution containing 4 mM Fe(CN)$_6^{3-}$ in an aqueous 0.1 M KCl solution at 25 °C. The diffusion coefficient for Fe(CN)$_6^{3-}$ is 7.62×10^{-6} cm^2/s in this electrolyte [11]. (Note that 1 M $= 1 \times 10^{-3}$ mol/cm^3). What is the value for the electrode surface area?

Answer: $i = \frac{nFAC_{Ox}D_{Ox}^{1/2}}{\pi^{1/2}t^{1/2}} = \frac{nFAC_{Ox}D_{Ox}^{1/2}}{\pi^{1/2}}(t^{-1/2})$ or slope $= \frac{nFAC_{Ox}D_{Ox}^{1/2}}{\pi^{1/2}}$

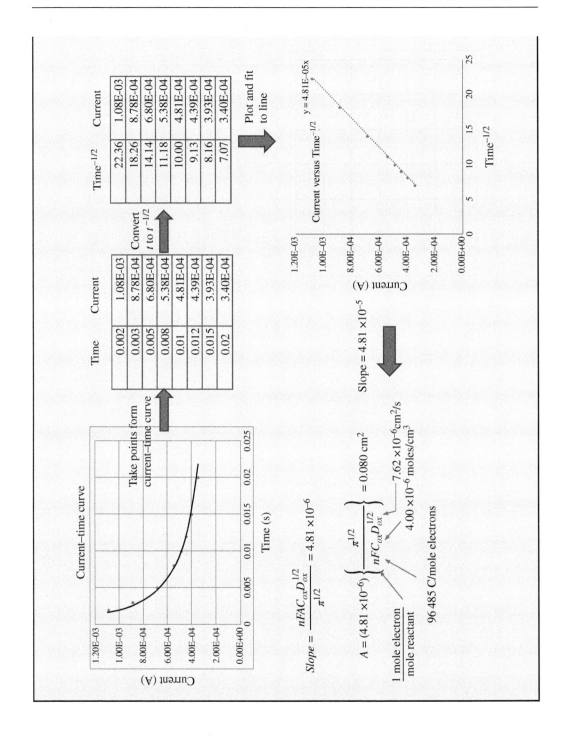

5.3.5.2. *Linear Voltage Scan: Cyclic Voltammetry (CV).* CV is the most widely used

controlled potential method employed in unstirred solutions. The name derives from the fact that the voltage applied to the working electrode is programmed in a cyclic pattern. The pattern is illustrated in Figure 5.11a. The potential is scanned in a linear manner from a chosen starting point, E_i, out to another, preselected value, $E_{\lambda 1}$, called the switching potential. At $E_{\lambda 1}$, the direction of the scan is reversed, and the voltage is ramped back toward the starting point at the same rate. Often the experiment is programmed to stop once the voltage returns to the start. However, it is also common to continue the scan past the starting voltage to some new value, $E_{\lambda 2}$. At $E_{\lambda 2}$, the scan direction is switched again, and the applied voltage drives back toward the original voltage, E_i. Hence, the applied voltage completes a cycle. Often, immediately repeating the cycle can reveal peaks associated with species formed by chemical reactions coupled to the electron transfer process.

How does the lack of stirring affect the shape of the voltammogram? The mass transport is similar to that governing a chronoamperometry experiment. Consider a quiet solution containing a modest concentration of a ruthenium complex in the oxidized form, $Ru(NH_3)_6^{3+}$. Even a hydrodynamic electrode cannot keep the concentration of the reactant species equal to the bulk concentration once the electron transfer rate begins climbing near $E^{o\prime}$ for the reaction. In a quiet solution, the only mechanism for moving the reactant to the electrode surface is diffusion. Compared to convection, diffusion is slow. Vigorous stirring can confine the depletion zone to a thin layer at the electrode surface, but in a quiet solution, the depletion layer continues to grow further and further out into the bulk

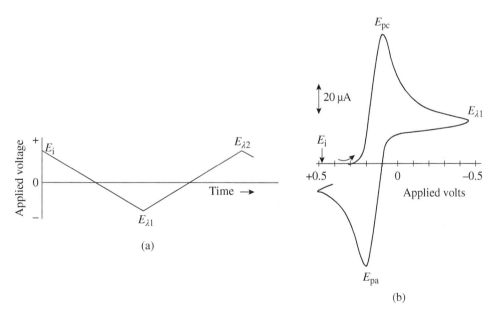

(a)

(b)

FIGURE 5.11 The CV experiment. (a) The applied potential starts at some initial value, E_i, and is scanned linearly to the first switching potential, $E_{\lambda 1}$, where the scan direction is reversed. The voltage is often stopped at a second switching point, $E_{\lambda 2}$. In some experiments, the voltage is cycled between $E_{\lambda 1}$ and $E_{\lambda 2}$ multiple times. (b) A current/voltage curve for the CV of 1 mM $K_3Fe(CN)_6$ at a platinum electrode in 0.1 M KCl solution versus a Ag/AgCl reference electrode in 0.1 M KCl. Source: Reproduced with permission from Mabbott 1983 [12]. Copyright 1983, American Chemical Society.

solution. As a result, the diffusion layer grows in thickness. In a sense, the reactant material has to travel further to get to the electrode. The mass transport step gets further behind. A more quantitative way of characterizing the situation is to describe the concentration gradient ($\partial C/\partial x$). In the hydrodynamic system, the concentration gradient is constant at applied voltages more negative than $E^{o\prime}$. In the quiet system, this gradient gets smaller with time at voltages more than 30 mV beyond $E^{o\prime}$. (The situation at this point matches the behavior of the changing concentration profile following a voltage step as in Figure 5.10c.) Consequently, the current begins to drop, even though the electron transfer rate constant is very fast at applied voltages well beyond $E^{o\prime}$. In fact, because the depletion zone is growing and the concentration near the electrode is decreasing with time, the current becomes independent of the applied voltage at this point. It actually begins to decline at a rate proportional to the square root of time. Thus, the current forms a peak in the voltammogram.

Although the concentration for the reactant species near the electrode is decreasing with time, the product is also building up at the electrode surface. Consequently, when the applied voltage is scanned back in the other direction, the $Ru(NH_3)_6^{2+}$ product is available to be oxidized back to the starting material. Once the applied voltage approaches the formal potential again, the current begins to increase rapidly in the anodic (downward) direction as the complex is oxidized back to $Ru(NH_3)_6^{3+}$. Of course, the supply of $Ru(NH_3)_6^{2+}$ at the surface is also limited, so the anodic current peaks out in the downward direction, as well. The fact that in a cyclic voltammogram, one obtains information about the oxidation as well as the reduction process has several advantages. First of all, it provides an excellent way of estimating the formal potential for the redox couple. The average of the peak potentials yields the same voltage as the half wave potential, $E_{1/2}$, from a hydrodynamic voltammogram, which is very close to the formal potential [12]. That is

$$\frac{E_{pa} + E_{pc}}{2} = E_{1/2} = E^{o\prime} - \frac{RT}{nF} \ln\left(\frac{D_O}{D_R}\right) \approx E^{o\prime} \qquad (5.29)$$

In Eq. (5.29), E_{pa} and E_{pc} are the potentials corresponding to the maxima of the peaks for the anodic and cathodic processes, respectively. CV is now the most commonly used method for determining the formal potential for electroactive species.

Unlike the current/voltage curve for a hydrodynamic electrode, the cyclic voltammogram provides a good estimate of $E^{o\prime}$, even when the electron transfer is quasi-reversible. Furthermore, the voltage separation between the peaks is also a good indicator of the reversibility of the electron transfer. If the system is reversible, then the difference in peak potentials will be close to 60 mV for a one-electron transfer process. More generally, for a reversible system,

$$E_{pa} - E_{pc} = \Delta E_p = \frac{57}{n}\,\text{mV} \qquad (5.30)$$

where the peak potentials are expressed in mV [13]. Systems where the peak separation is greater than 200 mV are considered irreversible in the electrochemical sense. The appearance of voltammograms for these two extremes is illustrated in Figure 5.12. It is worth noting that for a quasi-reversible system, scanning faster causes the peaks to separate even further. However, when peaks separate by more than 200 mV, one should be aware that the estimate for the formal potential based on the peak potential average is subject to errors. The anodic and cathodic peaks do not necessarily shift away from the $E^{o\prime}$ at the same rate.

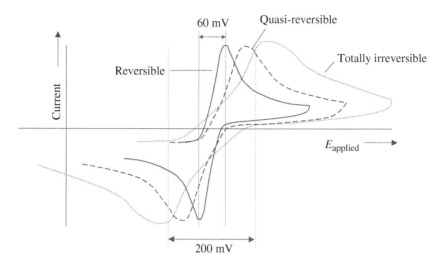

FIGURE 5.12   The shape of cyclic voltammograms for reversible, quasi-reversible, and electrochemically irreversible (slow electron transfer kinetics) systems.

EXAMPLE 5.2

Consider the cyclic voltammogram for an iron complex below (recorded with a Ag/AgCl reference electrode with 1 M KCl, E_{ref} = +0.222 V). Assuming that the electron transfer is a one-electron process, is this a reversible, quasi-reversible, or irreversible process in the electrochemical sense? Estimate the formal potential for this redox couple under these conditions. What is the value of the formal potential versus SHE? Answer:

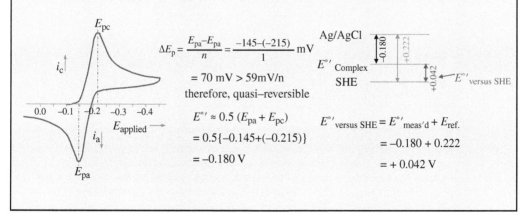

$$\Delta E_p = \frac{E_{pa} - E_{pa}}{n} = \frac{-145 - (-215)}{1} \, mV$$

$$= 70 \, mV > 59 mV/n$$

therefore, quasi-reversible

$$E^{\circ\prime} \approx 0.5 \, (E_{pa} + E_{pc})$$

$$= 0.5\{-0.145 + (-0.215)\}$$

$$= -0.180 \, V$$

$$E^{\circ\prime}_{versus \, SHE} = E^{\circ\prime}_{meas'd} + E_{ref.}$$

$$= -0.180 + 0.222$$

$$= +0.042 \, V$$

Changing the voltage scan rate can be helpful. First of all, if the reaction is associated with reactants and products that are dissolved in the surrounding solvent, and their mass transport is solely based on diffusion, then the magnitude of the peak current (in either direction) increases as the square root of the scan rate. The magnitude of the peak current is directly proportional to the concentration of the electroactive species as described by the Randles–Sevcik equation [4].

$$i_p = (2.69 \times 10^5) n^{3/2} A D^{1/2} v^{1/2} C^* \tag{5.31}$$

where n is the number of moles of electrons transferred per mole of reactant, A is the electrode surface area in cm^2, D is the diffusion coefficient of the reactant in cm^2/s, v is the voltage scan rate in V/s, and C^* is the bulk concentration of the reactant in mol/cm^3. The electrolysis of material that is bound to the electrode surface is easily distinguished,

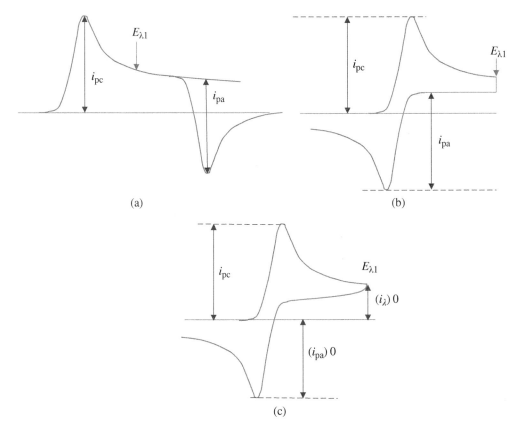

FIGURE 5.13 Three methods for estimating the peak current for both the anodic and cathodic processes from a cyclic voltammogram. Source: Adapted with permission Mabbott 1983 [12]. Copyright 1983, American Chemical Society. (a) The scan is stopped at the switching potential, E_{λ_1}, but the current plot is continued versus time in order to define the baseline for the anodic peak. Repeating the scan, a second time (plotting continuously versus time) and switching the scan direction at E_{λ_1}, allows one to measure the anodic peak, i_{pa}, with respect to the current at the same time on the first scan. (b) In this method, the scan is stopped at E_{λ_1}, and the voltage is held until the current decays to a steady state. Then, the scan is continued in the anodic direction. Both peak currents are measured with respect to the steady current that immediately preceded them during the scan. (c) A normal scan is recorded. If the switching potential, E_{λ_1}, is at least 35 mV beyond the cathodic peak, three current measurements with respect to the $i = 0$ baseline allow one to calculate the true peak current ratio from theory [14]:

$$\frac{i_b}{i_f} = \frac{i_{pa}}{i_{pc}} = \frac{(i_{pa})_0}{i_{pc}} + \frac{0.485(i_\lambda)_0}{i_{pc}} + 0.086$$

because it produces a current peak that grows linearly with scan rate. Because the background current is often a relatively large component of the total current in a CV (and it increases with scan rate), one needs to estimate the baseline for the background current in order to quantify the signal for the electrolysis of the analyte (the Faradaic current). Methods for approximating the baseline and measuring the current peak signals are illustrated in Figure 5.13.

Conventional CV is typically performed in solutions with analyte concentrations in the 0.1–200 mM range. Eq. (5.31) indicates that the peak current increases as the square root of scan rate. Why not scan faster in order to obtain better sensitivity? The problem is that the background current increases linearly with scan rate. The background can eventually overwhelm the desired signal at high scan rates. At lower concentrations, the signal looks small compared to the background current. Section 5.4 discusses special techniques that are better suited for quantitative analysis in dilute solutions.

A ratio for the peak currents that differs from unity can indicate a homogeneous reaction of the product after electron transfer. CV is particularly well suited to studying chemical reactions associated with an electron transfer step. For example, Figure 5.14A shows a cyclic voltammogram of aniline at a glassy carbon electrode in a buffer solution at

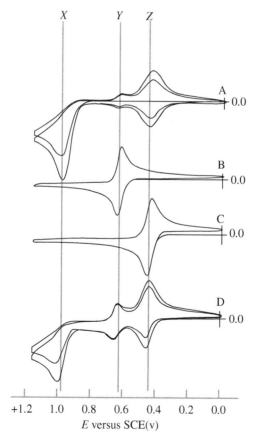

FIGURE 5.14 Cyclic voltammograms of aniline and related compounds at a carbon electrode in aqueous solution at pH 2.3. All scans started at 0 V in the positive direction. A: aniline alone; B: ben-zidine alone; C: 4-amino- diphenylamine alone; and D: mixture of all three. Source: Bacon and Adams 1968 [15]. Adapted with permission of American Chemical Society.

pH 2.3. The applied voltage started at 0.0 V (versus SCE [saturated calomel electrode]) and was scanned first in the positive direction. In the first positive-going segment, the current was essentially zero until a very large anodic current peak appeared near +1.0 V (shown by the vertical line marked X). The scan direction was reversed at +1.2 V. There was no return peak corresponding to the original anodic peak, but there was a reduction peak at about +0.65 V (marked as Y). A second reduction peak appeared at +0.45 V (marked as Z). The scans were repeated for two more complete cycles. In the second and third scans, two pairs of peaks appeared around the positions marked Y and Z even though no sign of the anodic peaks was evident for either pair in the original positive scan [15, 16].

The species associated with peaks at Y and Z appear to be products of the oxidation reaction at X. The size of the peak at Y decreased and the size of the peak at Z grew as the pH was increased (not shown). At pH values above 4, the peaks at Y are insignificant. A solution of the starting material at a pH above 4 was exhaustively oxidized (using a platinum screen electrode with a large area) while holding the working electrode at a voltage slightly more positive than 1 V. Using spectroscopy, the structure of the product was shown to be this imine [15, 16].

This structure looks like the oxidized form of an aniline dimer, p-aminodiphenylamine.

In fact, when a cyclic voltammogram of p-aminodiphenylamine was recorded under the same conditions, it looked exactly the same as the set of peaks at Z as shown in curve C of Figure 5.14. It was reasoned that since aniline ring can stabilize a cation radical by resonance, that the anodic

process at X likely represents two separate oxidation processes with a chemical step in between. That is, a single-electron oxidation forms the aniline cation radicals first, and these radicals quickly dimerize. At a pH above 4, the dimer is entirely p-aminodiphenylamine. From curve C in Figure 5.14, the formal potential appears to be about +0.45 V. The reduced form of the p-aminodiphenylamine is stable at potentials more negative than +0.45; therefore, at +1 V where it is first formed, it immediately oxidizes to the imine form. Consequently, the peak at X coincides with a series of three rapid steps: a one-electron oxidation followed by a dimerization followed by a two-electron oxidation. (Electrochemists refer to this as an ECE mechanism to indicate electron transfer followed by a chemical step followed by electron transfer). After the first scan, a pair of peaks for cycling the p-aminodiphenylamine between its oxidized and reduced forms appear at Z.

At pH below 4, another dimer is observed as well. This structure appears to form from the coupling of two species in the same resonance form, namely with the species with the radical in the para position.

benzidine

This dimer is benzidine and is also quickly oxidized as when it first forms (at +1 V). In a separate solution of the same buffer, a cyclic voltammogram (curve B of Figure 5.14) of benzidine gave a pair of peaks that corresponded closely with those observed at Y in the aniline solution. That reaction is also a two-electron, two-proton process.

This example is a brief introduction to the power of CV for probing electrode reaction mechanisms. Other examples, including strategies for measuring the rates of chemical steps and determining the order of steps and the number of electrons transferred per reactant molecule in electrode processes, are discussed in Chapter 6.

EXAMPLE 5.3

Imagine that one of your colleagues recorded the cyclic voltammogram for a complex that he was considering as a catalytic agent for the oxidation of a compound in a commercial synthetic scheme. He thought the redox potential was promising in that it was strong enough for oxidizing the other reagent, but not so strong that it would oxidize other sensitive functional groups on the target molecule. A second colleague said that she worried about the stability of the complex and asked for your input.

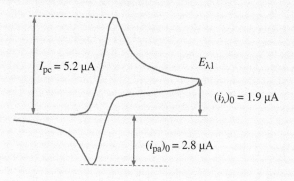

$$\frac{i_{pb}}{i_{pf}} = \frac{\left(i_{pb}\right)_0}{i_{pf}} + \frac{0.485(i_\lambda)_0}{i_{pf}} + 0.086$$

$$\frac{i_f}{i_b} = \frac{2.8}{5.2} + \frac{0.485(1.9)}{5.2} + 0.086 = 0.801$$

$I_{pc} = 5.2\ \mu A$

$E_{\lambda 1}$

$(i_\lambda)_0 = 1.9\ \mu A$

$(i_{pa})_0 = 2.8\ \mu A$

A current peak ratio < 1 indicates that the reduced form (product on the forward scan) is not stable.

Answer:
   Notice the return peak (for the oxidation of the complex back to its original form) seems a bit smaller than the forward peak. That suggests that the complex is not very stable in the reduced form. So, after it reacts with the target molecule, the reduced form of the catalyst may be decomposing leading to eventual loss of the catalyst. To check that hypothesis, one can check the current ratio of the forward to return peak, i_f/i_b. The calculation gives a ratio significantly less than 1.

5.4. METHODS FOR AVOIDING BACKGROUND CURRENT

Attempts to monitor current signals in CV for very dilute concentrations of electroactive species in solution can be limited by background current. At low concentrations, the Faradaic current can be very small compared to the background current. Amplifying the signal is not helpful because the background is also amplified in the process. However, understanding its source has led to several strategies for circumventing background current, and in some cases, these methods enable voltammetric determinations down to the nanomolar level.

   A large fraction of the background current is associated with charging the double layer. The double layer charging occurs quickly and generally decays to zero after the first few milliseconds. The current response to a 10 mV potential step at a carbon disk electrode in a clean electrolyte solution is shown in Figure 5.15.

   Because the biggest contribution to the background current is usually associated with charging the double layer, it is the key to lowering the detection limits in various forms of voltammetry discussed in the following.

   Whenever a voltage is applied to an electrode surface, an outside power source is being used to impose a difference in potential energy for an electron to move across the interface between the electrode material and some electroactive species in solution. For example, in order to make the electrode more negative, more electrons are pushed from the outside circuit to the electrode/solution interface raising the free energy on the electrons there. On the solution side of the interface, cations in the supporting electrolyte respond

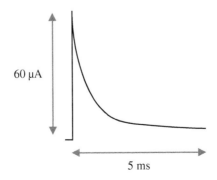

60 µA

5 ms

FIGURE 5.15 Double-layer charging current response of a 3 mm diameter carbon disk working electrode in 0.1 M KCl solution to a 10 mV step in applied voltage with no electroactive species present.

by moving from the bulk solution toward the OHP in order to balance the charge at the surface. The double layer is represented by the charge of electrons on the inside of the electrode surface versus the cations in the OHP (plus a thin neighboring "diffuse" region of solution where the cations outnumber the anions). The net excess of cations balances the charge on the electrode side. (For a positive electrode voltage, the electrode side has a deficiency of electrons while the solution side has an excess of anions in the OHP and diffuse charge region.)

For every applied potential, there is a different arrangement of charge. Consequently, when the applied voltage is changed, the charge must rearrange. Electrons move toward or away from the surface on the electrode side, and ions move into or out of the OHP on the solution side. Recall that the movement of charge constitutes a current. Hence, this is current that is required to charge the double layer in order to establish the new potential energy difference across the interface. Whenever the voltage changes, there will be an associated background current for charging the double layer.

Scanning the applied voltage linearly with time, as is done in a CV experiment, generates a continuous background current as shown in Figure 5.16. Effective ways of isolating the Faradaic component have been built around stepping or pulsing the voltage instead. An early strategy was a program to step the voltage in a staircase manner as shown in Figure 5.17a. There is an immediate surge in the current at the beginning of the step, but it dies out quickly in a few milliseconds. Measuring the current 10–15 ms after the step yields mostly faradaic current. If, instead of scanning the applied voltage continuously, the voltage is stepped in a staircase pattern, the current can be sampled after a short delay following each step and plotted versus the applied voltage to produce a voltammogram as in Figure 5.17a.

Some of the other popular voltage programs are shown in Figure 5.17. As with the staircase waveform, these techniques take advantage of the fact that the interesting (faradaic) current persists after the double layer charging current dies out following a step in the potential. These voltage "pulse" strategies circumvent charging current by sampling a few milliseconds after the voltage is stepped.

One of the most widely used methods for avoiding background current is called square-wave voltammetry (SWV). It was first introduced in the 1950s by Geoffrey Barker in England. It became more popular after innovations in semiconductor manufacturing

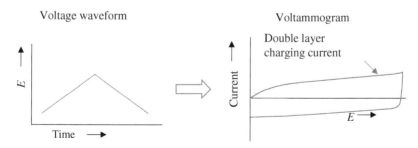

FIGURE 5.16 Scanning the voltage in a linear manner introduces charging current that can obscure the Faradaic current at low analyte concentrations.

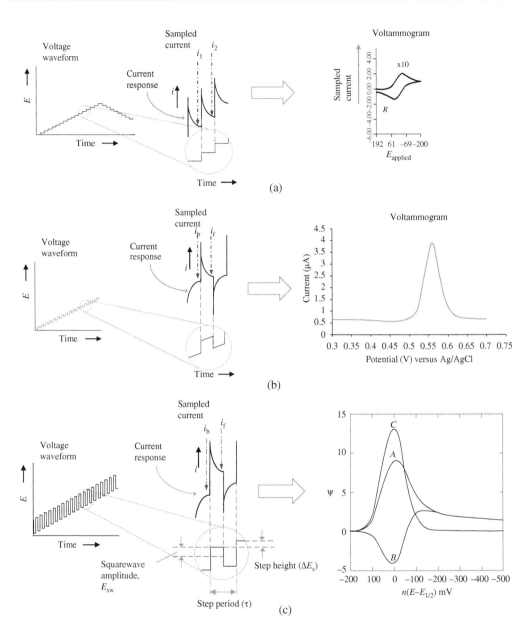

FIGURE 5.17 "Pulse voltammetry" techniques for overcoming background currents. (a) Staircase voltammetry. The current is sampled after a delay following each step in potential. Source: Adapted with permission from Miaw et al. 1978 [17]. Copyright 1978, American Chemical Society. (b) In differential pulse voltammetry, the applied potential is a linear ramp plus a periodic pulse in voltage. The current is measured immediately before the voltage pulse and again at the end of the pulse. The voltammogram is plotted as the difference in current measured before and after the pulse versus the applied potential at the top of the pulse. Source: Adapted with permission from Fan et al. 2012 [18]. Copyright 2012, American Chemical Society. (c) In Square wave voltammetry the potential waveform consists of a staircase with a potential pulse added to the beginning of each step. In the theoretical voltammogram on the right, Ψ stands for the current function. In curve A, the vertical axis represents sampled current, i_f, for the forward step. Curve B shows the reverse current, i_b, measured after the voltage is stepped back. For quantitative analysis, the difference current, $\Delta i = i_f - i_b$, as shown by curve C is normally used. Source: Adapted with permission from O'Dea et al. 1981 [19]. Copyright 1981, American Chemical Society.

made digital electronics common in the 1980s [20] and thanks to developments by Janet and Robert Osteryoung [19, 20].

In SWV, an additional enhancement of the faradaic signal is realized by recycling some of the electroactive material at the surface. Here is how that works. The SWV-applied voltage pattern appears in Figure 5.17c. It is helpful to think of the pattern as a pulse sitting on the rising edge of each step of a voltage staircase. The current is sampled twice during each step, once near the end of the voltage pulse and again after the voltage comes down off of the pulse (shortly before stepping up again). Imagine an experiment scanning from +1 V to −1 V using a staircase of 2 mV steps with a 20 mV pulse at the beginning of each step in a solution of the complex, $Ru(NH_3)_6^{3+}$. At voltages near the formal potential, the current sampled at the end of the negative going pulse captures current for the reduction of the $Ru(NH_3)_6^{3+}$. At the end of the pulse, the voltage becomes more positive by 20 mV, and the current is sampled again. Because this experiment is usually applied in an unstirred solution, some of the ruthenium(II) complex produced by the reduction process is available to be converted back to ruthenium(III) (producing some oxidation current). Consequently, there is more $Ru(NH_3)_6^{3+}$ available near the electrode to be reduced in the next negative-going part of the next step, and a bigger surge in reduction current is obtained than that would have been observed in a simple staircase voltammogram. This recycling of material produces a reduction current in the negative-going (cathodic) part of the pulse and current for the reoxidation following the positive-going (anodic) edge of the pulse. One can plot the current sampled following either the cathodic-going or the anodic-going edge of the pulse, but the difference in the current samples at each voltage step produces a voltammogram that is as much as a factor of two more sensitive than either of current voltage curves alone. The usual SWV plot displays a difference current, Δi, on the vertical axis. The peak current is proportional to the bulk concentration of the analyte and inversely proportional to the square root of the pulse width in seconds [20].

The peak current is also a function of both the voltage pulse amplitude, E_{sw}, (also called the square wave amplitude) and the voltage step size of the staircase waveform, ΔE_s. The pulse amplitude has an optimum value of about $50/n$ mV (where n is the number of electrons transferred for reactant species) [20]. The pulse width is chosen to be equal to half the cycle time between pulses (or $\tau/2$ in diagram at the bottom of Figure 5.17c). If τ is the time between the start of each pulse, then $1/\tau$ is the pulse frequency for the scan. Although the peak height increases with frequency, the optimum is about 200 Hz ($\tau = 5$ ms) for a reversible electrode process [20]. Beyond that rate, the current peak width increases making it more difficult to resolve peaks with similar $E^{o\prime}$ values. The peak is centered around the $E^{o\prime}$ value for a reversible process. For electron transfer of decreasing reversibility, the peak potential will shift away from the $E^{o\prime}$ value (toward more negative voltages for negative-going scans). Also, the peak height decreases in sensitivity as the electron transfer rate decreases. Nevertheless, the SWV peak is still many times more sensitive for quantitative determinations than CV. Furthermore, the time needed to record a scan is very short. For example, a 1 V scan can be completed within 0.5 s using a step size (ΔE_s) of 10 mV and a stepping frequency of 200 Hz. That allows one to take many scans during the elution of a solute from chromatographic column, for example, or to increase the precision of an analysis by recording and averaging many replicate scans quickly, such as during a flow-injection experiment.

Differential pulse voltammetry (illustrated in Figure 5.17b) is very similar to SWV in that the current is sampled both before and after a small voltage pulse. The difference current is plotted as the signal. This strategy also enhances the signal for reversible analytes in quiet solutions because of recycling.

5.5. WORKING ELECTRODES

5.5.1. Mercury Electrodes

In an ideal voltammetry experiment, the working electrode acts as an inert surface where electrons are exchanged between species in solution and the outside circuit. In actual practice, the nature of the electrode surface can influence the electron transfer process. As more is learned about electron transfer mechanisms, the design of working electrodes has evolved over time. Historically, mercury was widely used for working electrodes. In fact, modern techniques of controlled potential electroanalysis originated with the work of Czech chemist Jaroslav Heyrovsky and his students who applied mercury electrodes to the study of micromolar concentrations of metal analytes in the 1920s. Their work advanced the field of trace metal-analysis for which Heyrovsky won the Nobel Prize for Chemistry in 1959. Their apparatus used a reservoir that fed a continuous flow of liquid mercury through a capillary tube forming a tiny droplet of mercury that served as a working electrode. (A modern version of a dropping mercury electrode [DME] uses a piston mechanism to precisely control the lifetime and volume of each drop, which is shown in Figure 5.18.) The mercury flow rate was controlled so that a new droplet was formed every few seconds. The most attractive attribute of this DME was the fact that it provided a clean, reproducible electrode surface. Even if material from a sample solution adsorbed on to the electrode surface, a new droplet formed seconds later providing a clean surface again. In the early days,

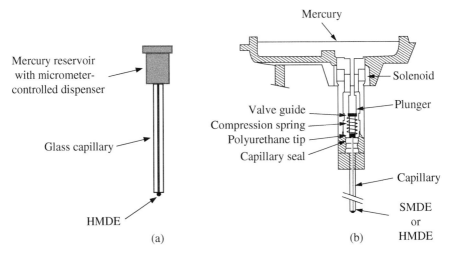

FIGURE 5.18 Mercury electrodes. (a) Hanging mercury drop electrode (HMDE). By rotating the micrometer mechanism at the top, a single droplet of mercury is formed at the bottom of the glass capillary. (b) Modern commercial electrodes have an enclosed reservoir and an electronically controlled plunger system that feeds mercury to a glass capillary. This static mercury drop electrode (SMDE) can hold a drop or form a new drop at a rapid rate mimicking a DME. Source: Adapted with permission form Strobel and Heinemann 1988 [2]. John Wiley and Sons.

voltammetry was called polarography. As solid electrodes began to be used in place of mercury as working electrodes, the term voltammetry also displaced polarography. One disadvantage of mercury is that it oxidizes relatively easily. The E° for the equilibrium between Hg_2^{2+} and metallic mercury is $+0.796\,V$ versus SHE.

$$Hg_2^{2+} + 2e^- \rightleftharpoons 2Hg \tag{5.32}$$

Furthermore, many common anions such as chloride and phosphate form insoluble salts with the mercuric ion stabilizing the Hg(I) form and limiting the positive range even more. Consequently, the useful potential range is limited mainly to negative potentials. Mercury is also toxic making it more of a concern for handling and disposal. These disadvantages have relegated the use of mercury working electrodes mainly to a few special applications today. (One of those applications is a very sensitive method called stripping analysis for determining trace concentrations of certain metal ions. Stripping voltammetry is discussed later in Section 5.7.)

5.5.2. Solid Working Electrodes

Solid working electrodes are much more common now. Platinum and gold are popular choices for working electrodes because they are relatively stable to oxidation and can be used for performing oxidations as well as reduction experiments over a window of approximately 2 V in aqueous solution depending on the pH (see Table 5.1). Oxidations of organic molecules generally involve the transfer of multiple electrons and hydrogen ions, and the breaking and forming of bonds. Gold and platinum have partially filled d-orbitals that can stabilize free radical intermediates at the surface [22]. This type of interaction can help oxidize the analyte, but it can also lead to the formation of interfering coatings on the electrode surface. Furthermore, these metals are not truly inert. They form surface oxides, depending on the applied potential and the solution conditions (such as pH). Although the electrode is still able to transfer electrons with solution species, the formation of these oxides creates another type of time-dependent background current that can complicate measurements of current signals for dilute analytes in solution.

There are also several types of carbon electrodes in use today. The lower cost of carbon compared to noble metal electrodes is one obvious attraction of carbon in developing voltammetric sensors. The useful potential window for carbon electrodes is also a bit wider than for Pt and Au because the kinetics of hydrogen evolution and surface oxidation are slower on carbon [21]. The voltage limits for carbon electrodes are associated with electrolysis of the solvent or supporting electrolyte in aqueous solution as shown in Table 5.1. These boundaries move by 59 mV per pH unit because the electrode reaction consumes hydrogen ions at the negative end ($2H^+ + 2e^- \rightarrow H_2$) and produces hydrogen ions at the positive end ($2H_2O \rightarrow 4H^+ + O_2 + 4e^-$). In some cases, part of the supporting electrolyte, such as chloride ($2Cl^- + 2e^- \rightarrow Cl_2$), is the more easily oxidized component and limits the scanning window.

5.5.2.1. Types of Carbon Electrode Surfaces.
The chemistry of carbon introduces more complexity to its uses as a working electrode. A little knowledge of this chemistry

TABLE 5.1 Working electrode materials and working voltage window

(A) Aqueous solutions[a]

Electrode material	Electrolyte	Positive limit (V versus SCE)	Negative limit (V versus SCE)
Glassy carbon[b]	0.1 M KCl	1.2	−1
	1 M HClO$_4$	1.6	0
[c]	0.1 M HClO$_4$	1.5	−1
Boron doped carbon[c]	0.1 M HClO$_4$	2	−1.2
Platinum	1 M H$_2$SO$_4$	1.3	−0.3
	pH 7 buffer	1	−0.7
	1 M NaOH	0.7	−0.9
Mercury	1 M H$_2$SO$_4$	0.5	−1
	1 M KCl	0.1	−1.8
	1 M NaOH	−0.1	−1.9
	0.1 M (CH$_3$CH$_2$)$_4$OH	0	−2.3

(B) Nonaqueous solutions

Electrode material	Electrolyte	Positive limit (V)	Negative limit (V)
Platinum	0.1 M TBABF$_4$/MeCN	2.5	−2.5
	0.1 M TBAP/DMF	1.5	−2.8
	0.1 M TBABF$_4$/BN	2.5	−1.8
	0.1 M TBAP/THF	1.4	−3
	0.1 M TEAP/PC	2.2	−2.5
	0.1 M TBAP/CH$_2$Cl$_2$	1.8	−1.7

[a]From Ref. [3], unless stated otherwise.
[b]Similar results for graphite.
[c]Ref. [21].

helps guide one in obtaining the best performance from carbon as an electrode. Many types of carbon have been used as electrodes. Humphrey Davy used graphite electrodes in the early nineteenth century for the production of alkali metals by electroplating [21]. Graphite is a good conductor of electricity. The structure of graphite consists of conjugated (sp^2) carbon atoms in two-dimensional sheets resembling layers of chicken wire on the molecular scale stacked on top of each other (see Figure 5.19). The individual layers are known as graphene. Natural graphite contains inorganic impurities that make it unsuitable as an electrode material for analytical purposes. Purified graphite is available in the form of rods, but it is somewhat porous permitting solution penetration into microscopic crevices and subsequently causing reproducibility problems [23]. The peeling back of layers and the insertion of solution change the active electrode area during the use of untreated graphite as an electrode, thus limiting the reproducibility of current measurements. Surprisingly, powdered graphite can be mixed with mineral oil, silicone oil, or other hydrophobic fluids to form a paste that exhibits good conductivity and exchanges electrons well with electroactive species in solution. The oil serves as a binder that not only keeps graphite particles in electrical contact with each other but also constrains the contact of solution to

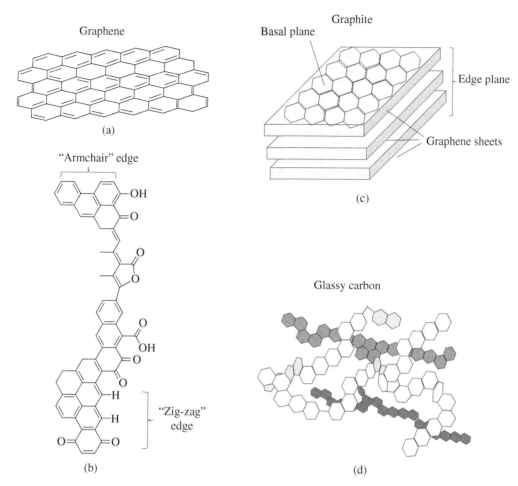

FIGURE 5.19 Structures of carbon electrode materials. (a) Graphene sheet. (b) Examples of surface oxides at edges of graphene sheets and glassy carbon ribbons. Source: Adapted with permission from McCreery 2008 [21]. Copyright 2008, American Chemical Society. (c) Stacked graphene/graphite particles showing edge planes and the basal plane. (d) Glassy carbon consists of tangled ribbons of graphene so that a glassy carbon surface exposes both graphene basal planes and oxidized edges.

particles on the outer surface of the paste. Furthermore, when packed into a small diameter tube, the material at the open end can provide a reasonably reproducible surface area in the shape of a disk [24] (see Figure 5.20). The particles in the exposed paste can be renewed by gently rubbing it against a piece of paper lying on a flat surface. The addition of reagents or catalysts to the paste has led to enhanced selectivity and/or sensitivity in the analysis trace metals [25], of drug molecules, and clinically relevant metabolites using immobilized enzymes [26, 27] and the simultaneous determination of similar hormones [28].

Highly ordered pyrolytic graphite (HOPG) is a synthetic form of graphite made from decomposition of acetylene or other gas on a hot surface. The process produces a material made of regularly oriented stacks of extensive graphene sheets with basal surfaces that behave differently from the edge planes.

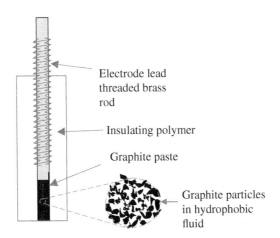

FIGURE 5.20 Carbon paste electrode. A purified graphite powder is mixed with a hydrophobic fluid, such as mineral oil, to form a paste. Source: Adapted with permission from Wang and Freiha 1984 [24]. Copyright 1984, American Chemical Society.

Another popular carbon electrode material is called glassy carbon (also known as glass-like or vitreous carbon). It is made by casting a polymer, such as polyacrylonitrile, in the desired shape using a mold and then heating it to between 1000 and 3000 °C under pressure in an inert atmosphere [21]. The polymer chains condense into highly conjugated, polyaromatic ribbons. Unlike graphite, the glassy carbon exhibits no crystalline planes in X-ray diffraction. The ribbons are thought to be tangled together with no long-range order. Further evidence that the material is amorphous is seen in the way it fractures. There are no flat planes found in the fragments. When struck, glassy carbon forms curved (often conical) pits in the surface in a manner similar to the way window glass pits when struck by a stone.

5.5.2.2. *The Role of Carbon Electrode Surface Chemistry.* In graphite, the two-dimensional graphene sheets stack in planes parallel to each other. The planar surface is often referred to as a basal plane. The edges of the sheet are often terminated with oxygen functional groups (see Figure 5.19b). These polar groups are sites for adsorption and exhibit greater electron transfer rates with some solution species [21]. Although the surfaces formed by collections of graphene edges are naturally not very smooth, they are called "edge planes" to distinguish these faces from basal planes. The tangled graphene ribbons found in glassy carbon present both areas resembling basal planes and edge planes at any electrode surface made from that material. Figure 5.19b shows a conceptual model for a section of a glassy carbon ribbon with some possible oxygen-containing groups, such as carbonyl, ether, phenol, lactone, and carboxylate groups that might appear along the edges [21].

The negative charge of carboxylates on the electrode surface influences the adsorption and electron transfer of charged species (repulsion of anions and attraction of cations). Why should adsorption matter? Consider an experiment where a positive voltage is applied to the working electrode. Recall that the molecules in direct contact with the electrode surface experience the full potential of the electrode. The potential drops very quickly moving away in a line perpendicular to the electrode surface because of the presence of counter ions that are drawn to the surface from the bulk solution. Because some anions can adsorb to the electrode (becoming part of the IHP), the next closest molecules to the surface experience a smaller applied potential than those in contact with the surface. Most counter ions that are recruited to form the double layer appear at the OHP. Solution species diffusing toward the electrode experience a steep potential transition at the OHP. It is at the OHP where electron transfer happens for many analytes. The closer an analyte approaches, the greater the applied voltage that the molecule experiences and the more likely an electron transfer will occur. A negatively charged functional group can attract or repel a charged analyte. Oxygen-containing surface groups can also draw analytes to the electrode surface by dipole–dipole interactions or hydrogen bonding. Interaction with analytes can enhance the electron transfer probability or decrease it when a nonelectroactive molecule is adsorbed and blocks access.

The redox processes of most organic compounds involve an exchange of hydrogen ions as well electron transfer. The structure of a carbon electrode surface may assist in the transfer of hydrogen ions. Whenever a specific chemical interaction with the electrode surface or a second solution species occurs that increases the electron transfer rate, the process is called electrocatalysis. There is evidence that the presence of hydroxyl and carbonyl groups catalyze these coupled chemical steps [21]. For example, catechols, such as the neurotransmitter dopamine (see Figure 5.21c), are oxidized in an overall process that removes two electrons and two protons from the phenolic groups. Oxidation of dopamine at a carbon electrode seems to depend on adsorption and interaction with these surface oxides. Surface oxides may provide a temporary parking place for the displaced hydrogen ions making the overall reaction more rapid.

The reduction of ferric ion (which is more appropriately represented as $Fe(H_2O)_6^{3+}$) in acid solution is an example of an inorganic species that is catalyzed by surface oxides at a carbon surface. A water molecule between the electrode surface and the iron atom is essentially a shared ligand through which an electron is transferred. (Figure 5.21b) Electron transfer by this path is called an "inner sphere" process. In this case, it has been demonstrated that carbonyl groups can play a catalytic roll in reducing $Fe(H_2O)_6^{3+}$. Richard McCreery showed that surface carbonyls on a carbon electrode could be blocked by reacting them with dinitrophenylhydrazine (DPNH) [21, 29]. Such treatment increased the peak separation for CV peaks (evidence that the rate constant, k^o, was decreased) for the iron aquo-complex. On the other hand, the electron transfer could be accelerated for $Fe(H_2O)_6^{3+}$ reduction by artificially increasing the number of carbonyls on the electrode surface through deliberately adsorbing a carbonyl-containing compound, anthraquinone, to the bare carbon electrode.

Some species appear to exchange electrons with relative ease at the OHP and are relatively insensitive to the presence of surface oxides on the electrode. These redox couples are called "outer sphere" electron transfer systems. The $Ru(NH_3)_6^{3+}/Ru(NH_3)_6^{2+}$ pair of ions is a good example of a redox couple that will give strong, reversible voltammograms regardless of the level of surface oxides (see Figure 5.21a). The ruthenium complexes are also insensitive to surface adsorption by other molecules. Because electrons can tunnel through a layer of nonelectroactive adsorbed molecules, $Ru(NH_3)_6^{3+}/Ru(NH_3)_6^{2+}$ gives

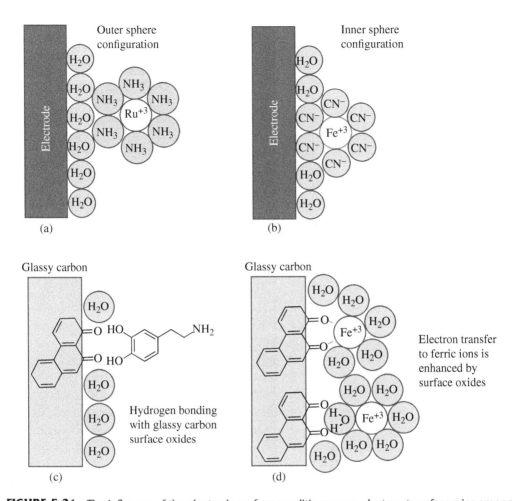

FIGURE 5.21 The influence of the electrode surface conditions upon electron transfer varies among analytes [21]. (a) The $Ru(NH_3)_6^{3+}/Ru(NH_3)_6^{2+}$ pair is an example of an outer sphere redox couple that transfers electrons at the OHP through the monolayer of solvent. Outer sphere electron transfer processes are indifferent to surface oxides or the presence of adsorbates. (b) Ferricyanide is an inner sphere electron transfer species that needs to adsorb to the electrode surface. Such complexes share a ligand with the electrode. Electron transfer of $Fe(CN)_6^{3-}/Fe(CN)_6^{4-}$ is indifferent to the presence of surface oxides, but is inhibited by competing adsorbates. (c) Hydrogen bonding with surface oxides is critical to electron transfer for dopamine. The presence of surface oxides may play a role in proton transfer coupled with the oxidation and reduction of the dopamine. (d) The reduction of $Fe(H_2O)_6^{3+}$ complex is strongly dependent on the presence of carbonyl groups. Pictured are two *hypothetical* ways that a ferric complex might interact with carbonyl groups [21].

essentially the same shaped cyclic voltammograms with a clean electrode as well as with an adsorbed monolayer of a nonelectroactive molecule such as bis(methyl styryl)benzene that blocks access to the surface. $Ru(NH_3)_6^{3+}/Ru(NH_3)_6^{2+}$ is one of the few redox systems that exhibit true outer sphere electron transfer at carbon electrodes. This behavior is in contrast to the $Fe(CN)_6^{3-}/Fe(CN)_6^{4-}$ redox couple. The ferricyanide/ferrocyanide pair also seems to be insensitive to the presence of oxygen species on the carbon electrode surface, but it appears these iron complexes themselves need to be adsorbed to the electrode to transfer electrons efficiently [21]. That is evident from the fact that an adsorbed layer of bis(methyl styryl)benzene dramatically decreases k^o. Consequently, $Fe(CN)_6^{3-}/Fe(CN)_6^{4-}$ is not a true outer sphere redox couple (as it is sometimes portrayed in the literature).

These common electroactive systems can be used to assess the nature of a carbon electrode surface when selecting carbon working electrodes and their surface preparation [21]. Dopamine and $Fe(H_2O)_6^{+3}$ seem to require hydrogen bonding with some group on the carbon surface. The $Fe(CN)_6^{-3}/Fe(CN)_6^{4-}$ redox couple is insensitive to surface oxides, but it is inhibited by adsorption by other molecules, and the $Ru(NH_3)_6^{3+}/Ru(NH_3)_6^{2+}$ couple is an outer sphere electron transfer system that should give peak currents that are dependent mainly on the active electrode surface area. Characterizing the nature of the carbon surface with these redox species can guide the user in creating a method (including a method for surface preparation) that will optimize the desired signal and sensitivity for a given analyte.

5.5.2.3. *Working Electrode Surface Preparation.*

The sensitivity of many analytes to the nature of the carbon electrode surface leads to a concern among electroanalytical chemists about the proper preparation of the working electrode. Polishing glassy carbon exposes new edges that react with air and water to form various oxygen-containing functional groups. Polishing yields an oxygen to carbon atom ratio of about 0.1–0.15. A fresh glassy carbon surface performs electron transfer well. This was demonstrated by fracturing a rod of glassy carbon while it was submerged in an electrolyte solution [29]. The new surface produced an electrode that gave unusually fast electron transfer. Presumably, the enhanced reactivity of a freshly formed glassy carbon surface is due to the cleanliness of the surface. This surface also had a relatively low oxygen content compared to a polished glassy carbon surface. Results similar to those caused by fracturing can be produced by polishing with course grained grit. However, it should be noted that polishing leaves behind carbon fragments and adsorbed contaminants from the polishing medium. For polishing, knowledgeable workers form a slurry of pure, dry alumina (usually with 1 μm or smaller particles) with purified (18 MΩ) water on a clean cloth to minimize contamination. Subsequent sonication in a volatile solvent is frequently recommended to remove alumina and carbon debris.

Electrochemical "activation" or "pretreatment" of carbon electrodes is also commonly used. For most analytes the beneficial effect of these procedures is partly one of removing unwanted adsorbed material. However, applying a voltage of +1 V to +2 V (versus SCE) – a value outside the useful analytical window – as a pretreatment step can roughen up and oxidize the surface (to yield an O/C mole ratio of >0.2) [21]. The extreme voltage produces a high level of current leading to a film of "electrogenerated graphitic oxide" (EGO) that is porous enough to allow small molecules to penetrate. This surface oxidation can dramatically improve performance for analytes, such as catechols, that undergo surface oxide-assisted electrocatalysis. At the same time, the process increases the effective electrode surface area (and, consequently, the sensitivity). However, it should be noted that if the pretreatment is performed in solutions above pH 10, then the EGO is hydrolyzed and does not adhere to the electrode surface [21].

When electrocatalysis of the analyte by surface oxides is not indicated, a simple, general surface preparation is to skip polishing and electrochemical pretreatment. A clean

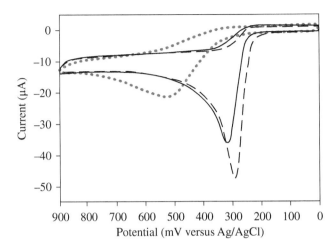

FIGURE 5.22 Cyclic voltammograms of ascorbic acid on glassy carbon electrode in 0.1 M H_2SO_4. Improved performance after briefly soaking the electrode in isopropanol treated with activated charcoal (dashed line). The solid line is the performance of a freshly polished electrode and the dotted curve shows the same electrode after treating it with isopropanol without activated charcoal. Source: Adapted with permission from Ranganathan et al. 1999 [30]. Copyright 1999, American Chemical Society.

electrode surface can be prepared by washing it in a clean, volatile organic solvent. The key is that the solvent must be clean. For isopropyl alcohol, this can be achieved by treating it with activated carbon [30]. The activated carbon adsorbs organic impurities from the solvent. Then, immediately before using, the electrode is briefly soaked in the solvent for a few seconds. It is removed and the solvent on the electrode is allowed to evaporate. Figure 5.22 shows a voltammogram of ascorbic acid for a glassy carbon electrode treated this way compared to a freshly polished surface. The shift in the oxidation peak toward 0 V and the stronger current signal for the electrode soaked in clean isopropanol is indicative of faster electron transfer (larger k^o) for that surface [30].

EXAMPLE 5.4

Imagine that you have been asked to develop a flow injection analysis for an electroactive molecule. This molecule is likely to be easily oxidized at a glassy carbon electrode. In addition to studying its redox behavior, how should one decide whether the electrode needs to be polished, cleaned, or electrochemically pretreated in order to get the most sensitive and reproducible performance?

Answer:

1. Start with a well-cleaned, polished electrode. Sonicate it after rinsing off the polishing powder in order to remove microscopic debris. If the electrode insulating material is not affected by propanol, rinse the electrode with propanol that has been previously cleaned with activated carbon. Run a cyclic voltammogram of the

analyte in an electrolyte buffer that is appropriate for the analytical samples. Note the peak separation, ΔE_p, and peak current height.

2. It is possible that the signal can be enhanced by introducing surface oxides on the electrode. Apply an electrochemical pretreatment that involves a few minutes of exposure to a positive voltage between +1 V and +2 V, such as 10 one-minute cycles of stepping between +1.5 V and −1.5 V. The success of the pretreatment process in terms of introducing surface oxides can be confirmed if a cyclic voltammogram of a dopamine solution has a smaller ΔE_p and higher peak current after pretreatment. Likewise, a smaller peak separation for the analyte suggests that surface oxides and pretreatment will be advisable for the flow injection method.

3. If no improvement in signal is observed with electrochemical pretreatment, it may also be true that the analyte undergoes outer sphere electron transfer. To test for this possibility, expose the polished, clean electrode briefly to an adsorbate, such as bis(methyl styryl)benzene in a separate solution. Transfer the electrode back to the analyte and buffer solution and record another cyclic voltammogram. An increase in ΔE_p indicates that the analyte must adsorb to the electrode for good electron transfer. On the other hand, no change in peak separation indicates an outer sphere process, and need for electrode cleaning will be minimal.

5.5.2.4. Carbon Fiber Electrodes.

Carbon fibers have been a popular working electrode material, especially for studies in living organisms. Commonly used fibers have diameters in the range of 5–50 μm. The carbon material in these fibers appears to vary structurally but have electrochemical properties that are similar to glassy carbon. In addition to their wide working voltage range and favorable redox behavior with organic molecules, the fiber electrodes' small size enables their precise placement near or within cells of interest, such as at the end of neurons. Carbon fibers are typically sealed with epoxy in a glass capillary exposing only the disk-shaped end to the sample solution.

5.5.2.5. Novel Carbon Electrode Materials.

There are also several other, advanced carbon materials being used as electrodes, such as carbon nanotubes [31], graphene [32], boron-doped diamond (BDD) [33], printable carbon inks [21], pyrolyzed photoresist films [34], and chemically modified carbon surfaces, which offer special advantages for voltammetric and amperometric sensors. For example, one of the attractive properties of BDD as an electrode material is its very wide potential window [21]. The usual constraint to scanning the voltage to more extreme values is the fact that the solvent or electrolyte oxidizes or reduces at some point contributing a large background current that obscures the current for the analyte. The electrolysis of water seems to be inhibited at BDD (see Figure 5.23). The slow kinetics for the oxidation of water at BDD means that some electron transfer reactions can be studied at the BDD electrode even when background current interferes with the Faradaic signal on other types of working electrodes.

5.5.3. Ultramicroelectrodes

Electrodes are often categorized based on their area because currents are proportional to the working electrode area. Disk electrodes with diameters on the order of 1 mm have

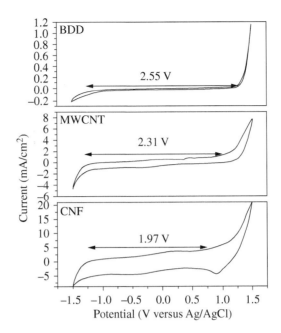

FIGURE 5.23 Background scans in aqueous pH 7 phosphate buffer showing the useful potential window. The limits are associated with electrolysis of the solvent or supporting electrolyte. The boron-doped microcrystalline diamond (BDD) electrode has the widest range. Multiwalled carbon nanotubes (MWCNTs), the carbon nanofiber (CNF) and glassy carbon (not shown) have somewhat narrower useful voltage ranges. Source: Adapted with permission from Poh et al. 2004 [33]. Copyright 2004, American Chemical Society.

areas on the order of 3×10^{-6} m^2 and were known as microelectrodes in the mid-1900s to distinguish them from much larger electrodes that were commonly used for plating out metals for gravimetric analysis. An electrode with a diameter of 10 μm has an area 10 000 times smaller than those of microelectrodes and can exhibit remarkable electroanalytical behavior. Because of their special utility, electrodes of this dimension (and smaller) have become known as "ultramicroelectrodes" in order to distinguish them from conventional working electrodes.

Most of the material that is transported to a conventional electrode in order to replenish reactant that has been oxidized or reduced diffuses from the bulk solution in a direction perpendicular to the surface. This pattern of linear (or planar) diffusion breaks down around the edges where electroactive material can diffuse to the electrode from the side (see Figure 5.24a). The contribution to the total current from the sideways diffusion along the edge is small at electrodes in the millimeter diameter range. However, at a disk with a 10 μm diameter, the mass transport is dominated by the material diffusing from around the edge. The movement is said to be radial rather than linear. This is illustrated in Figure 5.24b.

The tiny amount of material that is consumed by electrolysis at the electrode is rapidly replenished by the radial diffusion. Consequently, the depletion zone is confined to a small volume near the surface. This action creates a constant diffusion layer thickness

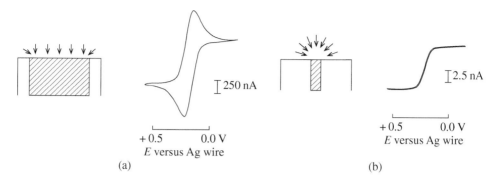

FIGURE 5.24 (a) Mass transport to a millimeter-size electrode follows a path that is mainly perpendicular to the surface and leads to a conventional CV current/voltage curve (2 mm diameter electrode). (b) Mass transport to an ultramicroelectrode is dominated by radial diffusion. Here, diffusion is rapid enough to keep the depletion zone caused by the tiny rate of turn-over of electroactive material confined to a constant volume near the electrode leading to a wave-shaped voltammogram. (10 μm diameter electrode). In both voltammograms, the solution was 1 mM ferrocene in CH_3CN; scan rate: 50 mV/s. Source: Reproduced with permission from Ching et al. 1994 [35]. Copyright 1994, American Chemical Society.

similar to the situation at a hydrodynamic working electrode. The current/voltage curve at an ultramicroelectrode disk takes a wave-like shape with a plateau current given by the following: [35]

$$i_{\lim} = 4nFrDC^* \tag{5.33}$$

where n is the number of moles of electrons transferred per mole of reactant species, F is Faraday's constant (96 485 C/mol), r is the electrode radius in cm, D is the diffusion coefficient for the reactant species in cm^2/s, and C^* is the analyte concentration in mol/cm^3.

Although the peak current at a conventional electrode is proportional to the electrode area, the limiting current of an ultramicroelectrode is proportional to the electrode radius. Even though the conventional electrode in Figure 5.24a has an area that is 40 000 times bigger than the second electrode, the peak current for the bigger electrode is no larger than about 150 times the limiting current of the ultramicroelectrode. The thin, stable depletion zone accelerates the mass transport. Still, the smaller electrode leads to a smaller current. Why is this approach appealing?

The major factor that makes smaller better on this scale is the improvement in the signal compared to the background current. Even though the total current diminishes with size, the faradaic current does not decrease as rapidly as the double layer charging current. The charging current remains proportional to the electrode area. Therefore, the ratio of the faradaic current to the charging current is bigger at an ultramicroelectrode. The signal-to-background current ratio is better at the tiny electrode. The problem of background current obscuring the faradaic signal is less of a problem. Amplifying the current enables one to measure faradaic signals for much lower analyte concentrations without the background charging current overwhelming the signal.

The voltammogram at an ultramicroelectrode reaches a plateau because the turnover of electroactive material and its replacement by diffusion reach a steady state. However,

FIGURE 5.25 The development of the reactant concentration depletion zone near the surface of an ultramicroelectrode (~10 μm diameter disk). (a) At very fast scans (≥10 V/s), the depletion zone is constrained to a thin disk-shaped region and the direction of mass transport is dominated by perpendicular diffusion. This produces a current–voltage curve with a shape similar to a voltammogram seen with larger electrodes. (b) At slower scan rates, the depletion zone has time to grow to a hemi-spherical shape and diffusion follows radial paths. The corresponding voltammogram looks like a current–voltage curve for a hydrodynamic system [36].

it is possible for the electron transfer step to get ahead of the mass transport during very fast scans (≥100 V/s). At fast scan rates, the voltammogram looks more like a CV at a conventional electrode with peaks corresponding to the forward and reverse electrode reactions. What causes that behavior? The wave-like profile depends upon a steady-state mass transport process associated with radial diffusion to the electrode surface, whereas the conventional current/voltage profile is associated with linear or planar diffusion leading to a continually growing depletion zone. The shape of the current/voltage curve for an ultramicroelectrode depends on the time required for the radial pattern of the depletion zone to form. Very fast scanning rates do not give sufficient time for the steady-state diffusion zone to establish itself (see Figure 5.25). It may help to think about this in terms of a simple voltage step experiment with a reversible redox species. About a microsecond after the voltage is stepped out to a value where the original reactant is oxidized or reduced, virtually all the electroactive starting material next to the surface reacts. At this point in time, the depletion zone is only a few hundred nanometers thick and takes the shape of the electrode [36]. A disk-shaped electrode produces a disk-shaped depletion zone in the neighboring solution. This is depicted in Figure 5.25a. At this point, the direction of the concentration gradient over most of the electrode is perpendicular to the surface [3].

Therefore, the flux of the reactant will be mainly perpendicular to the surface over most of the electrode, as observed for a conventional electrode. The depletion zone will also grow in the direction perpendicular to the surface as more material diffuses in to react. The exception to this is for the region around the electrode rim where mass transport is enhanced by diffusion from the side. As time increases, the depletion zone continues to grow deeper into the solution opposite the center of the disk, but less rapidly near the rim. Eventually, the depletion zone approaches a hemispherical shape [4] with a thickness of about six times that of the electrode radius [36]. At that point, the system reaches a steady state. The direction of mass transport includes all angles from 0° to 90° with respect to the surface. The depletion zone volume remains fixed, and the current becomes constant. Steady-state current voltammograms are generally observed for electrodes on the order of 10 μm in the smallest dimension at scan rates below 100 mV/s [36]. At scan rates above 200 V/s, the system does not have time to reach steady-state. While the depletion zone is growing, the concentration gradient (and, therefore, the rate of mass transport) is declining and the current peaks out. (Intermediate scan rates show mixed behavior.)

Another predictor of wave-shaped, rather than peak-shaped voltammograms is the ratio of Dt/d^2, where D is the solute diffusion coefficient, t is the time for a scan, and d is the smallest dimension of the electrode [35]. When the ratio is greater than 1, the voltammogram will look like a hydrodynamic system. Furthermore, in the faster scans, the product formed on the forward part of the voltage scan does not have sufficient time to diffuse away from the neighborhood of the electrode, and it can be converted back to the starting material on the reverse scan (producing the return current peak).

The amount of current required for charging the double layer is much smaller at an ultramicroelectrode. This is equivalent to saying its capacitance is smaller (on the order of 10 pF for a 10 μm disk compared to about 40 μF for a 3 mm disk) [37]. Consequently, the time that it takes to charge the double layer is much faster (about 5 μs compared to about 5 ms for the ultramicroelectrode and conventional electrode, respectively, depending on the solution resistance). The rapid response of the electrode makes scan rates up to 10^6 V/s possible. However, there is a practical limit. With electrodes of any size, the charging current eventually will obscure the faradaic current. Another factor limiting fast scanning at conventional electrodes is the ohmic loss to solution. The solution between the electrodes has a finite resistance (typically from 50 to 250 Ω). Any current passing through the cell will lead to voltage loss in overcoming the resistance according to Ohm's law. This loss is energy that does not get applied to the potential difference at the solution/working electrode interface and represents an error in the applied potential.

$$V_{error} = iR_{cell} \tag{5.34}$$

Only at very high scan rates and moderate concentrations, peak currents would reach 500 nA. Under those conditions, a cell resistance of 200 Ω would lead to an error of only 1 mV. More modest scan rates would make it possible to work in solutions of higher resistance. Therefore, an ultramicroelectrode can be appealing for studying the electrochemistry of a molecule that is soluble only in nonpolar solvents where the solubility of supporting electrolytes is low and the electrical resistance is high.

EXAMPLE 5.5

How does one know if Ohmic loss is an important factor in a voltammogram?

Answer:
Usually errors of 5–10 mV are negligible for voltammetry and amperometry experiments. For quantitative analysis, one usually chooses an applied potential that is well beyond the formal potential so that the current being measured is well out on the current limiting plateau where a 10 mV shift does not change the current. In order to estimate the voltage error caused by Ohmic loss, one needs a reasonable estimate of the solution resistance and the maximum current. Rough estimates are often sufficient. For example, a solution of 0.1 M KCl has a resistance of ~100 Ω. Consider a current level of 50 μA. The Ohmic voltage error would be:

$$iR = (50 \times 10^{-6} \text{ A})(100 \text{ Ω}) = 5 \times 10^{-3} \text{ V or 5 mV}$$

5.5.4. Fast Scan CV

One of the advantages of scanning the voltage very rapidly is the opportunity to explore very rapid processes such as the chemical reactions of transient species generated at an electrode surface. For example, bromoanthracene (ArBr) can be reduced to anthracene (ArH) using CV. At modest scan rates, the reaction (in acetonitrile) appears to be a two-electron reduction accompanied by the exchange of the halogen for a hydrogen ion all in a concerted process.

$$\text{ArBr} + 2e^- + H^+ \rightarrow \text{ArH} + Br^- \tag{5.35}$$

The intermediates disappear too rapidly to permit any evidence of a mechanism. At slow to moderately fast scan rates, the cyclic voltammogram for the reduction of 9-bromoanthracene appears as it does in Figure 5.26b [38, 39].

The cathodic peak at about $-1.75\,V$ appears to be a two-electron process as calculated from a rearranged form of the Randles–Sevcik equation:

$$n^{3/2} = \frac{i_p}{269AD^{1/2}v^{1/2}C^*} \tag{5.36}$$

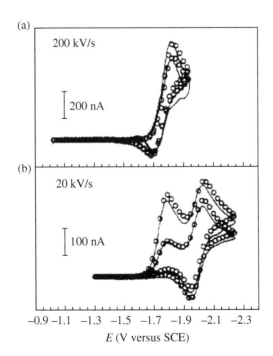

FIGURE 5.26 Reduction of 4.8 mM 9-bromoanthracene at a 3 μm Au disk electrode in acetonitrile. Only at very high scan rates (above 100 000 V/s) is it possible to catch the short-lived 1-electron intermediate, ArBr•⁻, as evidenced by the return peak in (a). At slow and moderately fast scan rates, the reduction peak at −1.75 V appears to be a concerted two-electron reduction with the loss of the bromine as in (b). This process appears chemically irreversible (no return peak). If the scan is extended to more negative voltages, a second reduction peak appears corresponding to the reversible, 1-electron reduction of anthracene. Circles represent current from digital simulation of the best-fit mechanism. Source: Adapted with permission from Wipf and Wightman 1989 [38]. Copyright 1989, American Chemical Society.

Scanning further in the negative direction produces a second reduction peak around -2.0 V (Figure 5.26b). This second peak appears to be a one-electron reduction of unsubstituted anthracene (Eq. (5.37)). (This assignment can be supported by recording a separate CV for authentic anthracene.) [38]

$$ArH + e^- \rightleftharpoons ArH^{\cdot-} \tag{5.37}$$

Notice that the reverse scan in Figure 5.26b shows a return peak associated with the oxidation of the anthracene radical anion back to anthracene (the reverse of Eq. (5.37)). However, no return peak associated with the first reduction peak is observed, even when switching the scan direction at a voltage closer to the first reduction peak and before the second. If, however, one increases the scan rate above $100\,000$ V/s, a return peak for the first process does appear (Figure 5.26a). Furthermore, at these fast scan rates, calculation of the number of electrons in the reduction step indicates it is now a one-electron transfer process. Both of these observations indicate that the fastest scans were able to catch some of the radical anion and reoxidize it back to the starting material [38]. The reduction of 9-bromoanthracene is not a concerted two-electron process, but rather proceeds in a series of steps. Using digital simulations of various mechanisms, the investigators were able to show that the data best fit a mechanism of a single electron reduction followed by two different chemical steps, one of which regenerated the starting material (shown in Eqs. (5.38–5.40)).

$$ArBr + e^- \rightleftharpoons ArBr^{\cdot-} \tag{5.38}$$

$$ArBr^{\cdot-} \rightarrow Ar^{\cdot} + Br^- \tag{5.39}$$

$$H^+ + ArBr^{\cdot-} + Ar^{\cdot} \rightleftharpoons ArH + ArBr \tag{5.40}$$

The three-step mechanism gives the same overall two-electron process as the concerted reaction in Eq. (5.35).

It should be noted that as scan rates increase, even at ultramicroelectrodes, the double layer charging current does eventually dominate the voltammogram. The most widely used strategy to extract the faradaic signal from the total current has been to subtract the background current. Prior to recording CVs for the analyte, a number of scans (10 is common) are recorded and averaged in the same electrolyte, under the same conditions except without the analyte [38]. This procedure has been reproducible enough to make quantitative analysis in challenging systems possible. (There are also hardware strategies for removing charging current.)

Another practical consideration associated with scanning faster is the fact that for analytes that exhibit sluggish electron transfer kinetics, CV peaks shift to more extreme potentials as the scan rate increases. Sometimes, this has the serendipitous effect of separating the peaks of two different compounds that have similar formal potentials but different electron transfer kinetics. But, it can also crowd peaks at the extreme end of the voltage window.

One fruitful application area for fast scan CV and ultramicroelectrodes is the field of *in vivo* analysis of neurotransmitters. In living systems, the concentration of signaling molecules varies rapidly in very specific locations. In fact, the original motivation for exploring ultramicroelectrodes and fast scanning voltammetry was the interest in studying brain chemistry. Mark Wightman and his students have pioneered the development in all these areas. They have shown how carbon fiber electrodes can be used to monitor transient dopamine signals in brain tissue of freely moving animals in response to various stimuli. The ability to scan voltammograms gives them the ability to discriminate dopamine from other electroactive constituents in the tissue, such as ascorbic acid. Fast scans at a high repetition rate provide them with temporal resolution on the sub-second time scale. One of the methods that they developed to deal with the huge amount of data that is generated in a short amount of time in these experiments is to display it in three-dimensional plots with voltage on the vertical axis, time on the horizontal axis, and the electrode current intensity in the third dimension represented by a color scale (shown in grayscale in Figure 5.27a,b). A very rapid voltage ramp permits a high repetition rate and very good time resolution. Figure 5.27 shows data recorded for experiments in which a carbon fiber ultramicroelectrode was implanted in the brain of a live anesthetized rat. A stimulus triggered the release

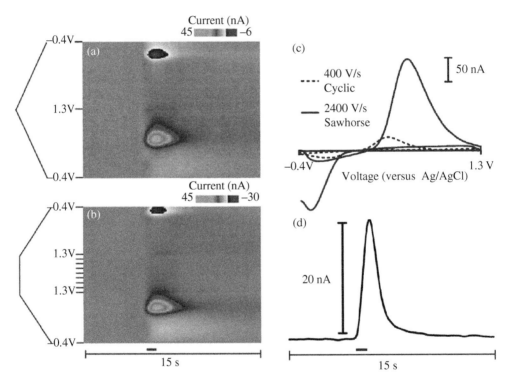

FIGURE 5.27 Fast scan CVs were recorded for a carbon fiber placed in the brain of an anesthetized rat. Periodical stimulation of the rat caused transient releases of dopamine. Multiple CVs are displayed in a three-dimensional format in (a) and (b). The vertical axis represents the applied voltage and the horizontal axis is time. The current intensity is indicated by a color scale (shown here in grayscale). The voltage was scanned between −0.4 V and +1.3 V in a triangular pattern in (a) at 400 V/s. In (b), the waveform was ramped at 2400 V/s between the same two extremes, but with a 0.55 ms pause at +1.3 V before the return scan. Vertical slices of the 3-D plot represent individual CV curves. Plots in (c) show multiple repetitive CVs for the faster scan rate overlaid on a CV for the slower scan rate. A significant boost in sensitivity was realized with the faster scan rate. Also, (d) shows a current/time curve for the oxidation of dopamine in an anesthetized rat. Source: Adapted with permission from Keithley et al. 2011 [40]. Copyright 2011, American Chemical Society.

of dopamine as detected by the sudden appearance of the oxidation peak for dopamine in cyclic voltammograms. The background subtracted CVs in Figure 5.27c are representative of the current profile that one would see by taking vertical slices through the 3D data near the middle of the plot in Figure 5.27a and b [40]. The stronger current signal in Figure 5.27c illustrates the advantage of using the faster scan rate. Figure 5.27d is a representative current/time curve for at the voltage where dopamine is oxidized in response to stimulating the rat brain.

5.6. PULSE AMPEROMETRIC DETECTION

From thermodynamic considerations, there are many organic functional groups that would seem to be either oxidizable or reducible at an electrode surface. Table 5.2 lists some common functional groups that have been shown to be electroactive. However, there are some kinetic factors that can inhibit electrode reactions making direct voltammetric or amperometric analysis impractical for some compounds. For example, aliphatic alcohols and aliphatic amines are two classes of compounds that can be oxidized at platinum or gold electrodes but tend to form products that stick to the electrode surface and block further oxidation. Research suggests that the reason that some organic compounds are difficult to oxidize is that they lack π-bonds that can stabilize intermediates in the

TABLE 5.2   Electroactive organic functional groups [41]

Reducible organic functional groups	Oxidizable organic functional groups
Acetylenes	Alcohols
Aldehydes	Aliphatic halides
Conjugated aromatics	Aromatic halides
Conjugated carboxylic acids	Aromatics
Conjugated double bonds	Carboxylic acids
Diazo compounds	Ethers
Disulfides	Heterocycles
Heterocycles	Nitroamines
Hydroquinones	Organometallics
Imines	Phenols
Ketones	
Nitriles	
Nitro compounds	
Nitroso compounds	
Organometallics	
Oximes	
Peroxides	
Quinones	
Sulfides	
Sulfones	
Sulfonium salts	
Thiocyanates	

electrode reaction [42]. As a result, these molecules have a high activation energy barrier for oxidation at an electrode. (The lack of π-bonding also makes these compounds unsuitable for analysis by absorption spectroscopy in the visible and ultraviolet range.) Both platinum and gold have partially unfilled d-orbitals that readily form covalent bonds with free radical intermediates when these compounds are oxidized at these metal surfaces. This stabilization of the initial products makes oxidation possible, but it creates a layer of organic material sticking to the surface that eventually blocks further oxidation. Dennis Johnson and his students [22, 42] demonstrated that platinum and gold electrode surfaces fouled by organic reaction products can be cleaned by pulsing the voltage to a positive value (such as +0.8 V versus Ag/AgCl for Pt or +1.4 V for Au) where the metal forms an oxide film, breaking the metal–carbon bonds in the process. A brief interval of 50–200 ms is adequate to eliminate the organic coating. The potential can then be stepped

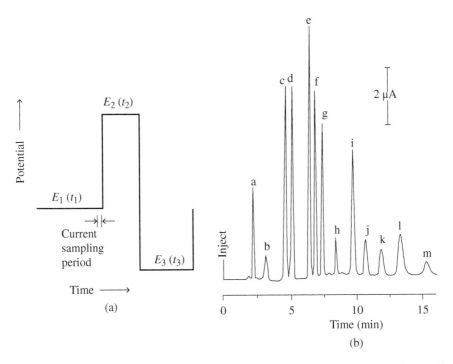

FIGURE 5.28 HPLC separation and pulse amperometric detection of a mixture of alcohols and glycols at a gold flow-through electrode. (a) Applied potential waveform. E_1 is the anodic detection voltage, +0.30 V, E_2 is the oxidative cleaning step (+1.4 V) and E_3 (−0.4 V) is the cathodic voltage where the gold oxide is reduced back to metallic Au. The time spent at E_1 is fixed within a range of 300 ms. The current is averaged during the last 16.7 ms of this step. The cleaning interval, t_2, was set to 120 ms and the reactivation time was 420 ms. (b) Chromatogram. OmniPAC PCX-500 column, 18–85.5% ACN gradient at 1.0 ml/min, postcolumn base addition of 0.3 M NaOH at 1.0 ml/min, 50-μl injection. Peaks (a) ethanol, 1840 ppm; (b) 1-propanol, 460 ppm; (c) 2-methyl-2-propen-1-ol, 46 ppm; (d) cyclopentanol, 460 ppm; (e) phenylmethanol, 69 ppm; (f) 1-phenylethanol, 115 ppm; (g) 3-phenyl-1-propanol, 115 ppm; (h) 2-ethyl-1-hexanol, 460 ppm; (i) 1-decanol, 460 ppm; (j) 1-undecanol, 920 ppm; (k) 1-dodecanol, 920 ppm; (l) 1-tridecanol, 920 ppm; (m) 1-tetradecanol, 920 ppm. Source: Adapted with permission from La Course et al. 1991 [42]. Copyright 1991, American Chemical Society.

back to a negative voltage (such as -0.4 V for Au) where the oxide is reduced creating a clean metal surface again. At that point, the voltage can be stepped to a positive value where the oxidation of the analyte occurs. The faradaic current is read after a short delay (100–400 ms) to let the background current decay. This voltage cycle of cleaning and reading the current at fixed voltages is called "pulse amperometric detection" (PAD). The whole cycle can be repeated in less than a second. Figure 5.28 shows the applied voltage pattern as it was used for a gold electrode serving as a detector at the end of a liquid chromatograph. The analytes were a mixture of aliphatic alcohols in this case. PAD is now available in commercial HPLC detectors and has been applied to the separation and detection of carbohydrates, aliphatic amines, and amino acids that are difficult to detect by UV-visible detectors.

5.7. STRIPPING VOLTAMMETRY

Quantitative analysis of extremely low analyte concentrations can be performed using a technique known as stripping analysis. The method owes its tremendous sensitivity to a step in the procedure that preconcentrates the analyte at the electrode surface before the voltage is scanned to produce a signal. Anodic stripping voltammetry (ASV) is the term used to describe the earliest variety of stripping analysis. The first applications were based on a hanging mercury drop working electrode (Figure 5.18a). When a metal, such as Pb^{2+}, is reduced to its metallic form at a mercury surface, the metal dissolves into the mercury. The concentration of lead atoms inside the drop increases with the time spent reducing lead ions. Because the volume of the liquid mercury is very small, the concentration of lead inside the drop grows quickly. In a short time, it will exceed the concentration of lead ions in the surrounding aqueous solution. After a predetermined interval (known as the deposition time), the voltage at the working electrode is scanned in the positive (anodic) direction, and the lead inside the drop is oxidized releasing Pb^{2+} ions back into the aqueous solution. The corresponding oxidation current provides a large signal peak whose height and area are proportional to the amount of lead inside the drop. The main features of an ASV experiment are diagramed in Figure 5.29. The lower part of the diagram indicates how the voltage changes during the course of the experiment. At the start, the potential is held at a negative voltage to deposit and preconcentrate the analyte. During this time, the solution is usually purged with nitrogen in order to remove dissolved oxygen. Oxygen is easily reduced at the mercury electrode and would contribute to the background current and complicate the voltammogram if it were not removed. The solution is often stirred as well during the deposition step in order to enhance the rate of analyte reduction and improve the sensitivity for the determination. A RDE or a flow through electrode system is particularly well-suited for providing reproducible stirring. After a predetermined period, the stirring and the nitrogen purge are stopped in order to let the solution become quiet. The quiet solution yields lower background current during the measurement step. The deposition time, rest time, and the rate of stirring are all carefully controlled in order to obtain the most reproducible results. When the voltage is scanned in the positive direction, the lead inside the mercury diffuses to the drop surface and is oxidized causing a surge of current. Of course, the amount of lead inside the drop is limited, so the current peaks

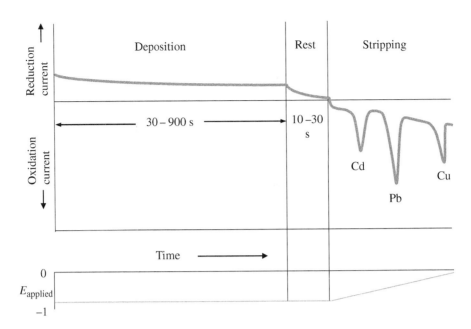

FIGURE 5.29 Diagram of an anodic stripping voltammetry experiment. The bottom graph shows that the applied potential is held constant during the deposition time during which the sample solution is often stirred in order to gather more analyte into the mercury drop electrode. The voltage is held for a brief time (fixed between 2 and 30 seconds) after turning off any stirring, allowing the solution to become quiet before scanning the voltage in the anodic direction. For optimum sensitivity, a square wave voltage scan is often used in place of a linear voltage ramp in the stripping step.

out. A SWV scan is often used instead of a linear voltage scan in the stripping step because significant enhancement is obtained by the effect of recycling the analyte as the potential steps back and forth. Stripping peaks for square wave scans also tend to be somewhat narrower, and, therefore, when multiple metals are present, the signals are better resolved.

Some other details of the experiment are worth noting. The position of the peak on the voltage axis can be used to identify the analyte. The peak potential is close to the standard reduction potential for the metal. The actual value is given by

$$E_p = E_{1/2} - (1.1)RT/nF \tag{5.41}$$

where $E_{1/2}$ is the half-wave potential for the metal analyte, R is the ideal gas constant, T is the temperature in Kelvins, n is the number of electrons in the half reaction, and F is Faraday's constant [43]. There are commercial instruments that use a device that hangs a mercury drop for the working electrode. These devices form very reproducible single droplets and hold them until the measurement is finished. A single run is performed on an individual drop, then it is removed and a fresh drop is created for the next run. The relative standard deviation in the drop size is very good, on the order of 0.5–0.9%.

In addition to lead, antimony, bismuth, cadmium, copper, indium, silver, thallium, and zinc can be determined by the method described here using a mercury drop. Arsenic, mercury, selenium, silver, and tellurium can be determined by anodic stripping from a solid gold electrode. Several other metals, as well as many organic molecules, can be determined in very dilute solutions after preconcentrating them by adsorption onto the surface of a mercury electrode, or in some cases, onto the surface of a solid working electrode [43].

Several anions can be determined by preconcentrating their mercury salts on the surface of a mercury electrode. In those cases, instead of holding the working electrode at a negative voltage, the electrode is poised at a deposition potential where the mercury will oxidize and form a stable salt if the sought-for anion is present. For example, chloride can be deposited as Hg_2Cl_2 at 0.4 V versus SCE. An insoluble salt forms a film on the electrode surface that can be stripped off by scanning the voltage in the negative direction, which reduces the mercury and releases the anion. This mode of stripping voltammetry is known as cathodic stripping analysis. Chloride, bromide, iodide, selenide, sulfide, chromate, molybdate (MoO_4^{2-}), vanadate (VO_3^{2-}), and tungstanate (WO_4^{2-}) have been determined by this sort of cathodic stripping [43]. These different anions form salts of varying stabilities (K_{sp} values) and, therefore, will be stripped at different voltages. Cobalt and nickel ions will form neutral complexes with dimethylglyoxime that adsorb on a mercury surface and can be determined by cathodic stripping. In this case, the cobalt or nickel ion is reduced rather than the mercury [44, 45].

There are also many organic compounds that can be absorbed (and thereby, concentrated) and subsequently stripped electrochemically. Thiols, such as glutathione and cysteine, as well as nucleic acids form insoluble salts with mercury that can be stripped in cathodic scans. Other interesting biologically active compounds can be determined by adsorptive stripping analysis. In these cases, the process of adsorption occurs spontaneously on the mercury surface as a function of time without the need for an applied voltage. For example, adsorptive stripping analysis has been applied to the determination of sedatives (such as diazepam and clozapine), antibiotics (such as penicillin and streptomycin), anti-tumor agents (such as chlorambucil and daunorubicin), anti-depressants (such as desipramine and trimipramine), cardiac agents (such as digitoxin and diltiazem), narcotics (such as cocaine and procaine), hormones (such as testosterone and progesterone), and vitamins (such as riboflavin and vitamin K1). A number of pesticides and other hazardous chlorinated compounds have been determined by adsorptive stripping analysis [46].

A mercury film deposited on a solid electrode (usually glassy carbon or pyrolytic graphite) is another form of mercury electrode that has been widely used for stripping analysis. Such a film can be deposited along with the analyte from the sample solution for ASV. Water samples can usually be spiked with a supporting electrolyte, such as KNO_3, and a stock solution of mercury nitrate in order to give a working solution of 5×10^{-5} Hg^{2+} (and usually about 0.1 M KNO_3). Typically, the working electrode is held at a negative potential for five minutes before scanning. Then, a second five-minute deposition and scan is carried out using the same film. The second scan yields a more reproducible signal [46]. There are three main advantages to this procedure compared to using a hanging mercury drop electrode. First of all, the film is easily prepared without elaborate equipment. After obtaining the signal for a given solution, the electrode is prepared for the next experiment by wiping it clean with a tissue. The second advantage is an improved sensitivity for anodic stripping work. This is the case because the surface-to-volume ratio of the mercury film is much greater than that of a typical hanging mercury drop electrode.

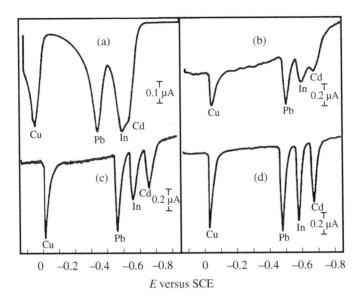

FIGURE 5.30   Anodic stripping experiments performed on a solution of four metals at 2×10^{-7} M for several different electrodes. (a) A hanging mercury drop, 30-minute deposition. (b) Mercury thin film electrode (MTFE) on a pyrolytic graphite rotated disk electrode, five-minute deposition. (c) MTFE on an unpolished glassy carbon disk electrode; five-minute deposition. (d) MTFE on a polished glassy carbon disk electrode; five-minute deposition. Source: Adapted with permission from Florence 1970 [47].

Consequently, the analyte concentration inside the mercury film builds up faster than it does for the hanging mercury drop. The film also has more surface area to reduce the analyte, so more analyte is collected, and the electrode stores that material temporarily in a smaller volume of mercury.

Both of these factors lead to a bigger stripping peak. Also, the analyte metal inside of the film electrode has a shorter distance to diffuse in order to get out of the film, so it is depleted quickly on the stripping step. That fact leads to thinner peaks and better resolution when several metals are present as can be seen in Figure 5.30 [47]. Using a rotated disk made of glassy carbon for a working electrode to enhance the deposition process makes very low detection limits possible. For example, nanomolar detection limits for lead (~0.2 ppb) have been reported for a five-minute deposition and a 2000 rpm rotation rate at a mercury film covered glassy carbon RDE [47].

5.8.   SPECIAL APPLICATIONS OF AMPEROMETRY

5.8.1.   Flow-Through Detectors

There are numerous methods of analysis in which the electrode voltage is fixed and the analyte is either oxidized or reduced producing a current that is measured as a function of time. These methods depend on some additional mechanism to provide some sort of selectivity for the analyte. For example, a flow-through electrode using a carbon or platinum surface to oxidize or reduce solutes exiting from a liquid chromatograph can be a very sensitive detector. The electrode relies on the chromatographic separation to provide the selectivity. A pulsed amperomeric detection scheme was discussed earlier for compounds that foul the surface, (Figure 5.28); however, it should be noted that excellent performance can be obtained for many easily oxidized or reduced analytes by holding the applied voltage at a fixed value with no need to change it during the chromatogram. The sensitivity for individual analytes will vary because their diffusion coefficients are different. Furthermore, if the potential is not positive enough to be out on the current limiting plateau for any of the solutes, then the calibration curves for these compounds will not be as steep as they could be. By the same reasoning, the choice of the applied potential can also be a way of increasing the selectivity of the detector. For example, if an important analyte is easily oxidized at a lower applied potential than another solute that elutes a similar time, it may be possible to avoid detecting the interfering solute by operating at a potential in between the $E_{1/2}$ values for the two compounds.

5.8.2. Dissolved Oxygen Sensors

A very popular electrochemical sensor for molecular oxygen relies on a gas permeable membrane for providing selectivity. In 1962, Leland Clark was interested in developing a heart-lung machine for aerating a patient's blood during surgery [48]. In order to demonstrate the effectiveness of his apparatus, he needed a sensor for molecular oxygen in blood. He took advantage of the fact that dissolved oxygen reduces readily at a platinum electrode at $-0.7\,V$ [48].

$$O_2 + 4H^+ + 4e^- \rightleftharpoons 2H_2O \tag{5.42}$$

Because there are other reducible components in blood and because of a concern that proteins or other material in the blood might adsorb to the platinum surface and block the reduction of the O_2, Clark made a crucial modification of the original idea. He placed a gas-permeable Teflon membrane between the sample and the solution containing the platinum working electrode and a reference electrode. Molecular oxygen is the only electroactive species in blood that can cross the membrane and be reduced at the working electrode.

The membrane usually consists of a thin film stretched tightly over the platinum working electrode and silver reference electrode trapping a very thin layer of electrolyte solution. In operating the oxygen probe, users assume that the current reaches a steady state that is proportional to the oxygen concentration and inversely proportional to the membrane thickness, b.

$$i = nFAP_mC/b \tag{5.43}$$

where n is the number of electrons transferred, F is Faraday's constant, A is the working electrode surface area, P_m is the coefficient for the membrane permeability, and C is the oxygen concentration of the test solution [49].

Equation (5.42) looks a lot like Eq. (5.24) that describes the limiting current for hydrodynamic voltammetry in general where the diffusion coefficient and diffusion layer thickness have been replaced by a membrane permeability coefficient and the membrane thickness. However, this model is an oversimplification of the real system. There is usually an induction period required for the system to come to a steady state after stepping the voltage to the working potential. The current actually decreases as the square root of time in accordance with the Cottrell equation (Eq. (5.26)) for the first few minutes before a relatively stable level is reached. The fact that the system eventually reaches a steady state is probably due to the rate of transport of oxygen through the membrane becoming the rate-determining step [49].

Among the variables that influence this device are the effects of temperature on oxygen solubility. A saturated sample of water contains 9.09 mg/l dissolved O_2 at 20 °C, but only 7.65 mg/l at 30 °C. The diffusion of the O_2 through the Teflon membrane is also temperature dependent. The signal for the steady state sensor increases by about 3% per °C because of the temperature effect on diffusion. The effects of atmospheric pressure and of water salinity on the solubility of oxygen and the stirring of the sample solution are also important. Teflon also allows some passage of water through the membrane causing slow evaporation and loss of response [49, 50].

A newer version of the sensor exploits chronocoulometry for better immunity from influences such as convection of the sample solution. The potential is stepped from a value near 0 V where no faradaic current is observed to −0.7 V where O_2 reduction occurs. Chronocoulometry is merely integrating the current–time curve for the diffusion-controlled analyte. The current–time signal is integrated for a fixed interval after the beginning of the voltage pulse giving the charge for the reduction of O_2 during that period [50]. The charge is still proportional to the concentration of dissolved oxygen. The voltage is pulsed for only a short period, so that very little of the oxygen near the electrode gets depleted. At the end of the pulse, the concentration near the surface recovers quickly permitting many replicate measurements to be made in a short period of time. Of course, careful timing of the current and charge measurement with respect to the start of the voltage step is required. (A three-electrode system, as described in Chapter 7, is also used. One of the benefits of the three-electrode system is greater longevity of the reference electrode.)

The amperometric oxygen sensor is particularly good at responding to relative changes in the concentration of oxygen in samples. However, calibration is necessary in order to determine absolute concentration levels. A single-point calibration of the device is often made by allowing the sensor to respond to a clean portion of water that has been equilibrated with oxygen in the air [50]. Measuring the temperature and barometric pressure allows one to estimate the dissolved oxygen level from a table of oxygen solubility versus temperature. Multipoint calibration procedures have normally depended upon reactions that consume oxygen. Many of these procedures have been based on enzyme-controlled reactions in which both the enzyme and substrate are added to solution. Of course, in order to make this a quantitative analysis, both the solution volume and the quantity of substrate added must be measured. An example reaction is the oxidation of hypoxanthine to form uric acid. Deliberately adding catalase and using

$$\text{(5.44)}$$

hypoxanthine as a substrate ensures that the overall stoichiometry is one mole of oxygen per mole of substrate [49, 51].

5.8.3. Enzyme Electrodes

As a biochemist, Clark saw the possibilities for combining the oxygen sensor with enzymes in order to create sensors for other molecules. For example,

$$(5.45)$$

glucose oxidase is an enzyme commonly isolated from a fungus that can oxidize the simple sugar, glucose, consuming oxygen in the process.

By covering an oxygen sensor with an immobilized form of the enzyme, the device registered a drop in current (corresponding to a drop in dissolved oxygen levels) whenever glucose was present. Clark compared that current with the current at a second O_2 electrode without glucose oxidase, in order to distinguish glucose signals from fluctuations in O_2 concentration that were not associated with the enzyme reaction [52]. The enzyme electrode was refined by Updike and Hicks [53] and was eventually commercialized by Yellow Springs Instruments in 1975 in a dedicated analyzer that could measure glucose in a 25 µl sample of blood without any sample treatment [52]. The use of an enzyme provides very high selectivity for the target compound making possible measurements in such a highly complicated matrix such as blood.

An alternative approach for exploiting oxidase enzymes is to monitor the H_2O_2 released. Hydrogen peroxide can be oxidized back to molecular oxygen at +0.6 V.

$$H_2O_2 \longrightarrow O_2 + 2H^+ + 2e^-$$
$$(5.46)$$

This approach has been demonstrated for a number of molecules that are substrates of oxidase enzymes. One of the disadvantages of using the measurement based on dissolved oxygen as an indirect indicator of glucose concentration is the challenge for miniaturizing the probe and gas-permeable membrane. Researchers have long sought to miniaturize the sensor in order to make possible the monitoring of glucose directly in the blood stream of a diabetic patient for use together with an insulin pump. Many researchers have thought that measuring hydrogen peroxide might be a simpler approach. Most strategies based on hydrogen peroxide still require membranes to immobilize the enzyme and to keep bigger molecules from fouling the electrode surface. In some cases, polymer films have been electrochemically generated at the working electrode making the preparation of tiny sensors easier [52].

There are two main disadvantages to using the oxidation of hydrogen peroxide to generate the current signal in enzyme electrodes. The first weakness is the fact that the voltage needed at the working electrode, +0.6 V, is positive enough to oxidize many other small molecules (such as ascorbic acid and acetaminophen) that are commonly found in the blood. Therefore, these other molecules can interfere. This problem has been addressed by coating a Pt working electrode with a film of Prussian blue – an insoluble salt formed from Fe^{3+} and $Fe(CN)_6^{4-}$. This salt is an electrocatalyst in that it can transfer electrons between a solution species and the electrode [27]. The film can reduce, rather than oxidize, hydrogen peroxide after which the iron(II) ions in the film transfer electrons directly with the electrode at a much lower voltage (−0.1 V versus Ag/AgCl) while remaining adsorbed to the platinum electrode surface. A robust approach was demonstrated in which a polymer film was generated by oxidizing 1,2-diaminobenzene at the electrode surface trapping both the enzyme and Prussian blue particles [27]. The process worked well enough that the authors successfully coated a carbon fiber making it into a glucose sensor of nanometer dimensions. Mixtures of carbon and Prussian blue particles and enzyme have also been screen-printed to create glucose biosensors [27, 54, 55].

One of the disadvantages of this and other approaches that rely on the formation of hydrogen peroxide is the fact that fluctuations in oxygen concentrations can cause errors. This is especially problematic in physiological fluids where the oxygen level can be as much as a factor of ten lower than the concentration of glucose. Thus, for *in vivo* monitoring, the response becomes oxygen-limited rather than glucose-limited, and the response curve for glucose is no longer linear at higher concentrations. This problem has led some researchers to develop sensors that use an electron transfer mediator in place of oxygen. The role of O_2 in oxidase reactions is to recycle the flavin cofactor bound to the enzyme.

$$GOx(FAD) + glucose \longrightarrow GOx(FADH_2) + gluconolactone \tag{5.47}$$

$$GOx(FADH_2) + O_2 \longrightarrow GOx(FAD) + H_2O_2 \tag{5.48}$$

where GOx(FAD) represents the glucose oxidase enzyme with the oxidized form of flavin, and $GOx(FADH_2)$ is the reduced form. The flavin moiety is held in the active site of the enzyme (through a linker at R in the following structure). Here is the flavin redox reaction.

$$\tag{5.49}$$

Oxidized flavin (FAD)   Reduced flavin ($FADH_2$)

Several redox agents, such as ferricenium and ferricyanide, can substitute for molecular oxygen and carry electrons between the enzyme and the electrode (see Figure 5.31b,c). These reagents are commonly referred to as mediators. Mediators have been essential to commercial glucose monitors used by millions of people in managing their diabetes.

$$GOx(FADH_2) + 2M_{Ox} \longrightarrow GOx(FAD) + 2M_{Red} + 2H^+ \tag{5.50}$$

$$2M_{Red} \longrightarrow 2M_{Ox} + 2e^- \tag{5.51}$$

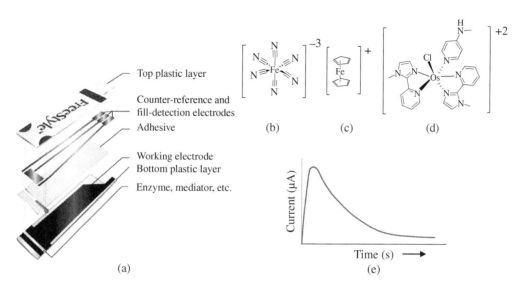

FIGURE 5.31 (a) Expanded view of a commercial glucose test strip. The blood sample is wicked into a narrow opening on the bottom right where enzyme and redox mediator are stored. The enzyme catalyzes the oxidation of glucose by the mediator. The bottom layer is the working electrode that is used to reoxidize the mediator. The second layer is an adhesive strip that defines the cavity that serves as the electrochemical cell. The next layer contains the Ag/AgCl reference/counter electrode. There are also two thin electrodes at the outer edges of the sample cell. Current between them indicate that the cell has been filled. Source: Reproduced with permission from Heller and Feldman 2010 [54]. Copyright 2010, American Chemical Society. (b) Ferricyanide, one of three different mediators used in commercial test strips, (c) ferricenium complex, and (d) an osmium bis(methyldiamidazolyl pyridine) complex. (e) Conceptual depiction of the current–time response of the glucose sensor.

In Eqs. (5.50, 5.51), M_{Ox} and M_{Red} are the oxidized and reduced forms of the mediator, respectively [52]. At least one company making single-use test strips for glucose monitors has replaced glucose oxidase with glucose dehydrogenase, but all schemes that have been used commercially oxidize glucose and reduce a mediator in the process. As long as the mediator reacts rapidly enough with the enzyme to out-compete oxygen, then it can replace O_2 in the analytical scheme. Ferricyanide and ferricenium complexes work well because they also are reoxidized at the electrode surface at a relatively low voltage (about +0.2 V versus Ag/AgCl).

Different companies have refined the design of glucose monitors over the last 20 years. Most designs use a carbon-working electrode and a combination Ag/AgCl reference/counter electrode (see Figure 5.31a). One company has decreased the analysis time to about five seconds, as well as decreasing the sample size to 300 nl (eightfold less volume than a mosquito bite!) and boosted the reproducibility by developing an osmium complex as a mediator [54] (see Figure 5.31d). After oxidizing (indirectly) the glucose, the reduced osmium mediator is reoxidized at the working electrode. In the tiny volume defined by the sample compartment of the test strip, there is some risk that the oxidized form of the mediator could diffuse to the counter electrode and be reduced. However, the osmium complex is too weak of an oxidizer to accept electrons from the Ag/AgCl reference electrode. That is, the $E^{o\prime}$ for the osmium complex is more negative than the

$E^{o\prime}$ for the Ag/AgCl couple. Hence, it is just strong enough to oxidize the glucose, but not strong enough to oxidize the Ag in the reference electrode. Test strips based on iron complexes do suffer from some mediator recycling or "shuttling" between the working and counter electrodes. Errors due to this background current mechanism need to be addressed in those test strips. The lack of mediator shuttling in the osmium mediator test strips means that the electrodes can be closely spaced (50 μm is the separation in these test strips) and placed directly opposite to each other making a very low working volume possible. The five-second analysis time is enough to permit nearly 100% conversion of the glucose in the sample. By integrating the current–time curve, the test strips yield signals that are less sensitive to temperature fluctuations and sources of noise. These carbon electrodes and reagents are screen-printed, single-use enzyme electrode systems [52].

Although a freely diffusing mediator has worked for test strips, loss of mediator in an implanted sensor or a device applied to intravenous measurements during surgery could be hazardous to the patient as well as lead to a loss of sensitivity. Remarkably stable and sensitive glucose sensors have been made using a conducting polymer to trap enzyme on the surface of electrodes. Poly(vinylpyridine) and poly(vinylimidazole) films covalently linked to osmium complexes have been demonstrated to provide lots of metal centers linked by long enough tethers to be able to reach into the active site of the enzyme for electron transfer with the flavin co-factor. Afterward, the osmium complex is able to relay the electron to the electrode surface [56]. The polymer also forms covalent links to the enzyme tying it to the electrode. This three-dimensional network is permeable to glucose and by-products turning over huge current densities (on the order of mA/cm^2) and making micron scale electrodes practical. Furthermore, the low operating voltage (+0.2 V versus SCE) minimizes the current from the oxidation of ascorbic acid, acetaminophen, cysteine, and uric acid [52].

The glucose sensor is the most successful example of a bioanalytical device based on electrochemistry. The utilization of biomolecules in order to provide selectivity for analytical sensors is an impressively active area of research today [57, 58]. There are many enzymes, such as oxidases and dehydrogenases, that control redox reactions and produce electroactive by-products. These can be coupled with electrodes to make sensors for the corresponding enzyme substrate. Other strategies have employed the molecular recognition abilities of DNA, RNA, immunoproteins, and other biomolecules to change the signal at a microelectrode in order to make implantable sensors and continuous monitoring possible. The ease of miniaturization and the relative mechanical simplicity of electrochemical methods give them an advantage over other types of instrumental analysis for sensing applications where operator attention to the device is not practical.

5.8.4. Karl Fisher Method for Moisture Determination

Another electrochemical method of analysis deserves special mention here because of its industrial importance. The automated coulometric version of the Karl Fischer method for determining trace levels of water has been in use for over 60 years [59]. Its high level of precision and accuracy has made it the method of choice for moisture analysis in the pharmaceutical industry, the petroleum industry, food science, material science, in the manufacture of paint and adhesive products, for quality control of transformer oil, cosmetics, and in many other industries where water can reduce the stability of a product or change its performance.

The underlying principle governing this method is based on a redox titration. The original method, published in 1935 by the German chemist, Karl Fischer, was performed using a buret [60]. The sample was dissolved in an alcohol solution (ethanol has been used as a solvent, but methanol or 2-methoxyethanol are commonly used today) of sulfur dioxide and a base, B (usually pyridine or imidazole and diethanolamine).

$$SO_2 + CH_3OH + B \ \rightleftarrows \ CH_3OSO_2^- + HB^+ \rightleftarrows (HB)CH_3OSO_2 \qquad (5.52)$$

The sample containing water is added to the alcohol solution of sulfur dioxide, imidazole, and diethanolamine. Then, the iodine titrant is added from a buret. The iodine is reduced in a 1 : 1 mole ratio with the sample water.

$$(HB)CH_3OSO_2 + H_2O + I_2 + 2B \rightleftarrows (HB)CH_3OSO_3 + 2I^- + 2HB^+ \qquad (5.53)$$

The quantity of I_2 added is carefully controlled, while the sulfur dioxide and base are present in excess. In the coulometric titration, the iodine is generated electrochemically at an electrode with a large surface area, usually a Pt screen (see Figure 5.32). Therefore, instead of iodine, an iodide salt is included in the solution with the sample. A porous glass frit or ceramic diaphragm is often used to minimize the loss of any I_2 to the cathode

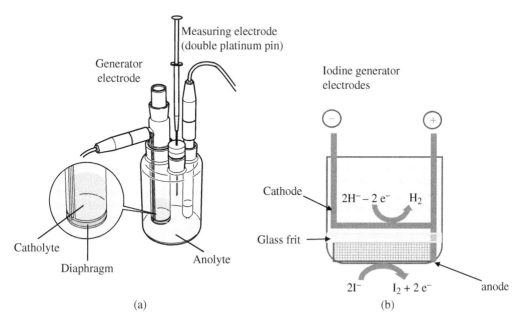

FIGURE 5.32 (a) KF coulometric titration cell. Source: Courtesy of Mettler-Toledo, Introduction to Karl Fisher Titration Guide. [61]. (b) Close-up view of the iodine generator. The lower platinum electrode oxidizes iodide in the titration solution to iodine. The anode process reduces hydrogen ions to hydrogen gas inside the glass tube. ˙The terms "generator electrode" and "measuring electrode" mean an arrangement of two electrodes (anode and cathode) to form an electrolytic cell.

compartment. The sample chamber is stirred thoroughly so that the iodine mixes quickly throughout the sample solution to react with the water. As with a titration using a buret, the generator stops producing iodine at the equivalence point. The amount of I_2 generated is directly proportional to the charge passed in the time between the start of the titration and the endpoint. This is a direct application of Faraday's law [62].

$$\text{moles } H_2O = \text{moles } I_2 \text{ generated} = \frac{Q}{nF} = \frac{\int i\,dt}{nF} = \frac{\int i\,dt}{(2)96\,485\,\text{C/mol}} \qquad (5.54)$$

A microcontroller circuit integrates the current throughout the experiment. These titrations are automated so that the generator stops at the equivalence point. How does the instrument recognize that the equivalence point has been reached? There are two common electrochemical approaches. Both use a second pair of small Pt wire electrodes to detect the endpoint. These detection electrodes are depicted in Figure 5.33a. The first of these detection schemes is called a biamperometric endpoint detection. In this mode, a modest voltage (about 500 mV) is applied between the two platinum wires, and the outside circuit monitors the current between them. The current at one electrode must match the magnitude of the current at the other electrode. Both of these electrodes are

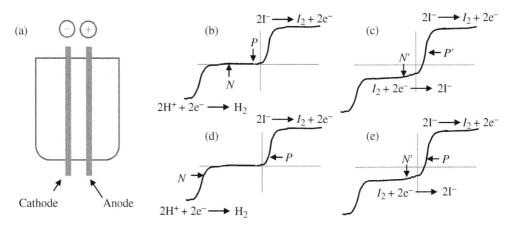

FIGURE 5.33 (a) Detection electrodes in Karl Fischer cell. (b) Voltammogram of sample solution before equivalence point. In the biamperometric detection mode, a 500 mV potential is applied between the electrodes. The negative electrode takes a voltage at N, and the positive electrode moves to P, at a value 500 mV more positive than N. The absolute voltage positions must correspond to currents of equal magnitude (in opposite direction). (c) After the equivalence point, a small excess of iodine appears in solution, and larger current is produced by I_2 reduction at the original value of N. Both electrode potentials, N′ and P′, shift to the right in order to provide enough oxidation current at P′ to balance the magnitude at N′. The sudden increase in the cell current signals the endpoint. (d) In the bipotentiometric detection mode, a constant current is forced between the electrodes. The potential for the negative electrode, N, must shift far to the left to reach a potential where the reduction of hydrogen ion can provide enough current. The voltage at the positive electrode will shift to the right until the current due to iodide oxidation at P matches the preselected current level. The measured potential between P and N is greater than 1 V before the equivalence point of the titration. (e) After the equivalence point, the presence of excess I_2 can provide reduction current at the negative electrode allowing N′ to shift in potential much closer to P. The sudden drop in the voltage between the electrodes signals the endpoint of the titration.

both polarizable. Neither is a reference electrode. That is, the absolute potential at the electrodes can shift (while maintaining the voltage difference at the pre-set value) as needed in order to match the oxidation current at the anode with the magnitude of the reduction current at the cathode. These wires are very close to each other so that the resistance between them is much smaller than the resistance between either and the generator electrodes. Consequently, they communicate with each other and ignore the processes occurring at the generator electrodes.

The voltage is not scanned during the experiment, but it helps to imagine what a voltammogram would look like for the sample solution in order to think through the processes that are occurring. Because the solution is being stirred, the current–voltage curve would look like a hydrodynamic voltammogram. Before the equivalence point, a voltammogram of the sample solution would look like diagram (b) in Figure 5.33. The potential for the positive platinum electrode would drift to whatever point, P, that would yield a current of the same magnitude, but of opposite sign to the current at point N (500 mV more negative than P). The only cathodic process available is for the reduction of hydrogen ions at the extreme negative end, far from the iodide oxidation wave. Consequently, N and P will lie somewhere in the horizontal region in between, where the background current is very small.

After the water is gone, there is no reactant left to consume the I_2. A tiny excess of I_2 forms at the generator and is quickly mixed throughout the cell by the stirrer. Although only a tiny excess in terms of moles, this excess makes the iodine concentration jump by several orders of magnitude. At this point, the negatively charged wire has a reducible species that can accept electrons, and current flows between the detector electrodes. The potential at the positive electrode will shift in a positive direction until the oxidation current, at P′, matches the reduction current at N′ (see Figure 5.33c). The outside circuitry will experience a surge in the current. When this current reaches a preset threshold (usually a few microamps), a microcontroller shuts off the power to the iodine generator, and the accumulated charge is displayed and/or sent to a computer. The higher that the threshold current is set, the greater the number of moles of excess iodine will be needed to signal the equivalence point. It is important to run a blank in order to calculate a correction for this error.

The same physical setup is used for the bipotentiometric mode of detection. In this case, a preset level of current is forced by the control circuit through the two detector electrodes. At the same time, the system monitors the difference in potential between the two detector electrodes. In Figure 5.35d, the current is supplied at the positive electrode by the oxidation of iodide ion (which is in great abundance) at point P. Before the equivalence point, there is virtually no I_2 in solution. The most easily reduced species is hydrogen ion. Consequently, the negative electrode potential will be found at a point N at the far negative end of the voltammogram. The voltage difference between the two electrodes will be greater than one volt. After the equivalence point, an excess in iodine provides a species that is much easier to reduce than hydrogen ion, as shown in Figure 5.33e. In response, the potential at the negative electrode will shift much closer to that of the positive electrode. This sudden decrease in voltage between the two detector electrodes triggers the system to stop the iodine generator. The microcontroller reads the integrated charge and displays the value or sends it to a computer.

Sample moisture levels of 1 ppm to 5% are most appropriate for coulometric Karl Fischer (KF) titrations. Higher concentrations take more reagent and are slow to perform coulometrically and are better done volumetrically with an automated buret. Samples with water content in the range of 10 μg to 100 mg are appropriate for automated coulometric titration. The "resolution" is ~0.1 μg water. Normal titration rate is ~2 mg H_2O/minute [64]. Coulometric titration is considered an "absolute" method, meaning the quantitative relationship between charge and moles of iodine generated needs no calibration. (Blanks still should be run for background correction.)

Because exclusion of moisture from air and from glass surfaces or reagents is important, the titration container is enclosed, and fresh reagents are pumped into the cell as needed. Before the first run, preliminary steps are required to eliminate moisture resulting from tiny leaks and accumulation of water over long idle periods. However, because the cell remains sealed between injections, there is little need for waiting between sample runs.

The Karl Fischer titration is quite specific for water. The major exception is interference from aldehydes and ketones that form acetals and ketals [64].

EXAMPLE 5.6

A 95.3 mg sample of olive oil was injected into a Karl Fisher coulometric titrator, and the endpoint was reached in 15.2 seconds at a current level of 100.0 mA. What was the weight % moisture in the original olive oil?

Answer:
    From Eq. (5.54),

$$\text{moles } H_2O = \text{moles } I_2 \text{ generated} = \frac{Q}{nF} = \frac{\int i\,dt}{nF} = \frac{\int i\,dt}{(2)96\,485\text{ C/mol}} = \frac{(0.100\text{ A})(15.2\text{ s})}{2(96\,485\text{ C/mol})}$$

$$= 7.87_6 \times 10^{-6}\text{ mol } H_2O$$

$$\text{wt\% water in oil sample} = \frac{(7.876 \times 10^{-6}\text{ mol})(18.015\text{ g/mol water})}{0.0953\text{ g oil}}(100\%) = 0.149\%$$

5.9. ION TRANSFER VOLTAMMETRY

Up to this point, the discussion in this chapter has focused on experiments in which the potential at a working electrode surface has been controlled for the purpose of forcing an electron transfer reaction to occur and measuring the resulting current. Recent work has demonstrated that there are advantages to be gained by doing similar experiments with ISEs. That is, an external power source can be used to control the voltage across the interface between a liquid membrane of an ISE and a sample solution in order to force ions to transfer between the sample solution and the membrane. In these experiments, known as ion transfer voltammetry, current–voltage curves are used to quantify the analyte concentration and to improve the discrimination against interfering ions.

There are many crucial details in making these experiments practical. However, the principles necessary to understanding the mechanisms and phenomena that are exploited in these experiments have already been laid out in this and previous chapters of this book. The discussion here focuses on two methods that are the ion transfer equivalent of stripping voltammetry and CV. Ion transfer stripping voltammetry (ITSV) is probably the easier of the two to understand and is presented first.

Recall that in an anodic stripping experiment, a huge gain in sensitivity results from reducing an analyte metal ion, such as Pb^{2+}, from a very dilute sample solution, to metallic Pb atoms dissolved in a tiny volume of liquid mercury, usually in the form of an extremely thin film, that serves as the working electrode. The small number of moles of analyte metal deposited into the film corresponds to a relatively big concentration inside the mercury because of the tiny volume of the film. This preconcentration step is followed by scanning the voltage from the negative deposition potential back in the positive direction. At a voltage near the formal potential for the analyte metal reduction, the lead is reoxidized and stripped out producing a current spike that has a height and area that are proportion to the lead ion concentration in the original aqueous sample solution.

In an ITSV experiment, the analyte ion is forced into a thin hydrophobic membrane on the order of 200 nm thick. Figure 5.34 shows a diagram of the membrane sandwich structure that was used for a stripping analysis of trace levels of perchlorate ions [63]. As with conventional liquid membrane ISEs, the membrane is made from a plasticized polymer, such as polyvinyl chloride (PVC) or polyurethane. This film is created by spin-casting the polymer solution from a volatile solvent onto a solid electrode, such as glassy carbon or gold, that has been previously coated with a conducting polymer. The purpose of the conducting polymer layer is to make the transition between electrons as current carriers in the electrode to ions as current carriers inside the membrane. Poly(3-octylthiophene) (POT) and poly(3,4-ethylenedioxythiophene) (PEDOT) are two popular redox polymers that can be deposited electrochemically directly onto a metal surface. The redox polymer can be oxidized or reduced by the underlying electrode. Each electron removed from the redox polymer introduces a cationic charge site on the polymer backbone. These charges attract anions, Y^-, from the hydrophobic salt, XY originally included in the PVC membrane. This creates a positive potential inside the PVC membrane which, in turn, extracts perchlorate ions from the aqueous sample solution. The sensor is held at a positive potential for a carefully controlled deposition time interval. During this time, the membrane preconcentrates the perchlorate analyte inside the PVC film. Then, the voltage on the sensor is scanned back toward zero. During the voltage scan, the positive charge on the redox layer is reduced, and the hydrophobic counter ions, Y^-, are pushed back into the PVC layer where the electrochemical potential drops. The perchlorate ions, in turn, are stripped out of the PVC membrane back into the aqueous solution, and the current is measured. Figure 5.35a shows current–voltage curves for the stripping step for nanomolar concentrations of perchlorate ion prepared in tap water.

During the preconcentration step, anions move from the sample solution into the membrane establishing a current analogous to the deposition of lead atoms into the mercury film electrode of an ASV experiment. This current is controlled by the mass transport of ions to the membrane surface, so it can be enhanced by stirring the sample. In the experiments that generated the data in Figure 5.35, the sensors were rotated using

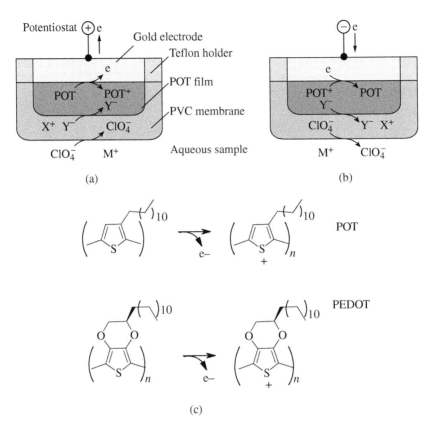

FIGURE 5.34   The structure of a liquid membrane sensor for ion transfer stripping analysis of perchlorate ions. Poly(3-octylthiophene) (POT) and poly(3,4-ethylenedioxythiophene) (PEDOT) are two redox polymers that act as electron to ion transducers. (a) Removal of an electron from either polymer leaves a cation charge site in the backbone. The new charge site pulls a hydrophobic ion, Y^-, from the PVC membrane increasing the positive charge there and attracting a perchlorate anion from the aqueous sample into the membrane. Thus, holding the sensor voltage at a positive value will preconcentrate perchlorate from the sample. (b) Scanning the voltage back toward zero reduces the charge in the POT layer pushing anions back into the PVC membrane and forces perchlorate ions back into the aqueous solution. Source: Adapted with permission from Kim and Amemiya 2008 [63]. Copyright 2008, American Chemical Society.

the same type of apparatus as is used for a rotated disk experiment in conventional voltammetry [63]. The deposition current follows the Levich equation for the limiting current as in a redox voltammogram at a rotated disk:

$$i_l = 0.62 z_i FAD\omega^{1/2} v^{-1/6} C_o \tag{5.55}$$

where z_i is the charge on the analyte ion, ω is the rotation rate in radians per second, v is kinematic viscosity (the solution viscosity divided by the density) of the sample solution, and C_o is the bulk concentration of the analyte ion.

   As with ASV, the amount of analyte extracted into the film depends on the conditions of the preconcentration step. The stripping signal increases with the time spent in the deposition process. However, in ITSV, the accumulated charge reaches a limiting value at a given sample concentration as shown in Figure 5.35d.

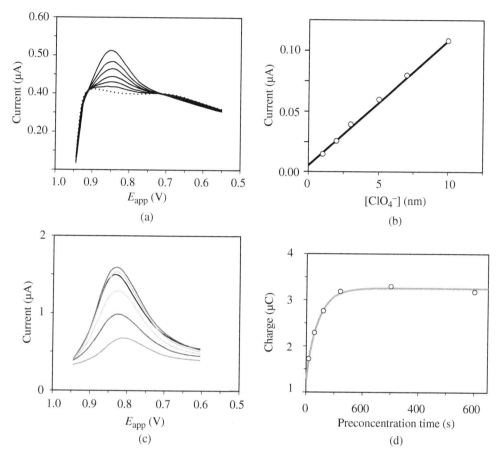

FIGURE 5.35 (a) Perchlorate stripping voltammograms of 10, 7, 5, 3, 2, 1, and 0 nM (from the top) in tap water at 0.1 V/s. (b) Peak current above background versus ClO_4^- concentration. The sensor was rotated at 4000 rpm. Deposition time was three minutes. (c) Peak current dependence on deposition time (from top: 600, 300, 120, 60, 30, and 10 s). (d) The accumulated charge versus deposition time indicates that the membrane saturates after about three minutes of deposition for a 100 nM standard solution. Source: Adapted with permission from Kim and Amemiya 2008 [63]. Copyright 2008, American Chemical Society.

The peak current for the stripping step is dependent on the square of the charge on the analyte:

$$i_p = \frac{z_i^2 F^2 v V_m C_m}{4RT} \tag{5.56}$$

where z_i is the charge on the analyte ion, F is Faraday's constant, v is the voltage scan rate, V_m is the membrane volume, C_m is the concentration of the analyte ion in the membrane, R is the ideal gas constant, and T is the temperature [65].

The limit of detection (LOD), for perchlorate was 0.5 nM in this work. That is comparable to the performance of the most sensitive of alternative methods, namely, liquid chromatography with a suppressed conductivity detector or electrospray ionization-mass spectroscopic methods. The LOD for a conventional potentiometric perchlorate ISE determination is 10 times better. However, the latter requires a few hours to reach equilibrium. Hence, this stripping technique provides a rapid alternative with good sensitivity.

In this example, selectivity depends solely on the relative solubility of the anion in a nonpolar environment. The relative selectivity of the membrane for anions follows the Hofmeister series. Although the Hofmeister series ranks common inorganic anions in the order of their ability to salt out proteins, there appears to be a correlation of the ranking in the series to the energy of solvation of the anions by water molecules [66]. Perchlorate appears at the end of the series among the most favorably extracted anions. (Strong salting out agents – $F^- \sim SO_4^{2-} > HPO_4^{2-} >$ acetate$^- >$ citrate$^- > Cl^- > NO_3^- > Br^- > ClO_3^- > I^- > ClO_4^- > SCN^-$ – favorably extracted.) [67] Interfering anions can be discriminated against by using a less positive applied potential for the preconcentration step. Furthermore, the stripping peak for *less* favored ions appears at a more positive voltage on the stripping voltammogram providing another characteristic for discrimination. Although clearly useful in some contexts, this sensor represents the least selective membrane system for an ISE.

Adding an ionophore to the PVC matrix provides another factor for stabilizing the target ion in the membrane and increasing the selectivity of the device. More recent work has demonstrated practical ITSV analysis for nanomolar levels of calcium, ammonium, and potassium ions. Figure 5.36 shows the membrane structure of a sensor for potassium ions based on the ionophore, valinomycin [68]. In addition to the use of an ionophore, the fact that the analyte was cationic meant that the composition was different from the perchlorate sensor in another way. When the redox polymer layer (PEDOT in this case) was formed, it was initially oxidized in an acetonitrile solution containing an electrolyte with hydrophobic ions. In the oxidized form, the charged sites in the polymer attract counter ions (in this case tetrakis(pentylfluoroborate) anions) to balance the charge. In the working device, the redox polymer is reduced to the neutral form during the preconcentration step. This forces the counter ions into the PVC layer giving it a negative charge. These ions are attractive to the potassium ions in the aqueous sample, so potassium ions transfer into the membrane and bind with the ionophore. The stripping step consists of scanning the voltage in the positive direction to reoxidize the PEDOT pulling the anions back into the PEDOT layer and expelling the potassium cations back into the water.

Stripping voltammograms for nanomolar potassium solutions are shown in Figure 5.37a. The voltage scan in the stripping step goes in the positive direction. More stable ion/ionophore complexes will produce stripping peaks at more positive voltage on the voltammogram. The K^+ peak appears at about $+0.54$ V; a second peak appears at about $+0.42$ V. The authors discovered that their lab water was contaminated with traces of ammonium ions that were hard to remove. The valinomycin ionophore is selective for potassium ion, but it will also bind ammonium ion. The valinomycin binding constant for potassium is about 60 times larger than that for ammonium. Consequently, during the stripping scan, the voltage must go to a more extreme positive potential (by about 100 mV) in order to strip out the potassium than for the ammonium peak. Despite the presence of approximately 100 nM levels of ammonium, the potassium signal was easily distinguishable, and potassium ions were detectable down to about 0.6 nM [68].

Figure 5.37c also shows the results for a stripping experiment for calcium ion using a hydrophobic ionophore. This experiment also used a rotated electrode system to achieve

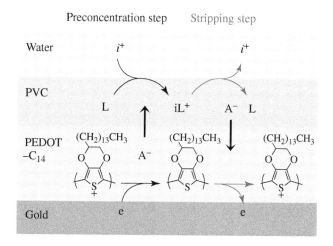

FIGURE 5.36 The diagram on the left shows the double polymer membrane ISE for potassium stripping voltammetry. When the PEDOT layer is reduced, the tetrakis(pentylfluorophenyl)borate counter ion, A⁻ (whose structure is shown on the far right side), is pushed into the PVC layer attracting the positive analyte ion into the PVC to bind with the valinomycin, L. The process is reversed during the stripping step producing a current peak. Source: Adapted with permission Kabagambe et al. 2012 [68]. Copyright 2012, American Chemical Society.

sub-nanomolar detection limits. The authors noted that working at these low calcium concentrations is a significant challenge due to the sensitivity to minute levels of contamination that are common in lab environments. It was necessary to prepare solutions and keep the electrochemical cell in a glovebox in order to minimize the contamination [65].

One concludes from these studies that trace-level analysis based on ITSV can be fast and effective. Analyte ions can be preconcentrated into a membrane device made from a combination of an ionophore-carrying hydrophobic thin polymer film on top of a redox polymer attached to a gold or carbon solid electrode. The stripping experiments are reasonably fast to perform (analysis time on the order of a few minutes) and the voltammogram enhances the discrimination of the analyte from possible interfering ions.

Not all applications of ISEs require such low detection limits. Ion transfer CV has some advantages to offer in the context of higher analyte concentrations. Using membranes

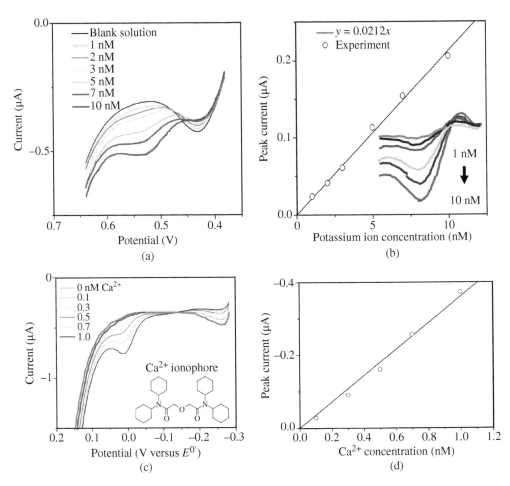

FIGURE 5.37 (a) ITSV of potassium ions in lab water after five-minute preconcentration step. The stripping peak near +0.42 V was shown to be ammonium contamination. Source: Adapted with permission from Kabagambe et al. 2012 [68]. Copyright 2012, American Chemical Society. (b) Calibration curve for potassium after voltammograms were corrected by subtracting the voltammogram for the blank (0 nM K$^+$). The corrected current-voltage curves are shown in the lower right corner of panel. (c) Calcium ITSV curves near the detection limit for Ca-ITSV device. Source: Adapted with permission from Kabagambe et al. 2014 [65]. Copyright 2014, American Chemical Society. (d) Calibration curve for sub-nanomolar calcium standard solutions using a 30-minute preconcentration step.

of ~200 nm thickness, current peaks in both extraction and depletion scans are directly observable in CV experiments with solutions in the range of 10 μM–0.1 M analyte. Figure 5.38 shows a cyclic ion transfer voltammogram of 10 mM Na$^+$ (in a stationary solution) for a polyurethane membrane on the redox polymer POT [69]. Since the volume of these thin membranes is so small, they saturate easily. Consequently, the membrane approaches the equilibrium conditions predicted by the Nernst equation (Eq. (5.46)) at each potential in the scan [65]. The integrated peak charge approaches a constant level. As a result, the peak height is not useful for quantitative analysis at these concentration levels.

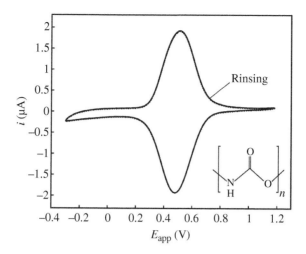

FIGURE 5.38 Multiple CV scans of 10 mM NaCl with a polyurethane membrane on the redox polymer POT with no loss of signal caused by rinsing in between scans. Good reproducibility indicates strong retention of ionophore in the sensor membrane. (Three-electrode system was used with a stationary working electrode.) Variation in peak potential <1.5% after 50 rinsings. Source: Adapted with permission from Cuartero et al. 2016 [69]. Copyright 2016, American Chemical Society.

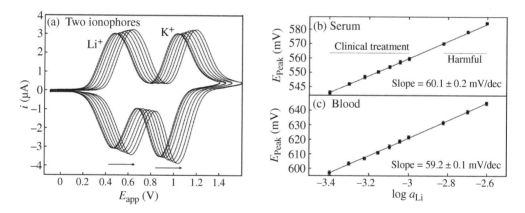

FIGURE 5.39 Two ions at once. (a) Separate ionophores for Li$^+$ and K$^+$ were used in the same polyurethane membrane for a set of cyclic voltammograms with increasing analyte concentrations (from left to right [Li$^+$] = [K$^+$] = 3.16×10^{-6}, 1×10^{-5}, 3.16×10^{-5}, 1×10^{-4}, 3.16×10^{-4}, 1×10^{-3}, 3.16×10^{-3}, 1×10^{-2} M). Supporting electrolyte: 10 mM NaCl. Scan rate: 100 mV/s. Each pair of peaks shifts to the right 60 mV per order of magnitude change in analyte concentration. Calibration curves for Li$^+$ ions prepared by spiking serum (b) and by spiking whole blood (c) showed good linearity over the clinical treatment range and into concentrations that would be harmful. Source: Adapted with permission from Cuartero et al. 2016 [69]. Copyright 2016, American Chemical Society.

However, the *peak position* shows a Nernstian dependence on concentration of the analyte. Both the forward and reverse peaks shift toward more positive potentials with increasing cation concentration. Figure 5.39a shows a series of ion transfer cyclic voltammograms for equimolar levels of both Li$^+$ and K$^+$ ions with a membrane containing separate ionophores selective for these two ions.

In this study, the authors used a polyurethane film in place of the PVC. Part of the motivation for using polyurethane was to find a more robust membrane polymer because PVC membranes exhibit limited life spans. Durability problems appear to be due to leaching of membrane components and/or polymer detachment. Polyurethane has greater adhesion, but similar performance. Polyurethane also has better biocompatibility [69]. The dual ion sensor tested in this work is robust enough to perform well in blood and serum as demonstrated by the calibration curves in Figure 5.39b. The sensor gave Nernstian response throughout the therapeutic concentration range for lithium as well as for higher toxic levels.

In addition to the speed of the analysis, the ability to quantify two ions with the same membrane offers the possibility of using one as an internal standard. The signals for both ions are subject to unwanted changes in the reference potential, junction potential, or ionic strength. However, if one of these analytes is present at a known concentration, its signal can be used to correct for these errors improving the accuracy and precision of the determination of the other analyte.

PROBLEMS

5.1 A chronoamperometry experiment was performed on a 4.00 mM solution of an iron(III) complex at 25 °C using a platinum electrode with an area of 0.075 cm². The potential was stepped from a voltage that was 100 mV more positive than the $E^{o\prime}$ out to a value that was 100 mV more negative than $E^{o\prime}$ where a one-electron reduction occurs. The data for the current–time curve are given below. Calculate the diffusion coefficient for the iron(III) complex under these conditions.

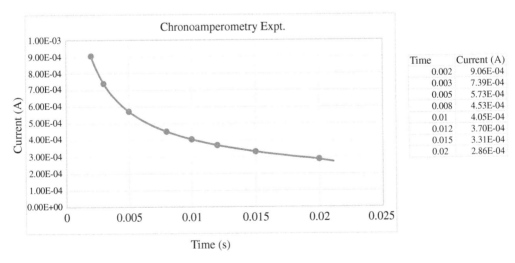

Time	Current (A)
0.002	9.06E-04
0.003	7.39E-04
0.005	5.73E-04
0.008	4.53E-04
0.01	4.05E-04
0.012	3.70E-04
0.015	3.31E-04
0.02	2.86E-04

5.2 A 33.6 mg sample of a freeze-dried enzyme preparation was injected into a Karl Fisher coulometric titrator and the endpoint was reached in 27.6 seconds at a current level of 150.7 mA. A blank run under the same conditions and current level require 1.9 seconds to react the endpoint. What was the wt% moisture in the enzyme?

5.3 From the CV below, estimate the formal potential for the redox couple.

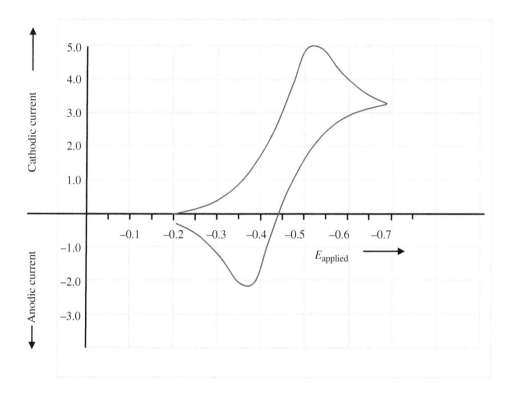

5.4 For the voltammogram in Problem 5.3, determine whether the one-electron transfer process is reversible in the electrochemical sense.

5.5 Does the voltammogram in Problem 5.3, indicate any possibility that the reduced or the oxidized form is unstable under the conditions in this solution? Support your answer with a calculation.

5.6 If the experiment in Problem 5.3 was recorded using a saturated calomel reference electrode at 25 °C, what is the value of $E^{\circ\prime}$ for the redox couple on the standard hydrogen electrode (SHE) scale?

5.7 The cyclic voltammogram below was recorded for a mixture of two different metal complexes, A and B, using a silver/silver chloride reference electrode (+0.197 V versus the normal hydrogen electrode [NHE]).

(a) What is the formal potential for complex B on the NHE scale?

(b) Comparing the oxidized forms of the two complexes, which complex is the stronger oxidizing agent?

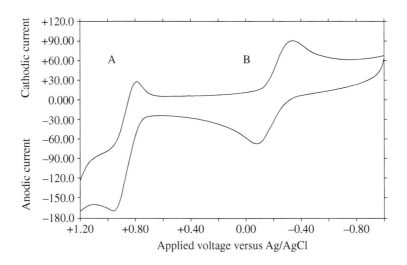

5.8 A staircase voltammogram was recorded for a 1×10^{-4} M solution of a reversible iron complex in which the voltage was stepped 5 mV every 10 ms sampling the current at 5 ms following the voltage step. The sampled current gave a well-defined peak when it was plotted versus voltage. An Osteryoung square wave voltammogram for the same solution with a 25 mV squarewave with a net 5 mV step at 100 Hz (current was sampled after 5 ms for each voltage transition). All of the other conditions were the same for both experiments. Give two plausible reasons why the square wave signal was much larger than the staircase current peak.

5.9 If for a cyclic voltammogram the $E^{o\prime}$ for quinone, Q, appears at a +0.220 V at a platinum electrode (versus silver/silver chloride ref. electrode) at pH 7.0, what direction would $E^{o\prime}$ shift at pH 6.0?

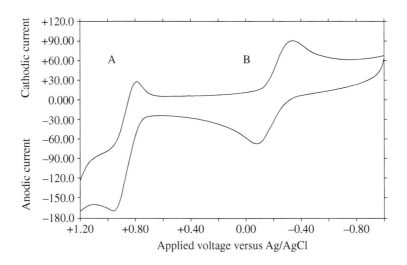

5.10 An organic compound produced the CV labeled A in an aqueous phosphate buffer on a glassy carbon electrode after polishing the electrode and cleaning it in a sonicating bath with deionized water. Curve B was the CV taken of the same solution on the same electrode after polishing, rinsing in the sonicating bath and electrochemically pretreating the electrode with five cycles of one minute at +1.3 V followed by one minute at −1.3 V. Give two plausible explanations for the difference in the appearance of the two curves and describe an experiment that could discriminate between these two hypotheses.

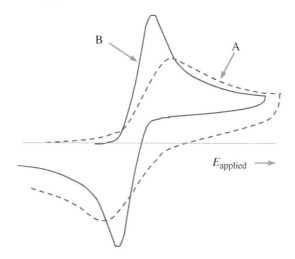

5.11   Imagine that your colleague is interested in performing anodic stripping analysis of Pb in drinking water and has been frustrated that he has not achieved the detection limits necessary to be able to quantify Pb at the low parts per billion level. His conditions have included a one-minute deposition time from a quiet solution at a hanging mercury drop electrode using a linear voltage scan for the stripping step. Make four suggestions for changes in his procedure that will each improve the sensitivity of the analysis.

5.12   Give a chemical explanation of why gold and platinum working electrodes are used as working electrodes for pulse amperometric detection (PAD) rather than glassy carbon.

5.13   One way to determine whether a current signal in a cyclic voltammogram is the result of a freely diffusing analyte or an electroactive compound attached to the electrode surface is to perform a scan rate study. Explain the distinguishing behavior of these two sources of current as a function of scan rate.

5.14   A cyclic voltammogram of p-aminophenol in $0.01\,M\,H_2SO_4$ was recorded with a carbon paste electrode scanning from $0\,V$ in the positive direction. The processes associated with each of the three current peaks are indicated on the voltammogram. Describe simple experiments that would help you confirm or refute the proposed products shown for each peak. (Source: Adapted with permission from H. A. Strobel and W. R. Heinemann 1988 [2])

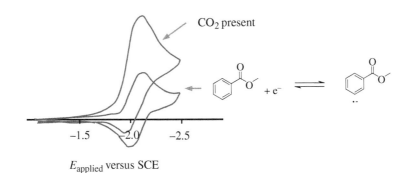

5.15 The cyclic voltammogram for methyl benzoate was recorded in dimethylformamide (DMF) using a platinum working electrode (lower curve). The product appears to be a singly-charged aryl radical. The solution was then exposed to CO_2 and a second voltammogram was scanned (upper curve).

Suggest a chemical explanation for the difference in behavior (in broad terms). Describe an experiment that would show support or refute your hypothesis.

5.16 For a RDE at 500 rpm a 0.1 μM solution of an iron complex gave a limiting current of 2.73 nA. What would be the limit of detection for this complex if the electrode were operated at 1000 rpm given a noise level of 0.5 nA?

REFERENCES

1. Kissinger, P.T. and Heineman, W.R. (1983). *J. Chem. Educ.* 60 (9): 702–706.
2. Strobel, H.A. and Heineman, W.R. (1988). *Chemical Instrumentation: A Systematic Approach*, 3e. New York: Wiley.
3. Bard, A.J. and Faulkner, L.R. (2001). *Electrochemical Methods*. New York: Wiley.
4. Wang, J. (2000). *Analytical Electrochemistry*, 2e. New York: Wiley-VCH.
5. Bioanalytical Systems, Inc. (2006). *BAS Epsilon LC Systems: Instruction manual*. West Lafayette, IN: Author.
6. Bioanalytical Systems, Inc. (2000). *BAS Rotating Disc Electrode (RDE-2): Instruction manual*. West Lafayette, IN: Author.
7. Blaedel, W.J. and Klatt, L.N. (1966). *Anal. Chem.* 38: 879–883.
8. Bard, A.J. and Faulkner, L.R. (1980). *Electrochemical Methods*. New York: Wiley, page 107.
9. Ahang, F. and Dryhurst, G. (1991). *J. Org. Chem.* 56: 7113–7121.
10. Saveant, J.-M. (2008). *Chem. Rev.* 108: 2348–2378.
11. Sawyer, D.T., Sobkawiak, A., and Robers, J. Jr. (1995). *Electrochemistry for Chemists*, 2e. New York: Wiley.
12. Mabbott, G.A. (1983). *J. Chem. Ed.* 60: 697–702.
13. Nicholson, R.S. and Shain, I. (1964). *Anal. Chem.* 36: 706–723.
14. Nicholson, R.S. (1966). *Anal. Chem.* 38: 1406–1406.
15. Bacon, J. and Adams, R.N. (1968). *J. Am. Chem. Soc.* 90 (24): 6596–6599.
16. Adams, R.N. (1969). *Acc. Chem. Res.* 2 (6): 175–180.
17. Miaw, L.-H.L., Boudreau, P.A., Pilcher, M.A., and Perone, S.P. (1978). *Anal. Chem.* 50: 1988–1996.
18. Fan, K., Luo, X., Ping, J. et al. (2012). *J. Agric. Food Chem.* 60: 6333–6340.
19. O'Dea, J.J., Osteryoung, J., and Osteryoung, R.A. (1981). *Anal. Chem.* 53: 695–701.
20. Osteryoung, J.G. and Osteryoung, R.A. (1985). *Anal. Chem.* 57: 101A–110A.
21. McCreery, R.L. (2008). *Chem. Rev.* 108 (7): 2646–2687.
22. Johnson, D.C. and LaCourse, W.R. (1990). *Anal. Chem.* 62 (10): 589A–597A.
23. Adams, R.N. (1958). *Anal. Chem.* 30 (9): 1576–1576.
24. Wang, J. and Freiha, B.A. (1984). *Anal. Chem.* 56 (4): 849–852.
25. Ping, J., Wu, J., Ying, Y. et al. (2011). *J. Agric. Food Chem.* 59 (9): 4418–4423.
26. Wang, J., Liu, J., and Cepra, G. (1997). *Anal. Chem.* 69 (15): 3124–3127.
27. Moscone, D., D'Ottavi, D., Compagnone, D. et al. (2001). *Anal. Chem.* 73 (11): 2529–2535.
28. Beitollahi, H., Karimi-Maleh, H., and Khabazzadeh, H. (2008). *Anal. Chem.* 80: 9848–9851.
29. Allred, C.D. and McCreery, R.L. (1992). *Anal. Chem.* 64 (4): 444–448.
30. Ranganathan, S., Kuo, T.-C., and McCreery, R.L. (1999). *Anal. Chem.* 71: 3574–3580.
31. Muguruma, H., Inoue, Y., Inoue, H., and Ohsawa, T. (2016). *J. Phys. Chem. C* 120 (22): 12284–12292.
32. Li, W., Tan, C., Lowe, M.A. et al. (2011). *ACS Nano* 5 (3): 2264–2270.
33. Poh, W.C., Loh, K.P., De Zhang, W. et al. (2004). *Langmuir* 20: 5484–5492.
34. Zachek, M.K., Takmakov, P., Moody, B. et al. (2009). *Anal. Chem.* 81 (15): 6258–6265.

35. Ching, S., Dudek, R., and Tabet, E. (1994). *J. Chem. Educ.* 71 (7): 602.

36. Wightman, R.M. (1981). *Anal. Chem.* 53: 1125A–1134A.

37. Howell, J.O. and Wightman, R.M. (1984). *Anal. Chem.* 56 (3): 524–529.

38. Wipf, D.O. and Wightman, R.M. (1989). *J. Phys. Chem.* 93 (10): 4286–4291.

39. Wightman, R.M. and Wipf, D.O. (1990). *Acc. Chem. Res.* 23: 64.

40. Keithley, R.B., Takmakov, P., Bucher, E.S. et al. (2011). *Anal. Chem.* 83 (9): 3563–3571.

41. EG&G (1990). *Princeton Applied Research Brochure on Polarographic Instrumentation*. NJ: Princeton.

42. La Course, W.R., Johnson, D.C., Rey, M.A., and Slingsby, R.W. (1991). *Anal. Chem.* 63 (2): 134–139.

43. Wang, J. (1985). *Stripping Analysis*. Deerfield Beach, FL: VCH.

44. Adeloju, S.B., Bond, A.M., and Briggs, M.H. (1985). *Anal. Chem.* 57: 1386.

45. Ostapczuk, P., Valenta, P., and Nürnberg, H.W. (1986). *J. Electroanal. Chem.* 214: 51.

46. Bott, A.W. (1993). *Current Separations* 12 (3): 141.

47. Florence, T.M. (1970). *J. Electroanal. Chem.* 27: 273–281.

48. Clark, L. and Lyons, C. (1962). *Ann. N. Y. Acad. Sci.* 102: 29–45.

49. Orna, M.V. (1989). Oxygen electrode. In: *Electrochemistry, Past and Present* (eds. J.T. Stock and M.V. Orna) Chapter 15, 196–210. ACS Symposium Series.

50. Yellow Springs, Inc. User manual for the series 6 multiparameter water quality sondes, https:// www.ysi.com/File%20Library/Documents/Manuals/069300-YSI-6-Series-Manual-RevJ.pdf (accessed 10 August 2019).

51. Holtzman, J.L. (1976). *Anal. Chem.* 48: 229–230.

52. Wang, J. (2007). *Chem. Rev.* 108 (2): 814–825.

53. Updike, S. and Hicks, G. (1967). *Nature* 214: 986.

54. Heller, A. and Feldman, B. (2010). *Accounts of Chemical Research* 43 (7): 963–973.

55. Komkova, M.A., Karyakina, E.E., and Karyakin, A.A. (2018). *J. Am. Chem. Soc.* 140 (36): 11302–11307.

56. Degani, Y. and Heller, A. (1987). *J. Phys. Chem.* 91 (6): 1285–1289.

57. Bollella, P. and Gorton, L. (2018). *Curr. Opin. Electrochem.* 10: 157–173.

58. Soto, R.J., Hall, J.R., Brown, M.D. et al. (2017). *Anal. Chem.* 89: 276–299.

59. Meyer, A.S. and Boyd, C.M. (1959). *Anal. Chem.* 31 (2): 215–219.

60. Fischer, K. (1935). *Angew. Chem.* 48: 394–396.

61. Mettler Toledo (2012). "Introduction to Karl Fisher Titration" Mettler Toledo KF Guide 1, https://www.mt.com/us/en/home/library/guides/laboratory-division/1/karl-fischer-titration-guide-principle.html (accessed 10 August 2019).

62. Dominguez, V.C., McDonald, C.R., Johnson, M. et al. (2010). *J. Chem. Ed.* 87: 987–991.

63. Kim, Y. and Amemiya, S. (2008). *Anal. Chem.* 80 (15): 6056–6065.

64. GPS Instrumentation Ltd. (2017) Measuring Principle Karl Fischer Titration, https://www.gpsil .co.uk/our-products/karl-fischer-titrators/measuring-principle/ (accessed 10 August 2019).

65. Kabagambe, B., Garada, M.B., Ishimatsu, R., and Amemiya, S. (2014). *Anal. Chem.* 86 (15): 7939–7946.

66. Adreev, M., de Pablo, J., Chremos, A., and Douglas, J.F. (2018). *J. Phys. Chem. B.* 122: 4029–4034.

67. Okur, H.I., Hladílková, J., Rembert, K.B. et al. (2017). *J. Phys. Chem. B* 121 (9): 1997–2014.

68. Kabagambe, B., Izadyar, A., and Amemiya, S. (2012). *Anal. Chem.* 84 (18): 7979–7986.

69. Cuartero, M., Crespo, G.A., and Bakker, E. (2016). *Anal. Chem.* 88 (11): 5649–5654.

CASE STUDIES IN CONTROLLED POTENTIAL METHODS

6

6.1. OVERVIEW

The purpose of this chapter is to demonstrate applications of voltammetry for investigating chemical reactions. These case studies highlight effective ways of garnering chemical information about a system for purposes other than the determination of the concentration of some analyte. If time is short, read the first case study (Section 6.2). It offers a brief introduction into the use of cyclic voltammetry (CV) as a tool for inorganic chemistry. It demonstrates the influence of a ligand on the formal potential of a metal complex. The formal potential of a redox complex is calculated from the average of the anodic and cathodic peak potentials in a cyclic voltammogram. In this case study, the replacement of single ligand on a ruthenium complex shifts the formal potential of the complex by roughly half a volt. The difference in formal potentials for the Ru(III) complexes is shown to be dependent on the relative binding strength of the ligands in question for the reduced and the oxidized forms of the complex. By applying Hess' law, one is able to calculate the equilibrium constants for both the formation of the Ru(II) and the Ru(III) complexes.

Case study II (Section 6.3) describes the use of CV as a tool in the design of redox catalysts. In this work, researchers manipulated the formal potential of iron and manganese complexes by changing the substituents on a single aromatic ligand. They demonstrated that the formal potential is linearly related to the electron-withdrawing effect of the substituent used on this ligand. Formal potentials of the catalysts were evaluated by CV.

Electrochemical studies have been useful for predicting intermediate compounds in cellular metabolism. Pharmaceutical manufacturers study the fate of a drug in the body because of the potential toxicity of its metabolites. Electrochemical studies are relevant because common pathways to biochemical detoxification and elimination of a substance in the liver use redox chemistry [1]. Case III (Section 6.4) describes the application of CV, chronoamperometry, and bulk electrolysis together with absorption spectroscopy and mass spectrometry in the identification of intermediates along the path toward the electrochemical oxidation of uric acid.

Several emerging technologies, such as the reduction of atmospheric carbon dioxide and efficient hydrogen fuel production from renewable energy sources, rely heavily on the use of electroanalytical tools to probe catalysis mechanisms. The final case study (Section 6.5) demonstrates the analysis of the shape of cyclic voltammograms for the purpose of probing mechanisms and measuring catalytic rate constants. This material represents an advanced topic and is intended to serve as a resource for the professional chemist. It looks at the methods used by two prominent research groups in this field.

Electroanalytical Chemistry: Principles, Best Practices, and Case Studies, First Edition. Gary A. Mabbott.
© 2020 John Wiley & Sons, Inc. Published 2020 by John Wiley & Sons, Inc.

The discussion builds on the principles presented in Chapter 5 and extends them to a technique called foot-of-wave analysis for estimating rate constants and turn-over efficiencies.

6.2. CASE I. EVALUATING THE FORMAL POTENTIAL AND RELATED PARAMETERS

One of the most important parameters for investigating redox reactions is the formal reduction potential, $E^{o\prime}$, for the reagent under study. The most commonly used method for evaluating $E^{o\prime}$ is CV. Consider, for example, the reduction of the ethylenediaminetetraacetic acid (EDTA) complex of ruthenium(III) in aqueous solution (shown here).

$$[Ru^{III}(EDTA)(H_2O)]^- + e^- \rightleftharpoons [Ru^{II}(EDTA)(H_2O)]^{2-}$$

$[Ru^{III}(EDTA)(H_2O)]^-$

CVs for three different concentrations of $[Ru(EDTA)(H_2O)]^-$ are shown in Figure 6.1A [2]. The dashed vertical line in Figure 6.1A appears at the average of the peak potentials and its position provides an estimate of the formal potential, $E^{o\prime}$ (about −0.01 V). It is useful

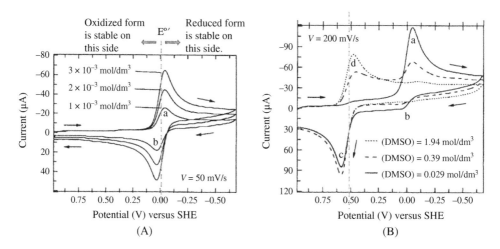

FIGURE 6.1 Cyclic voltammograms of $[Ru(EDTA)(H_2O)]^-$ and $[Ru(EDTA)(DMSO)]^-$. (A) Cyclic voltammograms of $[Ru(EDTA)(H_2O)]^-$ at different concentrations at a glassy carbon electrode. (B) Cyclic voltammograms of the same solution as in part (A) with the addition of DMSO to yield the indicated concentrations. Dashed vertical line indicates the formal potential of the corresponding complexes. Source: Adapted with permission from Toma et al. [2]. Copyright 2000, American Chemical Society.

to think of the formal potential in a manner similar to the pK_a for weak acids. Just as the pK_a represents a boundary between low pH conditions where the acid form is more stable and high pH conditions where the deprotonated form is more stable, the $E^{o\prime}$ represents a boundary between conditions where the oxidized and reduced species predominate. At potentials more negative than $E^{o\prime}$, the reduced form of the complex is stable; at potentials more positive than $E^{o\prime}$, the oxidized form is more stable. For this ruthenium complex, the oxidized and reduced forms occupy ranges of stability of similar size. The size of these zones of stability depends on the nature of binding of the metal to the ligands. It is interesting to note that the EDTA binds the ruthenium metal in five places rather than six sites as is common for most transition metals. This arrangement leaves the sixth coordination site available to bind with water. This water molecule can be replaced by other ligands. A very interesting ligand is nitric oxide, NO. The $[Ru^{III}(EDTA)(NO)]^-$ complex (shown below) has

dimethyl sulfoxide

$[Ru^{III}(EDTA)(NO)]^-$

$[Ru^{II}(EDTA)(DMSO)]^-$

been investigated as a possible pharmaceutical. Nitric oxide is a neurotransmitter, vasodilator (trigger for relaxing arterial blood vessels), as well as an antibacterial and antitumor agent [3]. The relative binding strength of these small ligands can have a big effect on the formal potential of the complex. This effect is demonstrated in Figure 6.1B, where the water in the ruthenium EDTA complex has been exchanged for dimethyl sulfoxide (DMSO) [2]. Additions of DMSO were added to the same solution that was used in Figure 6.1A in stages so that the partial displacement of water by the DMSO was also evident. After the first addition of DMSO, the reduction peak for the water-containing complex was still observed (peak (a) in the figure). The fact that peak (a) is unperturbed indicates that the $[Ru^{III}(EDTA)(H_2O)]^-$ complex remains intact. It is more stable than the corresponding $[Ru^{III}(EDTA)(DMSO)]^-$ complex under these conditions. However, the return peak (b) is almost entirely gone and a new return peak (c) appears near +0.6 V. This indicates that the reduced form of the water complex converts to the DMSO complex very rapidly. The DMSO form of the reduced complex is thermodynamically favored over the water form. At higher concentrations of DMSO, a new reduction peak (d) appears paired with the new oxidation peak (c). This pair of peaks represents the oxidation and reduction of the DMSO complex. A dashed line has been placed at the average of the peak potentials for peaks (c) and (d) representing the $E^{o\prime}$ for this couple. The $E^{o\prime}$ for the DMSO complex has shifted by over half a volt to more positive values (+0.53 V). The positive position of the $E^{o\prime}$ indicates that the reduced form of the complex is much more stable than the oxidized form in the DMSO complex. The DMSO stabilizes the Ru(II) form by binding more strongly to it than the ligand does the Ru(III). The oxidation state of the metal center is crucial to the binding because the bond between the metal center and

DMSO is primarily the result of electron density from a d-orbital on the metal donating to a π^*-orbital (anti-bonding orbital) on the ligand. The Ru(III) metal ion has one fewer electron to share [2, 4]. (The acceptor orbital on the DMSO happens to be an anti-bonding orbital. That may seem contradictory. Putting electrons into an anti-bonding orbital does lower the bond order for the S—O bond in the DMSO, but it provides an effective mode for sharing electrons between the metal and ligand.) Water does not have a π-orbital to provide that type of interaction. Water donates an electron pair to an empty d-orbital on the ruthenium. The important conclusion is that DMSO stabilizes the reduced form of the complex much more than the oxidized form and that causes a large shift in the formal potential of the complex. Just how much of a difference in the metal–ligand binding strength there is between the two redox forms of the DMSO-containing complex can actually be calculated from the data provided in the cyclic voltammograms in Figure 6.1B [2].

The equilibrium constant for the ligand exchange is a measure of the binding strength of the DMSO with the metal center. Consider ligand exchange for the Ru(III) complexes.

$$[Ru^{III}(EDTA)(H_2O)]^- + DMSO \rightleftarrows [Ru^{III}(EDTA)(DMSO)]^- + H_2O \tag{6.1}$$

The corresponding equilibrium constant for Reaction (6.1) can be written as follows:

$$K^{III} = \frac{[Ru^{III}(EDTA)(DMSO)^-][H_2O]}{[Ru^{III}(EDTA)(H_2O)^-][DMSO]} = \frac{[Ru^{III}(EDTA)(DMSO)^-]}{[Ru^{III}(EDTA)(H_2O)^-][DMSO]} \tag{6.2}$$

where the activity of H_2O is assumed to be unity because it is the solvent. With the use of Eq. (6.2) and current measurements from Figure 6.1B, it is possible to calculate an estimate of K^{III}. As DMSO is added to the system, more of the water complex converts to the DMSO complex. Assume that the Ru(III) is present in only as $[Ru^{III}(EDTA)(H_2O)]^-$ and $[Ru^{III}(EDTA)(DMSO)]^-$. Then the following is true:

$$[Ru^{III}(EDTA)(H_2O)^-] + [Ru^{III}(EDTA)(DMSO)^-] = [Ru^{III}]_{total} \tag{6.3}$$

In Section 5.3.5.2, it was noted that the peak current is proportional to the concentration of the electroactive species. The key relationship is the Randles–Sevcik equation (6.4). Because the background current is easiest to estimate for the peak closest to the start of the scan, we will work with the peak marked (d) for the reduction of the DMSO-containing complex.

$$i_p(d) = (2.69 \times 10^5)n^{\frac{3}{2}}AD^{\frac{1}{2}}v^{\frac{1}{2}}[Ru^{III}(EDTA)(DMSO)^-] \tag{6.4}$$

Assuming that the voltammogram for the highest concentration of DMSO represents the conditions where virtually all of the water in the complex has been displaced by DMSO, then:

$$i_p(d)_{max} = (2.69 \times 10^5)n^{\frac{3}{2}}AD^{\frac{1}{2}}v^{\frac{1}{2}}[Ru^{III}(EDTA)(DMSO)^-]_{max}$$

$$= (2.69 \times 10^5)n^{\frac{3}{2}}AD^{\frac{1}{2}}v^{\frac{1}{2}}[Ru^{III}]_{total} \tag{6.5}$$

These three equations can be used to express the fraction of oxidized ruthenium in the DMSO complex:

$$\text{Ru}^{III} \text{ fraction in DMSO form} = \frac{[\text{Ru}^{III}(\text{EDTA})(\text{DMSO})^-]}{[\text{Ru}^{III}]_{total}} = \frac{i_p(d)}{i_p(d)_{max}} \quad (6.6)$$

Because the fraction of the Ru^{III} in the DMSO form and the fraction in the H_2O form must sum to unity, the fraction for the H_2O form is given by

$$\text{Ru}^{III} \text{ fraction in } H_2O \text{ form} = \frac{[\text{Ru}^{III}(\text{EDTA})(H_2O)^-]}{[\text{Ru}^{III}]_{total}} = 1 - \text{Ru}^{III} \text{ fraction in DMSO form}$$

$$= 1 - \frac{i_p(d)}{i_p(d)_{max}} \quad (6.7)$$

Dividing Eq. (6.6) by Eq. (6.7) gives a useful ratio:

$$\frac{[\text{Ru}^{III}(\text{EDTA})(\text{DMSO})^-]/[\text{Ru}^{III}]_{total}}{[\text{Ru}^{III}(\text{EDTA})(H_2O)^-]/[\text{Ru}^{III}]_{total}} = \frac{i_p(d)/i_p(d)_{max}}{1 - \frac{i_p(d)}{i_p(d)_{max}}} = \frac{[\text{Ru}^{III}(\text{EDTA})(\text{DMSO})^-]}{[\text{Ru}^{III}(\text{EDTA})(H_2O)^-]}$$

$$= \frac{i_p(d)}{i_p(d)_{max} - i_p(d)} \quad (6.8)$$

Substituting into Eq. (6.2) yields an expression that can be used to calculate the equilibrium constant, K^{III}. Using estimates of the current at (d) for the middle concentration and the current at the maximum concentration (for $i_p(d)_{max}$) gives:

$$K^{III} = \frac{[\text{Ru}^{III}(\text{EDTA})(\text{DMSO})^-]}{[\text{Ru}^{III}(\text{EDTA})(H_2O)^-][\text{DMSO}]} = \frac{i_p(d)}{\{i_p(d)_{max} - i_p(d)\}[\text{DMSO}]}$$

$$= \frac{48 \, \mu A}{\{72 \, \mu A - 48 \, \mu A\}[0.39 \, M]} = 5.1 \quad (6.9)$$

A binding constant of 5 is not particularly strong. This number is a measure of the Ru^{III}–DMSO bond. It is interesting to compare that number with the binding constant for the reduced DMSO-containing complex. Fortunately, the voltammograms in Figure 6.1 provide the data for calculating that value, too.

This strategy takes advantage of Hess' Law. The ligand exchange process and the formal potentials for the two different complexes are related by the equations in Figure 6.2 [2].

There are conceptually two different paths that might be followed in order to go from $[\text{Ru}^{III}(\text{EDTA})(\text{DMSO})]^-$ as a reactant to the reduced product $[\text{Ru}^{II}(\text{EDTA})(\text{DMSO})]^-$. First of all, one can merely add an electron (step 4 in Figure 6.2). Alternatively, one can first exchange the DMSO ligand for H_2O, reduce that complex and then swap the water ligand with DMSO. Because the products and reactants are in equilibrium states, the energy spent to get from one to the other is independent of the path. Consequently, the energy spent going from $[\text{Ru}^{III}(\text{EDTA})(\text{DMSO})]^-$ to $[\text{Ru}^{II}(\text{EDTA})(\text{DMSO})]^-$ by steps 1, 2, and 3 is the

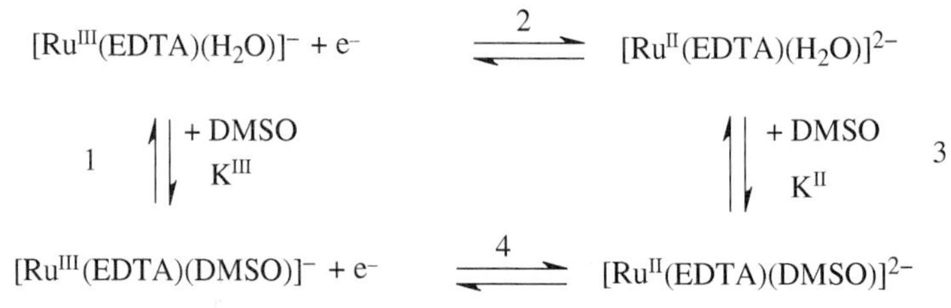

$$[Ru^{III}(EDTA)(H_2O)]^- + e^- \quad \xrightleftharpoons{\quad 2 \quad} \quad [Ru^{II}(EDTA)(H_2O)]^{2-}$$

$$1 \quad \Big\updownarrow \begin{array}{c} + DMSO \\ K^{III} \end{array} \qquad\qquad\qquad \Big\updownarrow \begin{array}{c} + DMSO \\ K^{II} \end{array} \quad 3$$

$$[Ru^{III}(EDTA)(DMSO)]^- + e^- \quad \xrightleftharpoons{\quad 4 \quad} \quad [Ru^{II}(EDTA)(DMSO)]^{2-}$$

FIGURE 6.2 Equilibria relating the different forms of some ruthenium–EDTA complexes. The energy required to go from $[Ru^{III}(EDTA)(DMSO)]^-$ to $[Ru^{II}(EDTA)(DMSO)]^-$ is the same following the path $1+2+3$ as amount of energy associated with the direct reduction of $[Ru^{III}(EDTA)(DMSO)]^-$ in a single, one-electron process (step 4). Source: Adapted from with permission from Toma et al. [2]. Copyright 2000, American Chemical Society.

same as the energy spent in the direct electron transfer, step 4. Expressed in terms of the Gibbs' free energy change:

$$\Delta G_1 + \Delta G_2 + \Delta G_3 = \Delta G_4 = RT\ln(K^{III}) - nFE_2^{o\prime} - RT\ln(K^{II}) = -nFE_4^{o\prime} \tag{6.10}$$

In Eq. (6.10), the free energy change associated with the formal potential for an electron transfer reaction is $-nFE^{o\prime}$ in joules. The free energy change associated with an equilibrium constant is $-RT\ln(K_{eq})$ in joules. The sign was changed for K^{III} because the direction for step 1 represents the inverse of the formation of the Ru^{III}–DMSO complex (opposite to the direction for which K^{III} was defined). All of the terms are defined or can be evaluated from the voltammograms in Figure 6.1 except for K^{II}, the formation constant for the Ru^{II}–DMSO complex. From Figure 6.1, it appears that $E_2^{o\prime} = -0.01$ V and $E_4^{o\prime} = +0.53$ V versus standard hydrogen electrode (SHE). The ideal gas constant, R, has a value of 8.314 J/K. The temperature was 298 K and Faraday's constant is 96 485 C/mol. Solving Eq. (6.10) for K^{II} gives:

$$RT\ln(K^{II}) = RT\ln(K^{III}) - nFE_2^{o\prime} + nFE_4^{o\prime} = (8.314)(298)\ln(5.1) + (1)(96\,485)\{0.53 - (-0.01)\}$$

$$= 5.25 \times 10^4 \tag{6.11}$$

$$\ln(K^{II}) = \frac{5.25 \times 10^4}{(8.314)(298)} = 21.2 \quad \text{and} \quad K^{II} = e^{21.2} = 1.6 \times 10^9 \tag{6.12}$$

The binding constant for the DMSO and Ru(II) is about 300 million times stronger than it is for the Ru(III) complex. Consequently, the DMSO ligand stabilizes the reduced form of the complex and that results in a much more positive $E^{o\prime}$. Another way of thinking about the outcome is that the $[Ru^{III}(EDTA)(DMSO)^-]$ complex is easier to reduce than the $[Ru^{III}(EDTA)(H_2O)^-]$. The DMSO complex is a stronger oxidizing agent than the water complex.

6.3. CASE II. EVALUATING CATALYSTS – THERMODYNAMIC CONSIDERATIONS

The case study with the ruthenium–EDTA complexes demonstrates that the strength of binding of ligands to the metal center can have a profound effect on the formal reduction

Net reaction:
$$A_{Ox} + B_{Red} \rightleftharpoons A_{Red} + B_{Ox}$$

FIGURE 6.3 The reduction of A_{Ox} by reagent, B_{Red}, as mediated by a catalyst, C. The thermodynamics of this cycling scheme require that the formal potential for C must lie between the formal potentials of A and B. In other words, $E_A^{o'} > E_C^{o'} > E_B^{o'}$.

potential of the complex. Because of this effect, inorganic chemists have studied the use of different ligand structures as a means of manipulating the redox potential of transition metal complexes for the purpose of making more efficient redox catalysts. A redox catalyst acts as a mediator in transferring electrons between two reactant molecules when a direct exchange is too slow. The general scheme is shown in Figure 6.3. Investigating ligand effects on the formal potential of complexes with structures similar to the active site of a redox enzyme is also an indirect method for studying how that enzyme functions [5]. By making subtle changes in the ligands, one can tune the formal potential of the complex to reach an optimum value.

In order for a redox catalyst to work, thermodynamics require that its formal potential lie between the formal potentials of the two redox couples in the net reaction that it is designed to catalyze. This situation is diagramed in Figure 6.3. In order for the catalyst, C, to cycle, C_{Ox} must be able to oxidize the reduced form of the reactant, B_{Red} and the reduced form of the catalyst, C_{Red}, must be easily oxidized by A_{Ox}. The formal potentials of these three redox pairs must follow the ranking $E_A^{o'} > E_C^{o'} > E_B^{o'}$. That is, A_{Ox} must be a stronger oxidizer than C_{Ox} which, in turn, must be a stronger oxidizer than B_{Ox}. Evaluating the formal potentials of individual complexes by cyclic voltammetry is an important step in testing complexes to see if they meet the thermodynamic criterion for catalyzing the reaction of interest.

The same criteria apply to redox enzymes. The influence of the ligands on the redox potential of the complex in the active site is a key to understanding how these enzymes work. Because the mass transport of proteins is relatively slow, direct voltammetry on enzymes is difficult. Instead, researchers have used complexes to model the environment of the active site of important enzymes. An example of this is a study of manganese and iron complexes that mimic the behavior of the active site of a group of enzymes called superoxide dismutases (SODs). Normal cell metabolism generates a small amount of superoxide, O_2^-, a very reactive free radical that can cause all sorts of damage, if left unchecked within the cell. The SOD enzymes protect the cell by hastening the disproportionation of the superoxide. The average oxidation state of the oxygen atoms in O_2^- is −0.5. It is an intermediate on the path between O_2 (oxidation state of 0) and hydrogen peroxide, H_2O_2, (at −1). Consequently, O_2^- can be oxidized back to O_2 or reduced to H_2O_2 [6].

$$O_2^- \rightleftharpoons O_2 + e^- \tag{6.13}$$

$$O_2^- + e^- + 2H^+ \rightleftharpoons H_2O_2 \tag{6.14}$$

In the enzyme system, O_2^- and O_2 play the role of the redox couple B in the diagram in Figure 6.3. Eq. (6.14) represents the reduction reaction for the redox couple A. The metal-containing active site of the SOD enzyme is represented by C in the diagram. For clarity, M_{SOD} represents the metal at the center of the enzyme's active site in the two different redox reactions in Eqs. (6.15, 6.16).

$$O_2^- + M_{SOD}^{III} \rightleftharpoons O_2 + M_{SOD}^{II} \tag{6.15}$$

$$O_2^- + M_{SOD}^{II} + 2H^+ \rightleftharpoons H_2O_2 + M_{SOD}^{III} \tag{6.16}$$

It has been estimated that the formal potential (at pH 7) for the half reaction in Eq. (6.13) is +0.89 V versus SHE and the formal potential for Eq. (6.14) is −0.16 V [6]. Consequently, the complex in the active site of the enzyme must have a formal potential between +0.89 and −0.16 V versus SHE at pH 7. Manganese(II) and iron(II) are commonly found in various forms of SOD. Small changes in the structure of the ligands that hold the metal in the active site are thought to be responsible for the apparent formal potential of various forms of the SOD enzymes. In order to explore this hypothesis, Sjödin et al. created a series of iron and manganese complexes with ligands that mimic the environment of some known SOD enzymes [6]. The structures of these model complexes are represented in Figure 6.4.

It was hypothesized that the tendency of the substituent to donate electron density to the aromatic ring and, indirectly, to the ligand site would influence the formal potential of the complex. The electron- donating/-withdrawing effect of substituents was quantified by the work of Hammet beginning in the late 1930s. Hammet measured the pK_a of para-substituted benzoic acids. He reasoned that an electron-withdrawing substituent would take electron density from the carboxylic acid group as well as the aromatic ring [7]. The effect would cause the acid to dissociate more easily. An electron-donating group would have the opposite effect. He quantified his comparisons by assigning each substituent a constant, σ, based on the difference between the pK_a of the substituted molecule and the

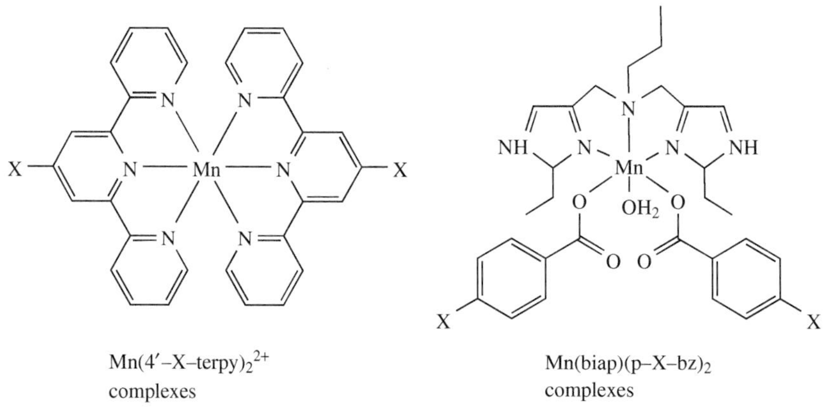

Mn(4′–X–terpy)$_2$$^{2+}$
complexes

Mn(biap)(p–X–bz)$_2$
complexes

FIGURE 6.4 General structures for a series of Mn(II) complexes that were studied as models of the active site for superoxide dismutase enzymes. (Similar Fe(II) complexes were also used.) The formal potential of the complex was measured for different ligand substituents, X. Source: Reproduced with permission from Sjödin et al. [6]. Copyright 2008, American Chemical Society.

pK_a of benzoic acid itself:

$$\sigma_x = pK_{a,x} - pK_{a,benzoic} \tag{6.17}$$

Positive values of σ are more electron-donating than a hydrogen atom, and negative values are more electron-withdrawing. Sjödin found that the formal potential of Fe and Mn complexes of the general form shown in Figure 6.4 could be tuned by changing the substituents, X, in two places [6]. Furthermore, the shift in the formal potential was a linear function of the Hammet constant. A summary is shown graphically in Figure 6.5.

Many of the complexes used in this work were poorly soluble in water. Consequently, the experiments were performed in acetonitrile (with an electrolyte of 0.1 M tetrabutyl-ammonium/hexafluorophosphate). The workers used an aqueous Ag/AgCl reference electrode. Because the difference in media can lead to an undefined and slowly drifting junction potential, the authors used a secondary reference technique. After recording voltammograms of each complex, they spiked the solutions with ferrocene in order to determine the $E^{o\prime}$ of the ferrocenium/ferrocene redox couple under the same conditions. The formal potential for the ferrocene provided a secondary reference that eliminated the need to know the absolute junction potential between the reference electrode and the working solution. All potentials were referenced to the $E^{o\prime}$ for the ferrocene. The data plotted in Figure 6.5 (and that for several other substituents) demonstrated that relatively small changes to parts of ligands that are distant from the coordinating site can influence the electron donating properties of the ligand and, in turn, shift the formal potential of these model complexes in a predictable manner. These results support the hypothesis that the reduction potential of the active site of a redox enzyme is also subject to subtle structural changes in the ligands that bind the metal.

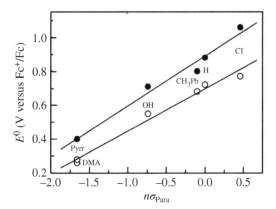

FIGURE 6.5 A plot of the formal potential for complexes versus Hammet constant, $n\sigma$, where n is the number of X-substituents ($n = 2$, in these cases). The experiments were carried out in acetonitrile, and the formal potential was measured with respect to the average of the ferrocenium/ferrocene peak potentials in order to account for any drift of the potential of the aqueous Ag/AgCl (saturated KCl) reference electrode or of the junction potential. Open circles represent the $[Fe(4\prime-X-terpy)_2]^{3+/2+}$ complexes, and the closed circles are data for the $[Mn(4\prime-X-terpy)_2]^{3+/2+}$ complexes. Abbreviations for substituents: Pyrr = pyrrolidine, DMA = dimethylamine, OH = hydroxyl, CH_3Ph = methyl phenyl, H = hydrogen, Cl = chlorine. Source: Adapted with permission from Sjödin et al. [6]. Copyright 2008, American Chemical Society.

6.4. CASE III. STUDYING THE OXIDATION OF ORGANIC MOLECULES

Organic electrode reaction mechanisms can provide insight into biochemical reaction pathways. This case study demonstrates the use of controlled potential methods to investigate possible intermediates in the decomposition of a common cellular constituent, uric acid. Figure 6.6 shows the cyclic voltammogram for uric acid recorded at a moderately fast scan rate of 500 mV/s in an aqueous solution at a glassy carbon electrode [8]. The scan started by going from 0 V in the negative direction. There was no sign of reduction of the starting material, even out as far as −1.5 V. However, scanning in the positive direction produced a large oxidation peak around +0.4 V. Scanning back in the negative direction again yielded a small reduction peak paired with the huge oxidation peak. A second reduction peak (that was not present in the first cathodic-going segment) was also observed. As the scan rate was increased the first reduction peak (I_c) grew with respect to the oxidation

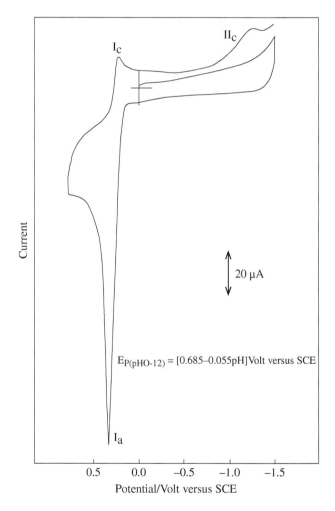

FIGURE 6.6 Cyclic voltammogram of 2 mM uric acid in pH 7.5 phosphate buffer recorded at 500 mV/s on a pyrolytic graphite electrode. **Source:** Reproduced with permission from Dryhurst et al. [8]. Copyright 1983, American Chemical Society.

peak height (data not shown) and the second reduction peak (II_c) decreased in size. This behavior suggests that a chemical step follows the initial electron transfer process (associated with peak I_a). This type of sequence is called an EC reaction mechanism (E = electron transfer, C = chemical step).

$$A \rightleftarrows B + ne^- \tag{6.18}$$

$$B + C \longrightarrow X + P \tag{6.19}$$

A and B are a redox couple, but B is unstable. It reacts to form X, leaving little B available near the electrode to be reduced to back to A on the cathodic scan. This scheme also includes the possibility that B decomposes on its own (that is, there is no second reactant, C and, perhaps no additional product, P). The new reduction peak, II_c, suggests that X can also be reduced.

What sort of information can one derive about the mechanism? Certainly, the identity of the products associated with each peak is of interest. Determining the number of electrons and protons involved in each process can help in making the identification and outlining a possible mechanism. Another important question is whether there are intermediates that can be identified. That is, are there transient species that appear on the way to forming compound X? If so, could these intermediates react with other molecules under different conditions? The probability of such side reactions is related, in part, to the rates of reaction for various chemical steps. Consequently, measuring the lifetimes of species under these experimental conditions is of interest.

Answering the question of the number electrons and number of protons transferred in the first oxidation step can be useful evidence for identifying the product of the first oxidation peak, that is, B. The best way of finding the number of electrons, n, is by coulometry. Holding the potential at +0.6 V (much more positive than the oxidation peak on the CV) and measuring the charge for oxidizing a known number of moles of reactant provide the data for calculating n from Faraday's law:

$$Q = \int idt = nFN \tag{6.20}$$

$$n = \frac{Q}{FN} = \frac{\int idt}{FN} \tag{6.21}$$

where N = the number of moles reacting and $F = 96\,485\,C/mol$.

One way of doing this is to electrolyze the entire quantity of a weighed amount of reactant added to the cell. This process usually means stirring the solution and using a working electrode with a large surface area (such as a Pt mesh) and recording the current until it drops to a tiny fraction of the original value. Such an experiment generally takes a few hours, but, in addition to providing a value for n, it also provides a supply of the final product, X, that can be identified from various forms of spectroscopy, including mass spectrometry. Dryhurst's group performed a bulk electrolysis at +0.6 V for the oxidation of uric acid and calculated a value of 2 for n [8].

It is interesting to think about the analysis of the electrolysis product for a couple of reasons. Product isolation can be a challenge and the authors employed a clever strategy. Furthermore, it was also an interesting example of electrochemical phenomena in action.

They freeze-dried the solution so that they could redissolve the material in a smaller volume of water and separate the electrolysis product from the salt that made up the original supporting electrolyte using liquid chromatography with water as the mobile phase. The authors' choice of column was a commercial gel permeation chromatography resin, Sephadex G-10™. It is a porous material in the form of beads made by cross-linking polysaccharide chains. Retention is based on molecular size. That material has a low molecular weight cut-off limit of only 700 Da. Molecules bigger than the cut-off limit do not penetrate the pores inside particles and are eluted by flushing the column with the volume of solvent equivalent to the space between particles. They are unretained. In contrast, the smallest solutes can explore the volume inside the channels of each particle and require more solvent to push them out of the column. They emerge last based on that mechanism. Curiously, the sodium phosphate, with a much smaller hydrated radius than uric acid, eluted first. Some mechanism other than size exclusion must have been operating. Earlier work with gel resins has demonstrated that the polysaccharide material contains a small fraction of carboxylic acid functional groups; some are naturally there in the polysaccharide and some are present because of residual oleic acid used in the manufacturing of the resin [9]. These acidic groups ionize even at slightly acidic pHs. Buffer solutions are most often used as the mobile phase with these resins and the ionic strength of these solutions are high enough to compress the electrical double layer on the surface of the resin (including within the channels inside the particles). However, in this work, pure water was the mobile phase. That means that the influence of the negative surface charge extended a much greater distance away from the surface. (Recall the Gouy–Chapman model discussed in Chapter 1.) This negative charge repels anions. Consequently, using the gel permeation column with pure water excluded the phosphate anions from the pores forcing the salt to elute well ahead of the organic molecules. This strategy worked well as long as the amount of salt in the sample was relatively small [9, 10].

Using mass spectrometry and infrared spectroscopy, the authors identified the electrolysis product, X, as allantoin [8]. Comparing the structures of uric acid and allantoin, it is apparent that a lot happens after two electrons are

Uric acid   Allantoin

removed from uric acid. Allantoin represents X in Eq. (6.2). There must be several intermediate steps in going from A to X. Identifying intermediate species would help define the pathway.

By repeating the cyclic voltammetry experiment with uric acid at different pHs, the authors observed the oxidation peak, I_a, and its partner, I_c, shift in potential by 55 mV toward 0 V for every increase of one pH unit. That is, in going between the oxidized and reduced forms of this redox pair, hydrogen ions are transferred as well as electrons. This reaction can be generalized as

$$Ox + ne^- + mH^+ \rightleftarrows Red \tag{6.22}$$

The corresponding Nernst equation reveals how the formal potential depends on pH.

$$
\begin{aligned}
E &= E^\circ - \frac{(2.303)RT}{nF} \log \left\{ \frac{a_{Red}}{a_{Ox}a_{H^+}^m} \right\} \\
&= \left[E^\circ - \frac{(2.303)RT}{nF} \log \left\{ \frac{1}{a_{H^+}^m} \right\} \right] - \frac{(2.303)RT}{nF} \log \left\{ \frac{a_{Red}}{a_{Ox}} \right\} \\
&= \left[E^\circ - \frac{(2.303)RTm}{nF} pH \right] - \frac{(2.303)RT}{nF} \log \left\{ \frac{a_{Red}}{a_{Ox}} \right\}
\end{aligned}
\tag{6.23}
$$

where term in the brackets, $\left[E^\circ - \frac{(2.303)RTm}{nF} pH \right]$, is the formal potential, $E^{\circ\prime}$, for the redox couple. The average of the peak potentials for the redox pair is a measure of the formal potential. Consequently, that average shifts with pH. The coefficient $(2.303)RT/F$ is equal to 59 mV at 25 °C. Then, in general the shift in $E^{\circ\prime}$ with pH is given by

$$\frac{\partial E^{\circ\prime}}{\partial(pH)} = \frac{\partial}{\partial(pH)} \left\{ \frac{-(2.303)RTm}{nF} pH \right\} = -0.059\,16 \left(\frac{m}{n} \right) \tag{6.24}$$

So, for a shift of 0.059 V per pH unit, $m = n$. In this case, $m = n = 2$. Therefore, the oxidation process at I_a is a two-electron, two-proton step [6].

$$A \rightleftarrows B + 2e^- + 2H^+ \tag{6.25}$$

A common structural change associated with the loss of two electrons and two hydrogen ions in an organic molecule is the formation of a double bond. This idea suggested a quinoid structure [11]. One of the resonance forms might be drawn as this:

Because this work was performed in buffers between pH 7 and 8, uric acid ($pK_a = 5.75$) would be in the anion form [11]. Therefore, the oxidation peak is more appropriately described with this first step:

$$\tag{6.26}$$

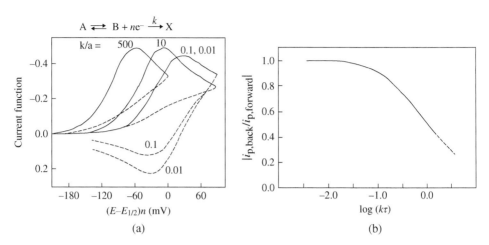

FIGURE 6.7 (a) Simulated cyclic voltammograms of an electron transfer step followed by an irreversible chemical reaction for different scan rates, v, where $k/a = (k/v)(RT/nF)$. The faster the scan rate compared to the rate of the chemical step (smaller values of k/a), the higher the concentration of the intermediate, Red, available near the electrode during the return scan and the bigger the return current peak. The term, $E_{1/2}$, on the x-axis represents the "half-wave" potential which is an estimate of the formal potential, $E^{o'}$. (b) A plot of the current ratio versus the log of the product of the rate constant, k, and the time, τ, required to scan between the formal potential and the switching potential. An estimate of k can be made, by finding the x-coordinate corresponding to the point on the curve associated with a measured ratio of peak currents. Source: Adapted with permission from Nicholson and Shain [12]. Copyright 1964, American Chemical Society.

What happens after the electron transfer? Apparently, structure B is unstable; an irreversible chemical reaction converts B to some other species, C. Just how fast is that reaction? There are two well-established ways of measuring the rate constant for the following chemical step. Both methods will be examined here.

The first approach is to scan at faster rates until the return peak grows in. The underlying idea is that the speed of the return scan must be comparable to the rate of the decomposition reaction in order to observe any current for the return peak. A small peak was observed for I_c in Figure 6.6, indicating that the time required to scan back from the switching potential was on the order of the lifetime of the intermediate, B. Figure 6.7a shows a set of computer-simulated current–voltage curves over a large range of scan rates.

This graph was adapted from a classic paper by Nicholson and Shain [12] and requires some explanation to make it digestible. (The curve for the anodic peak, I_a in Figure 6.6, points in the upward direction in this graph, but the oxidation step is shown here as the forward electron-transfer process to be consistent with the uric acid example.) Nicholson and Shain have tried to summarize a lot of information in one diagram for their readers. The current has been normalized (removing the square root dependence of the peak height on scan rate) and is represented here as a "current function" on the vertical axis. That maneuver merely keeps all of the voltammograms on the same current scale. The rate constant for the chemical step is represented by k. Cyclic voltammograms were plotted for different scan rates. The scan rate is embedded in the term, a. That is, $a = [(nF)/(RT)](vt)$, where t is the time in seconds from the start of the scan. $[(RT)/(nF)]$ has the units of volts; v is the scan rate in V/s; and k is in s^{-1}. Consequently, k/a is proportional to k/v, the rate constant of the chemical step compared to the voltage scan rate. The important idea to keep in mind is that the value of k/a decreases with faster scan rates. Another way of looking at it is that the larger values of k/a represent experiments at slower scan rates where the intermediate, B, has more time to decompose before the return scan starts. Consequently, there will be less B available on the return scan and a smaller return current peak will be observed. What is important about Figure 6.7a is the trend: as the scan rate increases (decreasing k/a), the shape of the CV approaches the ideal of a pair of peaks that are symmetric about the formal potential (represented by the $E_{1/2}$ in the x-axis). The panel on the right side in Figure 6.7 predicts the ratio of the peak currents for a wide range of scan rates. This plot can be used to calculate an estimate of the rate constant, k, for the chemical step. An example calculation should help to make this analysis clear.

Figure 6.8 is a cyclic voltammogram taken at a scan rate of 1200 mV/s. The return peak is smaller than expected for an ideal reversible electron transfer process. It decreases in size with decreasing scan rate suggesting an EC mechanism. A separate scan in the same electrolyte solution, but without the analyte was made in order to determine the baseline. The formal potential can be estimated from the average of the peak potentials. $E^{o\prime}$ appears to be 0.125 V versus normal hydrogen electrode (NHE). The switching potential was $=-0.270$ V. The time, τ, required to scan between the formal potential, $E^{o\prime}$ and the switching potential is

$$\tau = \frac{(E^{o\prime} - E_\lambda)}{v} = \frac{0.125\ \text{V} - (-0.270\ \text{V})}{1.2\ \text{V/s}} = 0.32_9\ \text{s} \tag{6.27}$$

The peak height ratio can be calculated from the method [13] that was illustrated in Figure 5.13 as shown in Eq. (6.28).

$$\frac{i_{p,back}}{i_{p,forward}} = \frac{(i_{p,back})_0}{i_{p,forward}} + \frac{0.485(i_\lambda)_0}{i_{p,forward}} + 0.086 = \frac{5.5}{10.4} + \frac{0.485(3.9)}{10.4} + 0.086 = 0.796 \tag{6.28}$$

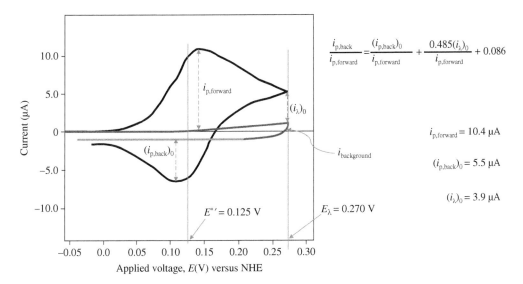

FIGURE 6.8 Example calculation for the rate constant for the following chemical step in an EC mechanism. The peak current ratio can be calculated from the formula shown as in Eq. (6.28).

From the curve in Figure 6.7b, the x-coordinate corresponding to a y-value of 0.796 is approximately equal to -0.70. That means $\log(k\tau) = -0.70$. Solving for the rate constant gives

$$\log(k\tau) = -0.70 \quad \text{or} \quad k = \frac{1}{\tau}\left(10^{-0.70}\right) = \frac{1}{0.329\ \text{s}}\left(10^{-0.70}\right) = 6.0_6\ \text{s}^{-1} \tag{6.29}$$

In summary, the pattern of a growing return peak (compared to the forward peak) with increasing scan rates suggested a mechanism of an electron transfer step followed by an irreversible chemical step. The scan rate study provided data for calculating the rate constant for the following chemical step. The average of the peak potentials provided an estimate of the formal potential. Table 6.1 is a list of other common electrode processes with coupled chemical steps that have been well studied (theoretically and experimentally) [12, 14] including some comments about distinguishing characteristics.

Scanning faster is not always practical. A second approach to finding rate constants for coupled chemical steps is to apply a double step chronoamperometry experiment [14, 16]. This is the method that Dryhurst and coworkers applied to the study of the uric acid oxidation. If the reaction is not too fast, one can determine whether the reaction with respect to the concentration of the product for the electron transfer step is first order (or pseudo-first order as might be the case, if the solvent or electrolyte were to react) or second order. One can also estimate the rate constant for the process of the chemical conversion of the intermediate to another compound. The fact that a small return peak (I_C) can be observed in the cyclic voltammogram in Figure 6.6 suggests that this chemical step is on the right time scale for such an investigation [8].

The basic idea underlying the double potential step experiment is that the first step brings the system to a potential where the intermediate is generated and the second step is used to observe how the intermediate behaves [16]. Figure 6.9a is a general diagram showing the applied potential and the current response as a function of time. In this case, the voltage was stepped from 0 V, a value where no electrode reaction occurs, out to +0.6 V, where the product B in Eq. (6.8), is formed. The voltage was held for a predetermined time, τ, and the current was measured. The current response to this "forward" step is an oxidation or anodic current, but for ease of reference it will be referred to here as the forward current, i_f. Next, the voltage was stepped back to 0 V again and the current was measured at a time, 2τ, or τ seconds from the start of the second potential step. The current response for this "back step" was for the reduction of compound B and was in the cathodic direction. It will be called the "back current," i_b, here for clarity. If the reaction in Eq. (6.8) was unperturbed by any following chemical step, the current following both the forward and back steps would follow the Cottrell equation (5.26) for the two species A and B, respectively. In an uncomplicated case (reversible electron transfer and no chemical reaction), the theory predicts that the ratio of the back current to the forward current, that is i_b(at $t = 2\tau$)/i_f(at $t = \tau$), would be equal to 0.293 for all values of τ [16]. (The ratio is not unity because most of B diffuses away from the electrode. Only a fraction is reduced back to A.)

However, when B takes part in a chemical reaction, even less B is available for the back current. Consequently, the current ratio decreases with τ and with increasing values of the rate constant, k. (Of course, for this one reaction, k is fixed.) If one performs the double step experiment multiple times at different values of τ, a plot of the current ratio versus τ

TABLE 6.1   Electrode mechanisms

I.   Reversible electron transfer

$$O + ne^- \rightarrow R$$

CV characteristics: CV peak height ratio ≈ 1; $\Delta E_p \approx 59/n$ mV.

II.   Irreversible electron transfer

$$O + ne^- \rightarrow R$$

CV characteristics: the current peak shifts in proportion to the square root of the scan rate. Totally irreversible in the electrochemical sense may still show a return peak, but $\Delta E_p > 200$ mV.

III.   Reversible chemical reaction preceding a reversible electron transfer – C_rE_r mechanism:

$$Z \rightleftharpoons O$$
$$O + ne^- \rightleftharpoons R$$

where the equilibrium for the chemical step, $K = k_1/k_{-1}$.

IV.   Reversible chemical reaction preceding an irreversible electron transfer – C_rE_i mechanism:

$$Z \rightleftharpoons O$$
$$O + ne^- \rightarrow R$$

where the equilibrium for the chemical step, $K = k_1/k_{-1}$.

V.   Reversible electron transfer followed by a reversible chemical reaction – E_rC_r mechanism:

$$O + ne^- \rightleftharpoons R$$
$$R \rightleftharpoons Z$$

where the equilibrium for the chemical step, $K = k_1/k_{-1}$.

VI.   Reversible electron transfer followed by an irreversible chemical reaction – E_rC_i mechanism:

$$O + ne^- \rightleftharpoons R$$
$$R \xrightarrow{k} Z$$

(a)   CV characteristics: small return peak for slow scan rates with increasing peak height ratio with increasing scan rates.
Chronoamperometry double step: $k = 0.406/\tau_{1/2}$ where $\tau_{1/2}$ = value of time from start of step for normalized current $R_I = 0.5$.

(*Continued*)

TABLE 6.1   (Continued)

(b) For product reacting with itself:

$$O + e^- \rightleftharpoons R$$

$$2R \overset{k}{\rightarrow} Z$$

Chronoamperometry double step: $k = 0.830[O]/\tau_{1/2}$ where $\tau_{1/2}$ = value of time from start of step for normalized current $R_I = 0.5$ and $[O]$ = original concentration of the starting material.

(c) For product reacting with starting material:

$$O + 2e^- \rightleftharpoons R^{-2}$$

$$R^{-2} + O \overset{k}{\rightarrow} Z$$

Chronoamperometry double step: $k = 0.922/\tau_{1/2}$ where $\tau_{1/2}$ = value of time from start of step for normalized current $R_I = 0.5$ and $[O]$ = original concentration of the starting material.

VII.   Reversible electron transfer followed by a regeneration of the starting material – catalytic mechanism:

$$O + ne^- \rightleftharpoons R$$

$$R + Z \overset{k}{\rightarrow} O$$

CVs vary in shape as a function of scan rate and concentrations of O and Z. See Case Study IV later in this chapter.

VIII.   Irreversible electron transfer followed by a regeneration of the starting material – catalytic mechanism:

$$O + ne^- \rightarrow R$$

$$R + Z \overset{k}{\rightarrow} O$$

IX.   Multiple electron transfer with intervening chemical reaction(s)
(a) ECE mechanism (first order):

$$O + n_1 e^- \rightleftharpoons R, \quad R \overset{k_1}{\rightarrow} X$$
$$X + n_2 e^- \rightleftharpoons Y$$

For double step chronoamperometry: $k = 0.273/\tau_{1/2}$ where $\tau_{1/2}$ = value of time from start of step for normalized current $R_I = 0.5$.

(b) ECE mechanism (second order):

$$O + n_1 e^- \rightleftharpoons R, \quad R + O \overset{k_1}{\rightarrow} X$$
$$X + n_2 e^- \rightleftharpoons Y,$$

When R reacts with O to form a dimer, double step chronoamperometry gives:
$k = 0.690/\tau_{1/2}$ where $\tau_{1/2}$ = value of time from start of step for normalized current $R_I = 0.5$.
Many systems have been reported in the literature for ECE and ECEC mechanisms. Their behavior is usually complicated but can be elucidated using digital simulation.

Source: Adapted with permission from Refs. [12], [14], and [15].

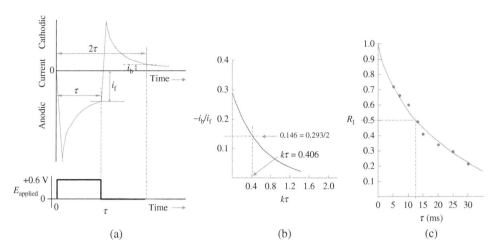

FIGURE 6.9 Double potential step chronoamperometry. (a) The general current response for a double potential step experiment. In the uric acid study, the potential was stepped from 0 V (versus saturated calomel electrode [SCE]) to +0.6 V, where oxidation occurred for τ milliseconds. At that time, the potential was stepped back to 0 V where reduction of the oxidized species occurred. The currents at τ (i_f in the diagram) and at 2τ (i_b in the diagram) were measured. Source: Adapted with permission from Dryhurst et al. [8]. Copyright 1983, American Chemical Society. (b) The ratio of currents, $(-i_b/i_f)$, for all values of τ is equal to 0.293 for an electron transfer process uncomplicated by a following chemical step. Theory shows that a following chemical step with a rate constant, k, will lead to a decline in the current ratio as plotted in the middle curve. When the current ratio falls to half the maximum value (a current ratio of $0.293/2 = 0.146$), the corresponding value of $k\tau = 0.406$. The only reason for displaying the middle curve here is to demonstrate that the assignment of $k\tau = 0.406$ comes from theory and applies to this mechanism when the current ratio falls to half of its initial value. Source: Adapted with permission from Schwarz and Shain [16]. Copyright 1965, American Chemical Society. (c) Multiple double step experiments were performed at various values of τ for the uric acid oxidation. In order to make the process of finding the point of interest (where the current ratio decreased to half its maximum value) easier, the current ratio was normalized by dividing each ratio by 0.293. In other words, one calculates $R_I = i_b/(i_f \cdot 0.293)$ for each trial (which pairs each R_I with a value of τ). Dryhurst calculated R_I for several trials and plotted R_I versus τ. Then, the data was fit with a smooth curve [6]. The point on the curve where R_I was equal to 0.5 corresponds to a value of 12.5 ms (or 0.0125 s) for $\tau_{0.5}$. This yields a value of $k = 0.406/\tau_{0.5} = 0.406/0.0125 = 32.5\,s^{-1}$.

will decrease in a curved fashion. (The exact relationship is not a simple function, but is a smooth, monotonic curve.) Figure 6.9b shows a theoretical plot of the current ratio as a function of $k\tau$ [16]. This plot is a convenient way of summarizing a lot of theoretical information. Obviously, this is not a practical way of plotting experimental data because one does not know the value of the rate constant, k, that would be needed for coordinates on the x-axis. Indeed, finding the rate constant is the objective of the analysis here. However, the graph does connect each current ratio to a corresponding value of $k\tau$. For example, consider the point where the current ratio has dropped to half of its maximum, that is at the value $(-i_b/i_f) = 0.293/2 = 0.146$. Graphically, the corresponding value of $k\tau$ is 0.406.

The Dryhurst group used this information to calculate k. Their strategy was to find the value of τ where the current ratio dropped to half the maximum value, namely a value of 0.146. In order to make a visual interpretation of the graph easier, they normalized the current ratio [8]. That is, they divided each current ratio by 0.293 to give a value called R_I. In this way, the maximum value became 1.0 and finding the point on the curve corresponding to $R_I = 0.5$ was straightforward. They plotted the current ratio for several different experiments as a function of τ. Then, a smooth curve was drawn through the data. This plot appears in Figure 6.9c. From the graph, the value of τ at $R_I = 0.5$ appears to be 12.5 ms or 0.0125 s. They calculated the value of the rate constant to be 32.5 s^{-1} using the value of τ, where $R_I = 0.5$ (see Eq. (6.30)).

$$k\tau = 0.406 \quad \text{or} \quad k = \frac{0.406}{\tau} = \frac{0.406}{0.0125} = 32.5 \, \text{s}^{-1} \tag{6.30}$$

The theoretical curve in Figure 6.9b is based on a first-order or pseudo-first order reaction model. The authors ruled out a second-order reaction by changing the initial concentration of the uric acid. A second-order reaction would proceed at faster rates at higher concentrations and the current ratio would fall off faster with increasing values of τ than at lower concentrations. Consequently, the data points from different concentrations would not follow the same curve when graphed as in Figure 6.9c. Clustering of the data along different curves would indicate that the reaction was not first order (or pseudo-first order). The uric acid data from different concentrations appeared to follow a single curve indicating that the reaction was first order with respect to compound B.

The theory has been worked out for double step chronoamperometry for other reaction mechanisms [14]. The work up of the data is similar, but the product of $k\tau$ at the point where $R_I = 0.5$ is different for each reaction model. The uric acid example just described can be called an EC mechanism (or more precisely, an E_rC_i mechanism, where C_i indicates that the chemical step was irreversible). Other reaction mechanisms are briefly described in Table 6.1.

There was reason to believe that product, C, of the relatively rapid chemical step described above was not the final product, allantoin (that was identified from bulk electrolysis). That implied that other chemical steps must follow the formation of C. To look for evidence of further reactions, the group turned to spectroelectrochemistry experiments [8]. Figure 6.10 is a diagram of the apparatus that was used to capture UV absorption spectra of reaction products. A porous, glassy carbon material called reticulated vitreous carbon (RVC) was cut into a thin section so that light could pass from one side to the other. This was used as the working electrode and placed in the beam path of a spectroscopic cell for a UV absorption instrument. The windows of the cell were made from quartz plates and made a sandwich with the RVC confining the solution in the beam to a thin layer. The bottom edge of the plates dipped into the sample solution and capillary action pushed solution up through the RVC electrode. Reference and counter electrodes were placed in a larger container of sample solution below the RVC.

The volume of sample solution surrounding the working electrode was small enough that the entire solution within the pores of the RVC could be electrolyzed in a few seconds and spectra could be recorded at different times. Figure 6.11a shows spectra recorded at 19-second intervals for the oxidation of 1 mM uric acid in the thin layer spectrochemical apparatus diagramed in Figure 6.10. Curve #1 in Figure 6.11a corresponds to the uric acid starting material and the successive curves show changes after the potential was stepped to +0.9 V (initiating the oxidation process associated with peak I_a in the voltammogram

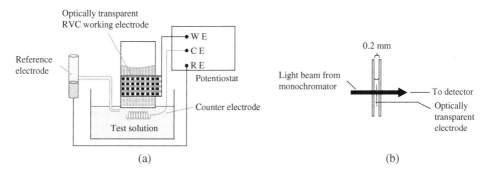

FIGURE 6.10 Spectroelectochemistry apparatus for observing UV spectra of electrode reaction products. (a) The working electrode appears as a screen-like rectangle in the diagram. It is made from a porous glassy carbon material called reticulated vitreous carbon (RVC) resembling a hardened sponge. (b) The material was cut into a very thin sheet and sandwiched between quartz plates. Sample solution was trapped between the plates by capillary action when the open end dipped into a container holding the sample solution, reference, and counter electrodes. The electrode was thin enough that light from the spectrometer was able to pass through the pores of the RVC. Source: Adapted with permission from Dryhurst et al. [8]. Copyright 1983, American Chemical Society.

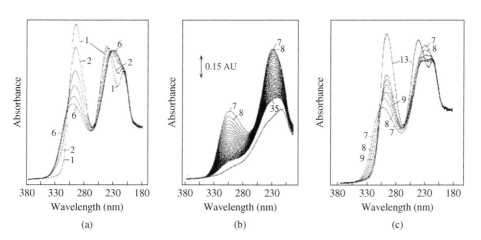

FIGURE 6.11 UV spectra recorded during oxidation of uric acid at an optically transparent electrode in a thin layer cell described as in Figure 6.10. (a) Spectra recorded after stepping to +0.9 V in 1 mM solution of uric acid at pH 8 at 19 second intervals. Curve 1 is of uric acid before application of oxidizing voltage. Curves 2–6 reveal the gradual formation of an oxidation product. (b) After curve 6, the electrochemical cell was disconnected and curves 7–35 were recorded. These spectra were taken on the same absorbance scale, but at intervals of 38 seconds. (c) In a separate experiment, the voltage was stepped to −1.2 V after scanning curve 6. Spectra were taken in 19 second intervals. Curiously, the spectrum gradually grows to match that of uric acid. Source: Adapted with permission from Dryhurst et al. [8]. Copyright 1983, American Chemical Society.

in Figure 6.6). After curve #6, the electrochemical cell was disconnected, and spectra were recorded at 38-second intervals. There appear to be some fast changes in Figure 6.11a and some much slower ones in Figure 6.11b. The authors inferred that, in addition to the uric acid starting material, there are at least two other distinct species represented here [8]. The more rapid changes are seen in the peak at 315 nm (Figure 6.11a). At the same time, the peak at 240 nm shifts to 225 nm. This most likely corresponds to the loss of uric acid. The new species is not likely to be compound B, the quinoid species in Eq. (6.26). Compound B has such a short half-life that it would not reach a significant concentration during this experiment. It would also rapidly decay once the cell was disconnected, but no sudden spectral changes occurred at that point. One assumes that this spectrum is associated with a compound that forms after B decomposes, call it compound C. Both peaks at 315 and 225 nm decay at a slower rate, suggesting that C reacts to form something else (compound D or E). An important outcome of this work was an estimate of the half-life for compound C of about 330 seconds. However, as interesting as these experiments were, they did not provide structural information about these intermediates.

Some mass spectral work was very helpful. The team repeated the spectroelectrochemistry work and collected the solution (~200 μl) surrounding the thin layer electrode around 100 seconds after applying +0.9 V, about the time when the spectrum appeared as in curve #6 in Figure 6.11a. They quickly froze the solution to quench the reaction at −78 °C (in a mixture of dry ice and acetone) [8]. This study of uric acid predates wide-spread access to electrospray ionization-Mass spectrometry instruments capable of liquid sample introduction. However, the sample was subsequently freeze-dried, and the authors made volatile derivatives of their intermediates using N,O-bis(trimethylsilyl)acetamide (BSA) and used gas chromatography-mass spectrometry (GC–MS) for structural analysis. BSA replaces active hydrogen atoms on alcohols, amines, and acids with trimethyl silyl (TMS) groups ((6.31, 6.32)). Substituting these bulky groups for active hydrogen atoms decreases hydrogen bonding and, consequently, makes a more volatile derivative.

$$(6.31)$$

$$(6.32)$$

From the GC–MS run, the workers obtained a reliable molecular weight for the derivative of the intermediate, C [8]. That mass was 472. Each TMS group increases the mass of the derivative by 72 Da above that of the original molecule. In order to find the mass of C itself, they needed to know how many TMS groups were on the derivative. For that reason, they repeated the reaction using a deuterated form of the BSA reagent. Each deuterated TMS group is 9 Da heavier than the normal form. This time the molecular weight of the derivative was 508 Da. The increase, $508 - 472 = 36$, indicated 4 (=36/9) TMS groups were present. Subtracting the additional mass of 4 TMS groups gave a molecular weight of 184 Da for the intermediate.

$$472 = MW + 4(73\,Da/TMS\,group) - 4(1Da/Hatom) \qquad (6.33)$$

$$MW = 472 - 4(72) = 184\,Da \qquad (6.34)$$

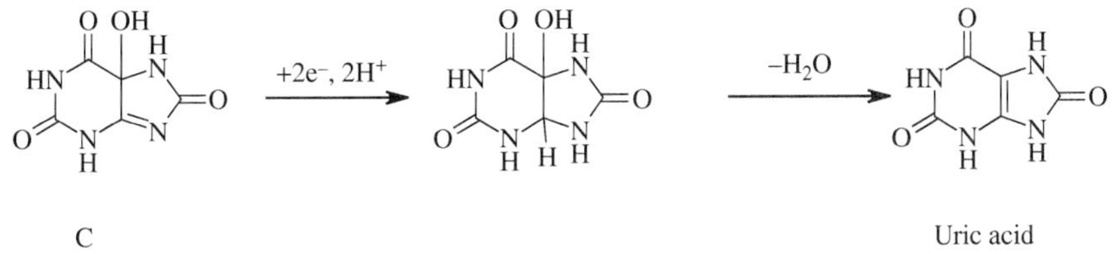

FIGURE 6.12 Proposed mechanism for the reduction process associated with peak II$_c$ in the cyclic voltammogram of uric acid, Figure 6.6 [6].

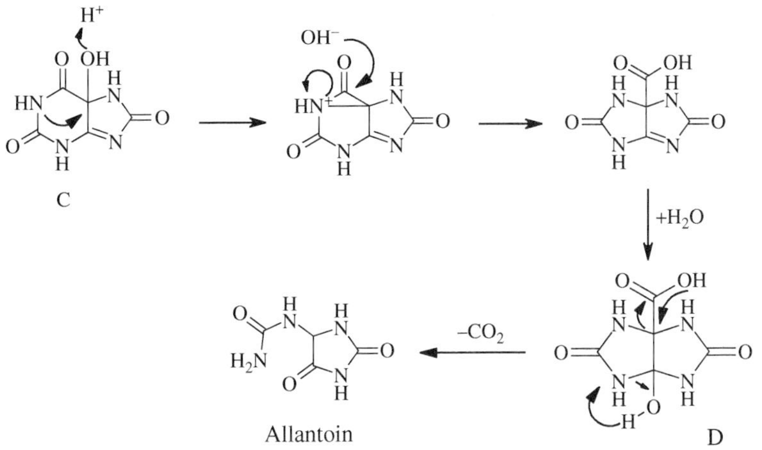

FIGURE 6.13 Proposed mechanism for the formation of allantoin from the intermediate, C [11].

The protonated form of intermediate B has a molecular weight of 166 Da. The difference of 18 Da suggests that B reacts with water to form C. It is this intermediate that is available for reduction at −1.2 V (peak II$_c$ in the CV in Figure 6.6). The spectra in Figure 6.11c monitor the reduction reaction at −1.2 V and indicate that uric acid is reformed in the process. The authors reasoned that the tertiary alcohol (shown here in Figure 6.12 in the protonated form) would be the most likely structure to lose water after a two-electron, two-proton reduction step.

Structure C has the formula, $C_5H_4N_4O_4$. The final structure, allantoin, has a chemical formula of $C_4H_6N_4O_3$. A reasonable way of accounting for the difference in the chemical formula would be a gain of one molecule of H_2O and a loss of CO_2. The authors proposed the steps shown in Figure 6.13 to get there [11].

The authors were candid in pointing out that electrochemical reaction mechanisms do not necessarily reveal the actual reaction pathways controlled by enzymes. However, they do provide insights about likely intermediates and their stability. In this particular case, Dryhurst et al. showed that the oxidation of uric acid catalyzed by the peroxidase enzyme produces an intermediate with the same spectrum as the intermediate described here as structure C [8].

6.5. CASE IV. EVALUATING CATALYSTS – KINETIC STUDIES

Of course, one of the most important attributes of a catalyst is the rate at which it turns over the substrate species to form products. In many redox catalysis systems, an electrode can replace one of the reactants. In other words, the electrode can donate or accept electrons to recycle the freely diffusing catalyst. The strategy of using an electrode to recycle the mediator is called electrocatalysis. This scheme is pictured for a net reduction of a substrate molecule in Figure 6.14. In addition to regenerating the catalyst, the electrode current is proportional to the turnover rate of the overall reaction. This fact makes voltammetry a powerful tool for studying the reaction kinetics. There are many promising technologies that are presently being investigated using voltammetric techniques such as hydrogen gas production for wind and solar energy storage [17], electrochemical incineration of organic wastes [5], and carbon dioxide reduction [18, 19].

The most widely used voltammetric method for investigating catalysis is cyclic voltammetry. Here the motivation for using a catalyst is usually the fact that the substrate molecule (A_{Ox} in Figure 6.14) does not exchange electrons easily with the electrode. In many cases, electron transfer between the substrate and the electrode does occur, but only at much more extreme voltages than the formal potential for the substrate. This problem is illustrated in Figure 6.15. Curve 4 is a CV for the substrate, A_{Ox}, alone. An electrode voltage that is slightly more negative than the formal potential for the A_{Ox}/A_{Red} redox pair ($E_A^{o'}$) should be enough to reduce the substrate directly at the electrode according to thermodynamics, but kinetic problems prevent that until much more negative voltages are applied. That leads to a large waste of energy. Adding a catalyst with a formal potential slightly more negative than $E_A^{o'}$ can be more efficient. Curve 1 is a CV for the catalyst alone and curve 2 is the voltammogram for a mixture of the substrate and catalyst. The catalyst mediates the turnover of the substrate at a voltage close to its own formal potential.

Figure 6.15 was created from a study of the catalytic reduction of CO_2 to CO by iron complexes. Thermodynamically, the formal potential for the reduction of CO_2 to CO is represented by the dotted vertical line in the lower left corner on the graph marked $E^o(CO_2/P)$. CO_2 reduction is much more favorable (has a more positive $E^{o'}$) than that of the catalyst. Therefore, carbon dioxide should be easier to reduce directly at the electrode, but the rate is insignificant until the voltage is quite extreme as is indicated by curve 4. Adding the catalyst leads to CO_2 reduction at a less negative voltage (curve 2). The result is that the reduction of CO_2 mediated by the catalyst saves a lot of energy.

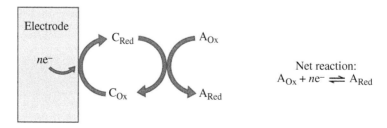

FIGURE 6.14   Redox catalysis with a freely diffusing catalyst is a variant of an EC electrode mechanism where the following chemical step regenerates the catalyst as well as forming the product, A_{Red}, from the substrate, A_{Ox}.

A lot can be learned about redox catalysis from observing how the shape of the current–voltage curve changes with different scan rates [20]. The catalytic mechanism gives rise to a variety of voltammogram shapes depending on conditions. The shape of the cyclic voltammogram is influenced by the scan rate and by the relative concentration of the substrate compared to the catalyst. Savéant investigated catalytic mechanisms with experiments and computer simulations and summarized the CV behavior over a wide range of conditions [20]. He used a diagram to organize the trends based on two dimensionless parameters. The first parameter, λ, is a kinetic term.

$$\lambda = \left(\frac{RT}{F}\right)\left(\frac{k[Cat_{Ox}]^o}{v}\right) \tag{6.35}$$

Equation (6.35) is written for the process depicted in Figure 6.15 where the reduced catalyst reacts with the substrate. Here k is the rate constant for the following chemical step of the catalyst reacting with the substrate in solution, v is the scan rate in V/s, $[Cat_{Ox}]^o$ is the concentration of the catalyst in bulk solution (present originally in the "unactivated" form), and the other terms have their usual meaning. The only other parameter needed is the ratio, γ, of the bulk concentration of the substrate, $[A_{Ox}]$, to that of the catalyst. This ratio is sometimes called the "excess factor" because catalysis reactions are normally operated at conditions where the concentration of substrate is in a stoichiometric excess compared to the catalyst.

$$\gamma = \frac{[A_{Ox}]^o}{[C_{Ox}]^o} \tag{6.36}$$

Dempsey and coworkers have refined Savéant's diagram by defining zones within which different CV shapes are found [21]. Her version appears in Figure 6.16.

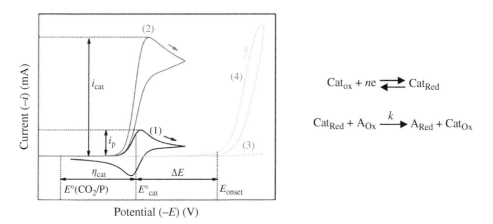

FIGURE 6.15 Theoretical cyclic voltammograms for the redox catalysis of carbon dioxide to carbon monoxide by an iron complex. In this diagram, the applied potential increases in the direction to the right. Cyclic voltammograms are overlaid for the catalyst alone (curve 1), for the substrate alone (curve 4), and for the mixture (curve 2). Curve 3 is a control run in the supporting electrolyte solution. Source: Adapted with permission from Francke et al. [18]. Copyright 2018, American Chemical Society.

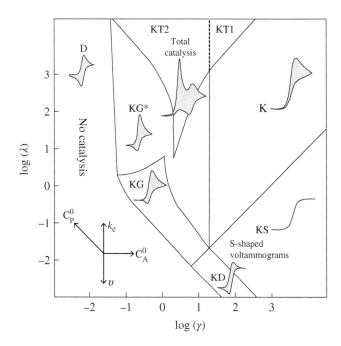

FIGURE 6.16 Zone diagram depicting the shape of cyclic voltammograms for the catalytic mechanism over a range of conditions. A log/log plot is needed to encompass the order of magnitude changes in scale. The parameters are defined in Eqs. (6.35, 6.36). The kinetic parameter, λ, of the y-axis increases with increasing values of the rate constant, k, for the chemical step and decreases with increasing scan rate, v. (Note the set of vectors in the lower left corner.) The excess factor, γ, increases with increasing concentration of the substrate, $[A_{Ox}]$. The concentration of the catalyst, $[C_{Ox}]°$, affects both γ and λ. Consequently, increasing $[C_{Ox}]°$ moves the conditions toward the upper left-hand corner of the diagram. Source: Adapted with permission from Rountree et al. [21]. Copyright 2014, American Chemical Society.

In experiments with a system that follows a catalytic mechanism, one can observe the different shapes in current–voltage curves by adjusting the scan rate and/or the substrate concentration, $[A_{Ox}]°$. By increasing the scan rate, conditions move vertically downward in the diagram. By increasing $[A_{Ox}]°$, the conditions move horizontally to the right. The one other adjustable variable is the bulk concentration of the catalyst, $[C_{Ox}]°$, but since it appears as a term in both γ and λ, increasing $[C_{Ox}]°$ moves conditions in a diagonal direction toward the upper left corner of the diagram.

The diagram is helpful in two ways. First of all, it can help confirm that the mechanism is a catalytic process. For example, imagine recording a voltammogram on a new system and observing a cyclic voltammogram with the appearance of the curve in zone K. This shape is also similar to a cyclic voltammogram for a noncatalytic E_rC_i process. However, faster scan rates lead to different behaviors in these two systems. In the noncatalytic case, faster scans can reverse the electron transfer process before the product has time to react chemically. When that happens, a return peak is observed in the voltammogram (recall Figure 6.2). In contrast, scanning faster for a catalytic system moves the conditions down from zone K to zone KS and leads to a current–voltage curve that looks more like

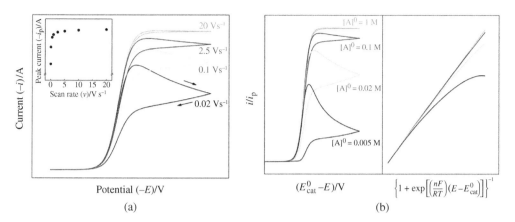

FIGURE 6.17 (a) Simulated curves at different scan rates (increasing from bottom to top). As scan rate increases, the system moves from kinetic zone K in Figure 6.16 vertically down to zone KS. The insert on the upper left of the diagram shows that the plateau current becomes independent of scan rate in this zone. (b) Simulated cyclic voltammograms for different substrate concentrations. As the substrate concentration increases (bottom to top), the cyclic voltammogram transforms from a curve with a shape representative of zone KG to the "S-shaped" curve of zone KS. Increasing the substrate concentration moves the conditions horizontally to the right in Figure 6.16. The plot on the right of box B is a "Foot-of-the-Wave" plot of the normalized current versus $\left\{1 + \exp\left[\left(\frac{nF}{RT}\right)\left(E - E_{cat}^o\right)\right]\right\}^{-1}$. The rate constant for the overall process can be extracted from the slope of the linear portion of this graph. Source: Adapted with permission from Francke et al. [18]. Copyright 2018, American Chemical Society.

a hydrodynamic voltammogram, such as might be seen with a rotated disk electrode. The current reaches a constant plateau and the curve retraces itself on the return scan. In fact, this happens because the starting material, C_{Ox}, is being replenished by the following chemical step rather than by convective mass transport. Likewise, scanning faster and moving vertically downward from zone KT2 (already a curious shape) or from zone KG* to zone KG both produce changes in shape that are not expected for a simple EC mechanism. Figure 6.17a shows a series of voltammograms generated from a computer simulation of a catalytic system and illustrates the changes in shape that one might see for a scan rate study. Figure 6.17b depicts the influence of the substrate concentration. At high substrate concentration, the curves take the "S-shape" indicative of total kinetic control as in zone KS.

There are useful equations that are applicable to different zones, and these can be used in order to extract an apparent rate constant. Of course, if the goal of using a catalyst is to speed up a reaction, then the overall rate is a figure of merit for evaluating the catalyst. Finding k is often a goal of these studies. In zone KS, the plateau current is related to the observed rate constant (also called the apparent rate constant), k_{obs}, for the chemical step as shown in Eq. (6.37) [21].

$$i_{cat} = \frac{nFA[C_{Ox}]^o \sqrt{D_C k_{obs}}}{1 + \exp\left\{\frac{nF}{RT}\left(E - E_C^o\right)\right\}} \tag{6.37}$$

In Eq. (6.37), the applied voltage is E and E_C^o stands for the formal potential for the catalyst and $[C_{Ox}]^o$ is the bulk concentration of the catalyst. D_C is the diffusion coefficient of the nonactivated form of the catalyst, C_{Ox}, k is the rate constant for the homogeneous reaction between C_{Red} and A_{Ox}, and A is the electrode area. (A is usually the geometric surface area, but the electrochemically active surface area is more rigorously correct. Recall that the active surface area can be obtained from a chronoamperometry experiment.) F is Faraday's constant and n is the number of moles of electrons transferred per mole of catalyst. When the current is measured on the plateau at a point at least 100 mV beyond E_C^o (that is, more negative in this case), then the denominator in Eq. (6.37) approaches unity and the equation simplifies to Eq. (6.38).

$$i_{plateau} = nFA[C_{Ox}]^o \sqrt{D_C k_{obs}} \tag{6.38}$$

Performing a cyclic voltammogram for the catalyst alone can helpful here. Because the catalyst by itself undergoes a simple, reversible electron transfer process, the peak current, i_{pc}, for this scan is given by the Randles–Sevcik equation:

$$i_{pc} = 0.446\,3nFA[C_{Ox}]^o \left(\frac{nFvD_c}{RT} \right)^{1/2} = 2.69 \times 10^5 n^{3/2} ACD_C^{1/2} v^{1/2} \text{ at } T = 298\,K \tag{6.39}$$

Dividing Eq. (6.38) by Eq. (6.39) eliminates some terms, such as the electrode area and the diffusion coefficient and provides a relationship that is easy to solve for the apparent rate constant, k_{obs} [21].

$$\frac{i_{plateau}}{i_{pc}} = \frac{1}{0.4463} \sqrt{ \left(\frac{RT}{nFv} \right) k_{obs} } \tag{6.40}$$

or

$$k_{obs} = \frac{(0.1991)nFv}{RT} \left(\frac{i_{plateau}}{i_{pc}} \right)^2 \tag{6.41}$$

In this context, the observed or apparent rate constant, k_{obs}, is defined as follows:

$$k_{obs} = k[A_{Ox}]^o \tag{6.42}$$

The apparent rate constant, k_{obs}, is also called the maximum turnover rate, TOF_{max}. It is the maximum number of moles of product generated per mole of catalyst per second. It should be pointed out that more complicated catalysis schemes have been observed [21]. In those cases, Eq. (6.42) is not an adequate definition of the observed rate constant. Fortunately, there are ways of handling these problems. In some reactions that involve multiple electron transfer steps and chemical steps, a single elementary step is rate-determining and the observed rate constant can be expressed as in Eq. (6.42).

Because Eqs. (6.37, 6.41) are applicable only in zones KS and KD, workers manipulate the substrate concentration and the scan rate to ensure that their conditions fall within one of those two zones. The prudent approach is to increase $[A_{Ox}]^o$ and the scan rate until the value of the limiting current (the plateau on the S-shaped curve) reaches a value that is no longer dependent on the scan rate. Then, the rate constant can be calculated from Eq. (6.41).

The catalytic mechanism that has been described here is a very simple one. Practical catalytic systems often involve multiple chemical and electron transfer steps. Furthermore, there are often complications such as instability of the catalyst, depletion of the substrate near the electrode, or inhibition as a result of product adsorbing to the electrode. However, Savéant has demonstrated that one can analyze the data at the foot of the current–voltage curve where the current is less sensitive to those problems [22]. The following example is a study by Savéant and coworkers with different iron–porphyrin complexes, represented here as [(por)Fe(I)], that can catalyze the reduction of CO_2 to CO [19]. Figure 6.18 shows the structure of an iron-TDHPP catalyst (one of the complexes represented by [(por)Fe(I)] in the reaction scheme in Figure 6.18) and the steps in the overall reaction sequence with CO_2. In order to evaluate the catalyst one would like to measure the rate constant, k, for the second step in Figure 6.18. The current that one observes for the electrocatalytic reduction is a function of an apparent rate constant, k_{obs}, for the overall process. In this case, the relationship between k and k_{obs} is fairly simple.

$$k_{obs} = 2k[CO_2] \tag{6.43}$$

So now the goal becomes finding k_{obs} from the voltammograms for this system. Figure 6.19A shows the cyclic voltammogram for the Fe(III) form of the complex being reduced in three separate one-electron steps to the corresponding Fe(II), Fe(I), and Fe(0) complexes in dimethyl formamide (DMF) in the absence of CO_2. Figure 6.19B is an overlay of the cyclic voltammogram from panel A with a cyclic voltammogram of the same solution after the addition of CO_2. The 60-fold increase in current coincides with the third peak in the reduction of the complex alone indicating that the Fe(0) (oxidation state 0) form is the active state of the catalyst. Panel C is the cyclic voltammogram for CO_2

FeTDHPP

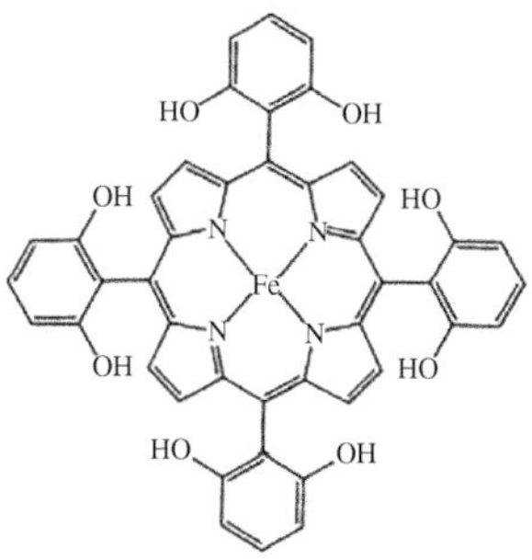

Iron 5, 10, 15, 20-
tetrakis (2′, 6′-
dihydroxylphenyl)-
porphyrin

$$[(por)Fe(I)]^- + e^- \rightleftharpoons [(por)Fe(0)]^{2-}$$

$$[(por)Fe(0)]^{2-} + CO_2 + 2AH \xrightarrow{k} [(por)Fe(II)CO] + H_2O + 2A^-$$

$$[(por)Fe(II)CO] + [(por)Fe(0)]^{2-} \xrightarrow{k' \gg k} 2\,[(por)Fe(I)]^- + CO$$

$$CO_2 + 2AH + 2e^- \longrightarrow CO + H_2O + 2A^-$$

FIGURE 6.18 Structure of the iron–porphyrin complex used in this case study. In the reaction scheme on the right [(por)Fe(0)]$^{2-}$ represents the activated form of the catalyst. The experiment described used 0.23 M CO_2 dissolved in dimethylformamide with 0.1 M tetrabutylammonium/phophorushexafluoride electrolyte. Water was added to a level of 2 M to act as the acid, AH. Source: Costentin et al. 2012 [19]. Adapted with permission of The American Association for the Advancement of Science.

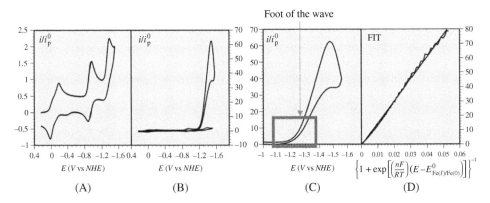

FIGURE 6.19 Cyclic voltammograms recorded in the presence of 2 M water in DMF with 0.1 M tetra-butylammonium/phophorushexafluoride electrolyte supporting electrolyte at a scan rate of 0.1 V/s. (A) FeTDHPP catalyst and water only. (B) Cyclic voltammogram for the same solution as in (A) with 0.23 M CO_2 overlaid on the cyclic voltammogram of the catalyst without CO_2. (C) Cyclic voltammo-gram of the catalytic peak from (B) showing the region for the foot-of-the-wave analysis. (D) Plot of the ratio of the current, i/ipo in (C), normalized to the peak current for the catalyst alone versus the $\left\{1 + \exp\left[nF(E - E_{\text{Fe(I)/Fe(0)}})/RT\right]\right\}^{-1}$. The slope of this plot yields the apparent rate constant, k_{obs} from Eq. (6.46). Source: Costentin et al. 2012 [19]. Adapted with permission of The American Association for the Advancement of Science.

and catalyst together with a rectangle framing the foot of the wave. Data from this region was used to construct the plot in panel D.

Savéant chose to work with data from the foot of the wave for the catalytic current in Figure 6.19C because it was less affected by undesirable complications. In this region, the current is also modelled by Eq. (6.37). Here is that equation again written with the appropriate terms for this cyclic voltammogram.

$$i_{\text{cat}} = \frac{nFA[\text{Fe}^{\text{III}}\text{TDHPP}]^{\circ}\sqrt{Dk_{\text{obs}}}}{1 + \exp\left\{\frac{nF}{RT}\left(E - E_{\text{Fe(I)/Fe(0)}}^{\circ}\right)\right\}} \tag{6.44}$$

In Eq. (6.44), [Fe^{III}TDHPP] represents the concentration of the starting form of the catalyst added to the solution, and [CO_2] was 0.23 M for this experiment. The diffusion coefficient, D, also is for the catalyst. The applied potential is E and the formal potential, $E_{\text{Fe(I)/Fe(0)}}^{\circ}$, is for the reduction of the Fe(I) state of the complex to the Fe(0) form. That formal potential can be obtained from the cyclic voltammogram of the catalyst alone in Figure 6.19A. The return wave for the third reduction peak is difficult to see in this figure, but the authors say that cyclic voltammograms at faster scans (using a much smaller elec-trode) showed a more prominent peak for the return scan. The average of the peaks gave a value of −1.333 V versus SHE for the formal potential.

Savéant applied a strategy that is often a rewarding approach to finding useful infor-mation from experiments, namely, he manipulated the mathematical model describing the signal until he found a way of plotting the data that put the desired quantity in the slope or intercept. One can follow his logic (and contributions of others) in the original papers

[18–22], but here is the basic pathway and his endpoint. Savéant divided equation (6.44) by the Randles–Sevick equation (6.39) to arrive at Eq. (6.45).

$$\frac{i_{cat}}{i_{pc}} = \frac{2.24\sqrt{\left(\frac{RT}{nFv}\right)k_{obs}}}{1 + \exp\left[\frac{nF}{RT}\left(E - E^{o}_{Fe(I)/Fe(0)}\right)\right]} \tag{6.45}$$

That removes several terms, such as the diffusion coefficient of the catalyst and the electrode area, and introduces others, such as the scan rate, v, and the peak current, i_{pc}, for the catalyst alone. Fortunately, these new terms are readily obtainable. A plot of the current ratio, i_{cat}/i_{pc}, versus $\left\{1 + \exp\left[\frac{nF}{RT}\left(E - E^{o}_{Fe(I)/Fe(0)}\right)\right]\right\}^{-1}$ is a straight line (see Figure 6.19D) with a slope containing k_{obs} [19].

$$\text{Slope of graph} = 2.24\sqrt{\left(\frac{RT}{nFv}\right)k_{obs}} \tag{6.46}$$

The slope of the best fit line to the data in panel D of Figure 6.19 appears to have a numerical value of $80/0.052 = 1.5 \times 10^2$. Given that the scan rate, v, was 0.1 V/s, and the number of electrons, n, in the electron transfer step at the electrode was 1, and the experiment was performed at 21 °C or 294 K, Eq. (6.45) can be solved for k_{obs}:

$$\text{Slope} = \frac{80}{0.052} = 1.5_3 \times 10^3 = 2.24\sqrt{\frac{(8.314)(294)k_{obs}}{(1)(96\,485)(0.1)}} = (1.13)\sqrt{k_{obs}} \tag{6.47}$$

$$\sqrt{k_{obs}} = \frac{1.5_3 \times 10^3}{1.13} = 1.3_5 \times 10^3 \quad \text{or} \quad k_{obs} = (1.3_5 \times 10^3)^2 = 1.8 \times 10^6 \text{ M/s} \tag{6.48}$$

For the experiment in Figure 6.19, the CO_2 substrate concentration was 0.23 M, solving Eq. (6.43) for k gives a rate constant for the homogeneous reaction between the catalyst and substrate a value of

$$k = \frac{1.8 \times 10^6 \text{ M/s}}{(2)(0.23 \text{ M})} = 3.9 \times 10^6 \text{ s}^{-1} \tag{6.49}$$

This value has been used as a figure of merit for comparing this catalyst with others that control this reaction. Of course, a full analysis of this system must also address the role of the acid, side reactions, catalyst stability, and product inhibition as well. However, those considerations are beyond the scope of the discussion here. Ultimately, a good electrocatalyst should have a high turnover frequency and a small overpotential (difference between the formal potential of the substrate and the operating potential for the electrode reaction) [19].

Evaluation of more complicated catalytic systems can be approached by using computer simulations to predict the shapes of cyclic voltammograms for various conditions based on a proposed model (sequence of reaction steps). Commercial programs are available that run on small computers and are flexible and powerful enough to handle complicated mechanisms. They enabled the chemist to predict voltammograms based on a hypothetical model and compare them with experimental current–voltage curves for a variety of conditions. Because rate constants for the coupled chemicals steps are adjustable parameters in these simulations, a good match between theoretical and experimental voltammograms also yield numerical estimates of the rate constants. (Two examples of popular programs for digital simulation of voltammograms are DigiSim and DigiElch [23, 24].)

REFERENCES

1. Szultka-Młyńska, M., Bajkacz, S., Kaca, M. et al. (2018). *J. Chromatogr. B* 1093–1094: 100–112.
2. Toma, H.E., Araki, K., and Dovidauskas, S. (2000). *J. Chem. Educ.* 77 (10): 1351–1353.
3. Fricker, S.P. (1995). *Platinum Met. Rev.* 39 (4): 150. https://www.technology.matthey.com/article/39/4/150-159/.
4. Miessler, G.L., Fischer, P.J., and Tarr, D.A. (2014). *Inorganic Chemistry*, 5e. Boston: Pearson.
5. Panizza, M. and Cerisola, G. (2009). *Chem. Rev.* 109 (12): 6541–6569.
6. Sjödin, M., Gätjens, J., Tabares, L.C. et al. (2008). *Inorg. Chem.* 47 (7): 2897–2908.
7. Hansch, C., Leo, A., and Taft, R.W. (1991). *Chem. Rev.* 91: 165–195.
8. Dryhurst, G., Nguyen, N.T., Wrona, M.Z. et al. (1983). *J. Chem. Educ.* 60 (4): 315–319.
9. Neddermeyer, P.A. and Rogers, L.B. (1968). *Anal. Chem.* 40 (4): 755–762.
10. Owens, J.L., Thomas, H.H., and Dryhurst, G. (1978). *Anal. Chim. Acta* 96 (1): 89–97.
11. Dryhurst, G., Kadish, K.M., Scheller, F., and Renneberg, R. (1982). *Biological Electrochemistry*. New York, NY: Academic Press.
12. Nicholson, R.S. and Shain, I. (1964). *Anal. Chem.* 36: 706–723.
13. Nicholson, R.S. (1966). *Anal. Chem.* 38: 1406–1406.
14. Childs, W.V., Maloy, J.-T., Keszthelyi, C.P., and Bard, A.J. (1971). *J. Electrochem. Soc.* 118: 874–880.
15. Mabbott, G.A. (1983). *J. Chem. Educ.* 60: 697–702.
16. Schwarz, W.M. and Shain, I. (1965). *J. Phys. Chem.* 69: 30–40.
17. Elgrishi, N., McCarthy, B.D., Rountree, E.S., and Dempsey, J.L. (2016). *ACS Catal.* 6 (6): 3644–3659.
18. Francke, R., Schille, B., and Roemelt, M. (2018). *Chem. Rev.* 118 (9): 4631–4701.
19. Costentin, C., Drouet, S., Robert, M., and Savéant, J.-M. (2012). *Science* 338: 90–93.
20. Savéant, J.-M. (2008). *Chem. Rev.* 108: 2348–2378.
21. Rountree, E.S., McCarthy, B.D., Eisenhart, T.T., and Dempsey, J.L. (2014). *Inorg. Chem.* 53 (19): 9983–10002.
22. Costentin, C., Drouet, S., Robert, M., and Savéant, J.-M. (2012). *J. Am. Chem. Soc.* 134: 11249–11242.
23. Bioanalytical Systems, Inc., Digisim® simulation software for cyclic voltammetry. https://www.basinc.com/products/ec/digisim (accessed 10 August 2019).
24. Gamry Instruments, DigiElch electrochemical simulation software. https://www.gamry.com/digielch-electrochemical-simulation-software/ (accessed 10 August 2019).

INSTRUMENTATION

7.1. OVERVIEW

Most of the apparatus associated with electrochemical experiments is electrical in nature. Consequently, it is appropriate to discuss briefly some basic circuits that are used for electrochemistry. Although the presentation here incorporates equations and simple quantitative reasoning, the emphasis is on a conceptual analysis. One does not need to be an electrical engineer to understand the important relationships and appreciate the operating principles of the measurement system. All of the material in this chapter is appropriate for coverage in a college course in instrumental analysis.

The basic principles of direct component (DC) circuitry that are normally described in introductory physics courses are applicable in the design and operation of electrochemical instruments. For example, Ohm's law is particularly useful. ($V = iR$, where a current, i, experiences a voltage drop, V, in crossing a resistance, R.) Other rules are almost intuitive. Kirchhoff's current law states that the sum of all current leading into any point on a circuit must equal the sum of all current leading away from that point. Kirchhoff's voltage law is equivalent to saying that electric potential energy is a state function. That is, the electrical potential energy spent by current going between any two points in a circuit is the same, regardless of the path between those two points. Thevenin's theorem allows one to analyze a circuit in terms of current and voltage behavior by replacing portions of that circuit containing multiple voltage sources and/or resistances by an equivalent circuit consisting of a single voltage source and a single resistance.

Much of the control circuitry, as well as the signal amplification and manipulation, are performed by operational amplifiers (op amps) in modern instrumentation. A short introduction to op amps is included here in order to make understandable the principles underlying a three-electrode system for performing voltammetry. Operational amplifiers are critical to most instrumentation that deals with analog (continuously variable) electrical signals. Some background in electronics can be extremely valuable to an analytical chemist, but a good coverage is beyond the scope of this book.

Because modern electrochemical instruments work with very small current signals, the electronic noise generated by electrical equipment in contemporary lab environments can be a serious problem for electrochemical methods. Strategies for shielding instrumentation and signals from noise are discussed. Finally, although working and reference electrodes are available commercially, they are also relatively easy to make. Construction of

Electroanalytical Chemistry: Principles, Best Practices, and Case Studies, First Edition. Gary A. Mabbott.
© 2020 John Wiley & Sons, Inc. Published 2020 by John Wiley & Sons, Inc.

platinum and carbon working electrodes as well as the preparation of silver/silver chloride reference electrodes are described in Section 7.5.

7.2. A BRIEF REVIEW OF PASSIVE CIRCUITS

In deciphering the behavior of circuits from schematic diagrams, it is often useful to look at an isolated portion of a bigger circuit. In these instances, it is helpful to think about the current or voltage at one point of the circuit in order to reason what the current or voltage will be at some other point. It is a bit like finding one's way on a road map. One focuses on the pathways between one's present location and another point of interest and the obstacles in between. There is actually a rule known as Thevenin's theorem that supports this sort of simplification [1]. Thevenin's theorem states that for the purpose of analysis, any combination of voltage sources and resistors can be replaced by a single voltage source and one resistor in series. This theorem will be applied later in the discussion of the circuit in Figure 7.1.

There are a few other rules or laws that help to simplify this task even further [1]. The first of these ideas is that at any junction of wires in a circuit diagram, the total current going into the junction must equal the total current leaving the junction. This rule is sometimes referred to as Kirchhoff's current law. Figure 7.1 shows a simple circuit with a battery and three resistors. Current flows from the positive end of the battery through the resistors to the negative end of the battery. Kirchhoff's current law indicates that the current coming into point A from resistor, R_1, equals the sum of the two currents leaving that point through resistors R_2 and R_3. Kirchhoff also stated a rule about voltages. That idea is actually a declaration that voltage is a state function. That is, the rule states that the voltage or energy spent in going between any two points in a loop is the same regardless of which path is

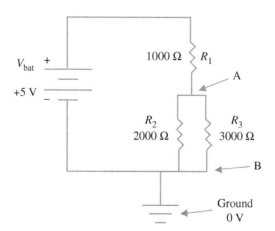

FIGURE 7.1 Simple circuit for demonstrating Kirchhoff's and Ohm's laws and Thevenin's theorem. These rules state: (i) The sum of currents going into any point, such as point A, is equal to the sum of the current leaving that point. (ii) The voltage between points A and B is the same regardless of the path taken between those points. (iii) The voltage difference between the leads of a resistor is equal to the product of the resistance and the current passing through the resistor. (iv) For the purpose of analysis, any combination of voltage sources and resistors can be replaced by a single voltage source and one resistor in series.

followed. Once again, this idea is almost intuitive. It can be illustrated with Figure 7.1. The battery provides a voltage to force current through the resistors. All of the energy provided by the battery is spent in pushing current through the loop from R_1 and then R_2 or R_3. Furthermore, the energy spent going from point A to point B is the same whether the current goes through R_2 or R_3. The potential energy between A and B is also equal to the energy (voltage) of the battery minus the energy spent in passing current through the resistor R_1.

In order to quantify voltages at various points in a circuit, probably the most useful rule is Ohm's law that states the energy spent in pushing current through a resistor is the product of the current times the resistance.

$$V_{\text{resistor}} = iR \tag{7.1}$$

The calculation of the voltage at point A with respect to the voltage of the ground and currents passing in different parts of the circuit in Figure 7.1 provides a good review of Ohm's law. First of all, one can simplify the diagram by applying Thevenin's theorem to replace R_2 and R_3 with an effective resistor with resistance R^*. These two resistors provide parallel paths for current to flow between points A and B. The effective resistance, in ohms, of two resistors, R_2 and R_3, in parallel can be shown to be equal to

$$R^* = \frac{R_2 \cdot R_3}{R_2 + R_3} = \frac{(2000)(3000)}{2000 + 3000} = \frac{6\,000\,000}{5000} = 1200 \tag{7.2}$$

This expression can be derived using Kirchhoff's laws and Ohm's law. The equivalent circuit is shown in Figure 7.2.

The current, i_1, that goes through R_1 also goes through R^*. The energy spent in pushing this current through these resistors is given by Ohm's law:

$$V_{R_1} = i_1 R_1 \tag{7.3}$$

$$V_{R*} = i_1 R^* \tag{7.4}$$

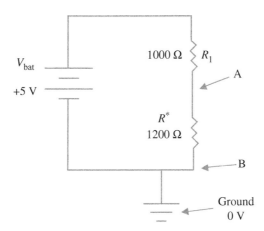

FIGURE 7.2 The equivalent circuit for Figure 7.1 where R^* represents the equivalent resistance for the parallel resistors R_2 and R_3.

The sum of the voltages through the resistors must be equal to the voltage produced by the battery.

$$V_{bat} = 5.0\,V = V_{R_1} + V_R^* = i_1 R_1 + i_1 R^* = i_1(R_1 + R^*) = i_1(1000 + 1200) \quad (7.5)$$

Rearranging gives the current through the first resistor.

$$i_1 = \frac{V_{bat}}{R_1 + R^*} = \frac{5.0\,V}{2200\,\Omega} = 2.27 \times 10^{-3}\,A \quad (7.6)$$

Now applying Ohm's law provides the voltage drop across resistor, R_1:

$$V_{R_1} = i_1 R_1 = (2.27 \times 10^{-3}\,A)(1000\,\Omega) = 2.27\,V \quad (7.7)$$

The voltage V_A, is also the voltage dropped across the equivalent resistor, R^*.

$$V_A = V_R^* = i_1 R^* = \left(\frac{V_{bat}}{R_1 + R^*}\right) R^* = (2.27 \times 10^{-3}\,A)(1200) = 2.72\,V \quad (7.8)$$

It is worthwhile noting that the voltage, V_A, is a fraction of the battery voltage, V_{bat}. The fraction is equal to the ratio of R^* and the total resistance, $R_1 + R^*$. That is, the fraction of the total resistance between the battery and the ground represented by R^* is equal to the fraction of the battery voltage that appears between point A and ground. Using two resistors in series such as shown in Figure 7.3a is a frequent strategy for dividing a voltage in a predictable way. In fact, this part of the circuit is called a voltage divider. A valuable way of implementing a voltage divider is to use a variable resistor, also known as a potentiometer, in place of the two resistors with fixed values. The potentiometer shown in Figure 7.3b has

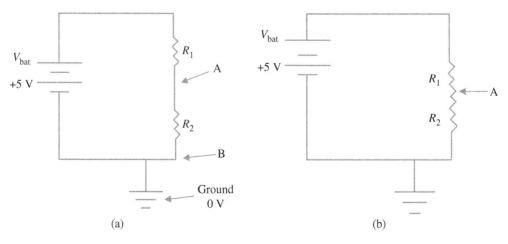

FIGURE 7.3   (a) A voltage divider made from two resistors provides a voltage at point A that is a fraction of the supply voltage, V_{bat}. (b) A voltage divider based on a variable resistor. The arrow represents a moveable contact that can be placed anywhere along the length of the device to create complementary resistances, R_1 and R_2, in order to provide any voltage between V_{bat} and 0.0 at the moveable contact.

a movable contact represented by the arrow that can be positioned anywhere along the path between the other two leads. The resistance between the moveable contact and the two separate ends defines two resistances, R_1 and R_2, in the voltage divider. Consequently, applying a voltage, V_{bat}, to one end of the potentiometer and connecting the other end to ground permits one to select any fraction of the voltage V_{bat} at the connection to the moveable contact (point A).

EXAMPLE 7.1

For the circuit in Figure 7.3a, choose resistors that will provide a voltage at point A of 1.00 V.

Equation (7.8) indicates that the voltage at point A compared to the ground (at point B) is equal to the battery voltage times the fraction of the total resistance $(R_1 + R_2)$ represented by the resistance between A and B. That is,

$$V_A = (5\,V) \left(\frac{R_2}{R_1 + R_2} \right) = 1.00 \text{ or } \left(\frac{R_2}{R_1 + R_2} \right) = 0.2$$

Lots of choices would work to give this ratio. A couple of practical considerations are helpful here. First of all, the choices for resistors above 1 MΩ are limited. Second, one might keep the total resistance at or above 1000 Ω in order to keep the current level small (that is at or below a few milliamps) in order to keep from draining the battery. That means the resistors should be in the range of 10^3 to 10^6 Ω. Let $R_2 = 10\,000\,Ω$. Then,

$$\left(\frac{10\,000}{R_1 + 10\,000} \right) = 0.2 \text{ or } R_1 = \frac{10\,000}{0.2} - 10\,000 = 40\,000\,Ω$$

7.3. OPERATIONAL AMPLIFIERS

The electronic instrumentation needed to perform basic electrochemical experiments is relatively simple. Even if one never needs to improvise an apparatus for one's self, understanding the fundamentals of the rudimentary circuitry helps in recognizing when the equipment is operating properly and what limitations the instrument places on the measurements. This section introduces a group of versatile devices known as operational amplifiers or "op amps" and demonstrates how to apply some simple concepts, such as Ohm's law, in order to explain how the op amps function in different circuits. Op amps are essential to performing modern electrochemical experiments.

An op amp is a semiconductor integrated circuit composed of many (often dozens of) transistors that can be employed in many different ways to manipulate an analog signal. The device will produce a voltage at its output connector that is a mathematical function of the voltage or current that appears at one of the two input connectors. In this context, the term "analog" means that the signal is continuously varying, as opposed to a digital

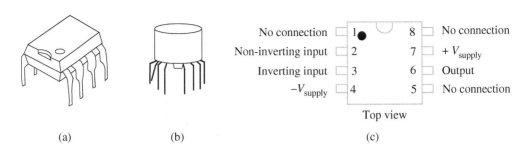

No connection   1   8   No connection

Non-inverting input   2   7 $+V_{supply}$

Inverting input   3   6   Output

$-V_{supply}$ 4   5   No connection

Top view

(a)   (b)   (c)

FIGURE 7.4 (a) Sketch of 8-pin dual in-line op amp package, (b) sketch of op amp cannister package, and (c) pin numbering system and function for many common op amps. Note the position of the circle identifies pin 1 [4].

signal that can have only a value that is one of a finite number of values arranged in a staircase manner, where each level increases by a fixed increment over the preceding one. One can think of a digital scale as being quantized. Values in between two adjacent levels are not allowed. Digital signals are in a format that computers can manipulate. Most analytical sensors, spectrometers, and chromatographic detectors generate an analog signal that is subsequently converted to a digital form before it is transferred to a computer. This sort of signal source-to-computer interface and much of the control circuitry that is common to modern instrumentation are generally covered in courses on digital electronics. These subjects are very useful for an analytical chemist, but they are beyond the scope of this book. Fortunately, the behavior of op amps that are at the heart of performing electrochemical measurements are fairly easy to use and comprehend. With the exception of design engineers in the big circuit-chip-manufacturing companies, no one thinks about the arrangement of transistors inside the op amp. Rather, the behavior of the device follows a few simple operating rules that lead to several useful circuits that are common to most analytical instrumentation.

Discrete operational amplifiers are commonly available in two different packages as shown in Figure 7.4. The rectangular package is the most common because its connecting pins form two rows that can be quickly placed into standard sockets or into predrilled holes in printed circuit boards. (With advances in circuit board manufacturing, a third type of package is growing in popularity. In those devices, the leg-shaped pins are replaced by small metal rectangles that all lie in the same plane with the bottom of the chip. This type of package is called a "surface mount device" or SMD. Special soldering techniques are used to connect SMD pins to appropriate contact points on the circuit board. Because holes do not need to be drilled for SMDs, these devices save labor in the assembly process.)

The function of each pin is indicated in Figure 7.4c. There are two connections for external power supplies, one positive and one negative. These provide the current and voltage that the device needs to operate. The upper voltage limit for the power supplies is typically 18 V [1]. Many devices can work with a range of power supply voltages, but the maximum output voltage of the op amp is always less than the supply voltage. Usually, the negative power supply is of the same magnitude as the positive power supply. Some op amps operate with a single power supply, but they are constrained to the positive or negative voltage range [1].

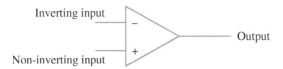

FIGURE 7.5 Schematic representation of an op amp for circuit diagrams. Positive voltage signals applied to the inverting input normally lead to negative voltages at the output.

The schematic representation of an op amp appears in Figure 7.5. It is common practice to omit drawing the connections to the power supplies in order to reduce the clutter in circuit diagrams. Of course, the power supplies must be connected for the circuit to work. Op amps have two inputs for receiving a signal. They are significantly different in their function. One of those is called the "inverting input" and is marked in diagrams with a negative sign as in Figure 7.5. The term "inverting" means that the output will be negative, if a positive voltage is applied to the inverting input and *vice versa*. The other input is marked with a plus sign and is cleverly called the "noninverting" input.

7.3.1. Properties of an Ideal Operational Amplifier

The behavior of an op amp can be predicted from the following two simple rules along with Ohm's law, Kirchhoff's laws, Thevenin's theorem and the properties of other circuit components connected to the output and inputs [1].

1. No current flows into the op amp at the inputs
2. The op amp drives its output until it establishes the same voltage at both inputs – usually through a feedback loop between the output and the inverting input.

A few common circuit examples will demonstrate how these rules are used.

7.3.2. The Voltage Follower

Figure 7.6 shows one of the simplest and most widely used op amp circuits. This configuration is called a voltage follower. On the left side of the op amp, some signal, V_{in}, is

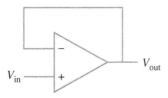

FIGURE 7.6 The voltage follower. The signal, V_{in}, is applied to the noninverting input. The output voltage drives the inverting input until its voltage matches the voltage at the non-inverting input, thus $V_{out} = V_{in}$.

applied to the noninverting input. By applying the properties of an ideal op amp, one can predict the output voltage, V_{out}.

The first rule states that no current will flow from the signal source into the op amp at the input. That statement is very important. It means that the op amp will not distort the signal by drawing current in the process of measuring the signal. However, it does not shed any light on the relationship of V_{out} as a function of V_{in}. Because the signal is applied to the noninverting input, the amplifier's output will have the same sign. The second rule of an ideal op amp indicates that it drives the output voltage until the voltages at the two inputs match. The output is connected to the inverting input. A connection between the output and one of the inputs is known as a feedback loop. In this case, the amplifier is "happy" when the output, V_{out}, matches the input, V_{in}. Although the amplifier does not change the voltage signal, this is a very important application for op amps. Because the op amp does not allow current to flow into the device at the input, the amplifier will not distort the signal by drawing current from the source. Another way of putting it is to say that the op amp has a large input impedance. Yet, the amplifier output can provide current, if necessary, to drive other, current drawing devices down the line.

7.3.3. Current Follower or Current-to-Voltage Converter

In a voltammetry experiment, the signal is a current. Because voltages are easier to manipulate, the standard practice is to convert the current to a voltage that is proportional to the original signal. The op amp configuration in Figure 7.7 is known as a current follower or current-to-voltage converter.

The noninverting input is attached to the ground setting its potential at 0 V. The current signal is applied to the inverting input. In addition, the inverting input is connected to the output through a feedback resistor, R_f. Consider a signal current moving in the direction of the inverting input as shown in Figure 7.7. Because the amplifier does not allow current to enter the device at the input, all of the current coming in is forced through R_f, in the feedback loop.

$$i_f = i_{in} \tag{7.9}$$

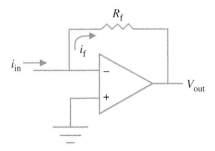

FIGURE 7.7 Current follower op amp configuration. The signal current cannot enter the op amp at the inverting input and must flow through the feedback loop. The voltage at the inverting input must equal the voltage at the noninverting input, namely zero. Ohm's law indicates that the difference in voltage between the input and output is equal to the energy lost in pushing a current through the feedback resistor. Hence, $V_{out} = -i_{in}R_f$.

Because, the current is moving in the direction of the input toward the output in the feedback loop, the voltage at the output must be at a lower value than the voltage at the input in this case. Furthermore, the second property of an ideal op amp indicates that the amplifier will drive the output to a voltage where the voltage at the inverting input matches the noninverting input. Because the noninverting input is grounded (it is fixed at 0 V), the voltage at the inverting input is also zero. Clearly, the current is moving through the feedback loop toward the output. Consequently, the output voltage is less than zero. How much less than zero? The difference in voltage between the input and output pins is the amount of energy lost in the feedback loop. Because the feedback loop contains only a resistor, the voltage "drop" can be calculated from Ohm's law ($V = iR$).

$$V_{resistor} = V_{in} - V_{out} = i_f R_f = i_{in} R_f \tag{7.10}$$

$$V_{out} = V_{in} - i_{in} R_f = 0 - i_{in} R_f \tag{7.11}$$

or

$$V_{out} = -i_{in} R_f \tag{7.12}$$

Equation (7.12) indicates that the output voltage is proportional to the signal current, i_{in}. Furthermore, the magnitude of the proportionality constant is the feedback resistance, R_f. That means that one can increase the magnitude of the output by using a bigger resistance. Because of that relationship, $-R_f$ is called the "gain factor" for the current follower. It is common practice to use bigger resistances to amplify weaker current signals with $10^7 \, \Omega$ as an upper limit for the gain factor.

EXAMPLE 7.2

What value of feedback resistance is needed to amplify a current signal of 2.0 μA to a voltage level of 0.30 V?

$$V_{out} = -i_f \cdot R_f = -2.0 \times 10^{-6} \cdot R_f = 0.30 \text{ V}$$
$$R_f = \frac{0.30 \text{ V}}{2.0 \times 10^{-6}} = 1.5 \times 10^5 \, \Omega$$

7.3.4. Inverter or Simple Gain Amplifier

Voltage signals can be amplified by using the circuit diagramed in Figure 7.8. The combination of components is very similar to that of a current follower, except that the voltage signal is connected to the inverting input through a resistor, R_{in}. The relationship between the input voltage and the output voltage is easy to derive using the properties of an ideal operational amplifier and Ohm's law. Because there is a feedback loop, the amplifier is able to drive the output voltage to a value to match the voltage of the inverting input to

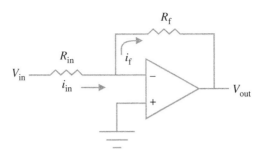

FIGURE 7.8 A simple gain amplifier multiplies the input voltage, V_{in}, by a gain factor of $-(R_f/R_{in})$.

the voltage at the noninverting input. The noninverting input is grounded. Consequently, the voltage at the inverting input is also 0.0 V. For the sake of discussion, assume that the input signal, V_{in}, is a positive voltage. While V_{in} is the voltage on the left-hand side of the resistor, R_{in}, the voltage on right side of R_{in} is 0. Ohm's law dictates that whenever a voltage difference occurs between the leads of a resistor, a current must flow. The more positive voltage appears on the left-hand side, so the direction of the current is from left to right through the resistor, R_{in}. Because the op amp does not allow any current to enter its input, the current must flow through R_f to the op amp's output. The direction of the current indicates that the output is at a lower potential than the input. The difference in voltage between the input, V_{input}, and the output, V_{out}, is the voltage dropped through the feedback resistor, R_f. Again, Ohm's law indicates what that difference is

$$V_{out} - V_{input} = -i_f R_f \quad \text{or} \quad V_{out} = V_{input} - i_f R_f = 0 - i_f R_f = -i_f R_f \tag{7.13}$$

The current in the feedback loop is equal to the current through R_{in}.

$$V_{out} = -i_f R_f = -i_{in} R_f \tag{7.14}$$

The current, i_{in}, through R_{in} can be calculated using Ohm's law.

$$V_{in} = i_{in} R_{in} \quad \text{or} \quad i_{in} = \frac{V_{in}}{R_{in}} \tag{7.15}$$

where V_{in} is the signal voltage. Combining Eqs. (7.14) and (7.15) gives

$$V_{out} = -V_{in} \left(\frac{R_f}{R_{in}} \right) \tag{7.16}$$

In conclusion, the output is proportional to the input voltage and the proportionality constant, or gain factor, is $-(R_f/R_{in})$. Because the sign of the signal is changed by this type of circuit, this configuration is also known as an inverter amplifier. It is also called a simple gain amplifier. The maximum resistance that one can use in a real operational amplifier circuit is limited to around $1 \times 10^7 \, \Omega$. Also, the input resistance is generally kept above $100 \, \Omega$ in order to keep the current and power consumption down. These constraints limit the gain factor for a single stage of amplifier, but it is possible to achieve a bigger boost in the signal by amplifying the output of one op amp with a second stage.

7.3.5. A Potentiostat for a Three-Electrode Experiment

A circuit used to apply a voltage to an electrochemical cell is sometimes called a potentiostat. When currents are small, 100 nA or less, concerns about errors resulting from
iR loss or current shifting the reference electrode from its rest potential are minimal
and a two-electrode cell consisting of the working electrode and a reference electrode
is adequate. Tiny working electrodes, dilute analyte concentrations, and high solution
conductivity and slow scan rates are conditions where two-electrode systems are practical. However, most voltammetry experiments employ three-electrode systems where an
auxiliary or counterelectrode is introduced to carry current for the reference electrode.

What sort of circuit is needed to make a three-electrode system work? Figure 7.9 shows
a collection of op amps configured for a three-electrode voltammetry experiment. The circle represents the electrochemical cell. Notice that the working electrode on the right side
of the cell is wired to the input of a current follower op amp (OA2). This op amp ampli-
fies the current that is, subsequently, plotted on the y-axis of the voltammogram. Op amp,
OA1, looks vaguely similar to a voltage follower. Indeed, its function is to apply the voltage that is sent to its noninverting input to the cell. The output pin is connected to the
auxiliary electrode, and the inverting input is connected to the reference electrode. The op
amp drives the voltage on the auxiliary electrode to whatever level is needed to force the
voltage at the inverting input to match the noninverting input. Thinking about a concrete
set of conditions can help. Imagine that $V_1 = +1$ V. The output voltage of OA1 must have
the same sign as V_1, so the solution side of the counter electrode must be positive also. In
fact, the solution potential, $\varnothing_{soln} \geq V_1 = +1.0$. The solution potential must also be different from the potential at the inverting input by the voltage across the reference electrode,
namely, E_{ref}. That is, $\varnothing_{soln} = (V_1 + E_{ref})$.

This potential represents the electrochemical potential energy of the solution. The factor that is important for driving an electron transfer reaction is the potential energy difference across the solution/working electrode interface. What is the voltage across that
boundary in this case? Notice that the voltage at the noninverting input of the current

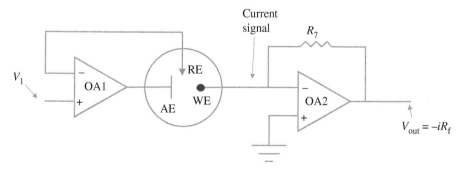

FIGURE 7.9 An op amp circuit for applying a potential in a three-electrode voltammetry experiment and measuring the current. WE, RE, and AE (auxiliary electrode) stand for working electrode,
reference electrode, and auxiliary electrode, respectively. The voltage that appears across the working
electrode/solution surface is equal to $-(V_1 + E_{ref})$. The output from OA2 is a voltage proportional to
the current signal.

follower (OA2) is zero because it is attached to ground. Consequently, the voltage at the inverting input and the working electrode must also be zero. The potential, E_{app}, that one plots in a voltammogram represents the potential energy difference across the working electrode/solution interface from the point of view of the working electrode. That is, if the potential of the solution is a positive value, the working electrode potential appears to be at a more negative voltage. Then,

$$E_{app} = \phi_{WE} - \phi_{soln} = \phi_{WE} - (V_1 + E_{ref}) = 0 - (V_1 + E_{ref}) = -(V_1 + E_{ref}) \qquad (7.17)$$

where ϕ_{WE} and ϕ_{soln} are the potential of the working electrode and the potential of the solution, respectively.

Another way of thinking about this is to say that the op amp circuit controls the potential on the electrochemical cell by raising or lowering the solution potential with respect to the working electrode. Because the circuit fixes the working electrode at 0 V, the applied voltage controls the solution potential, ϕ_{soln}. Then, $E_{app} = \phi_{WE} - \phi_{soln} = 0 - \phi_{soln} = -\phi_{soln}$. If one wants to start a cyclic voltammogram by scanning in the positive direction, the circuit must scan V_1 in the negative direction. As a result, the x-axis that the instrument plots for a cyclic voltammogram is $-V_1$.

Commercial instruments use a combination of operational amplifiers and digital components to generate the voltage, V_1, that is applied to the noninverting input of OA1. Other components control the scan rate, scan direction, starting potential, and current sensitivity as well as features for performing pulsed and square wave voltammetry experiments. Nevertheless, these more sophisticated instruments incorporate operational amplifiers for applying the cell potential via a three-electrode system, and a current follower to convert the current signal to a voltage as shown in Figure 7.7.

Op amps and digital circuits are valuable subjects for analytical chemists who need to maintain equipment or improvise simple instruments. Operational amplifiers perform many other functions commonly used in modern chemical instrumentation, including integrating signals, generating linear voltage ramps or other waveforms, adding voltages, subtracting signals, adjusting scales, and comparing voltage levels [1, 2]. Furthermore, real op amps have practical limits that one should be aware of when applying instruments to new situations. These applications and characteristics are beyond the scope of this book. However, Refs. [1, 2] are excellent resources and are readily available.

7.4. NOISE AND SHIELDING

A lab environment is full of electromagnetic waves that can induce tiny electrical currents in sensitive circuits. In experiments where sub-microamp currents are measured, these induced currents can be significant compared to the magnitude of the signal of interest and can cause interference [3]. Voltammetry experiments with ultramicroelectrodes are particularly susceptible to environmental noise pick-up because the signal levels are often on the order of only 10^{-12} A. An effective strategy for eliminating environmental noise is to place the electrochemical cell inside a metal box that is also connected to ground. The box intercepts electromagnetic fields in the room and provides a low resistance path

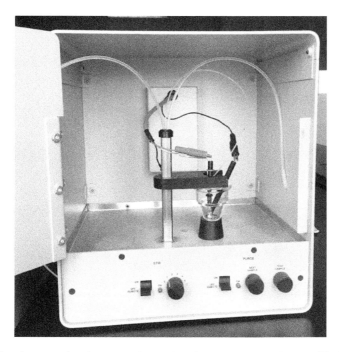

FIGURE 7.10 Faraday cage for ultramicroelectrode voltammetry experiments. The box is connected to the grounding pin in the lab so that electromagnetic fields in the environment are directed to ground and do not interfere with low current measurements.

to ground for any induced current, thereby protecting the electrical leads attached to the cell on the inside of the box. A tight screen, or even aluminum foil, around the apparatus is also effective. This type of structure is often called a "Faraday cage." Figure 7.10 shows a commercial example of a Faraday cage designed for voltammetry experiments with ultra-microelectrodes. In commercial equipment, the first stage of an amplifier system is often placed inside the Faraday cage as well, in order to boost the signal before external fields have a chance to effect it.

A combination pH electrode is another device that operates on extremely low current levels. It is rarely necessary to operate a pH electrode inside a Faraday cage. However, the wires between the electrode and the pH meter can act as an antenna for electrical noise and must be shielded. The most common strategy uses a coaxial cable to bring the signal from the electrodes to the meter. The signal from one of the two silver/silver chloride reference electrodes in the pH probe is connected to the center copper wire and the other is attached to the shield [3]. Figure 7.11d shows a cross-sectional view of a coaxial cable and a "bayonet" type connector. Coaxial cable consists of a copper wire surrounded by a polymer insulator. On the outside of the polymer is a sleeve made from many thin wires braided together to resemble a metal fabric (Figure 7.11c). Finally, another polymer layer covers the cable. The braided sleeve is eventually connected to the grounded frame of the instrument through the outer parts of the connector assembly. The signal is carried on the central copper wire that is soldered to a round connector pin, slightly less than a millimeter in diameter. That pin makes contact with the complementary connector by sliding into a

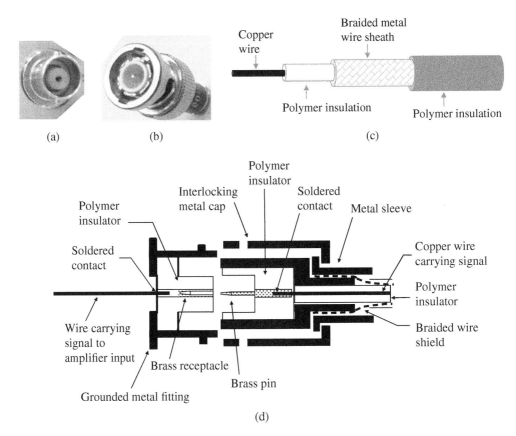

FIGURE 7.11 (a) BNC connector for mounting on an instrument frame. (b) BNC connector for coaxial cable. (c) Shielded coaxial cable has a central wire for sensitive signals and a braided wire sheath that is grounded to the instrument chassis. The second electrode in a combination pH electrode is connected to the braided sheath. (d) Cross-sectional diagram of complementary BNC connectors. The fitting on the left is fixed to the metal housing of the instrument containing the amplifier circuit. The portion on the right is attached to the coaxial cable carrying the signal from the electrode. The braided wire shield from the coaxial cable makes electrical contact with the outer metal housing of the connector on the right and subsequently with the grounding point on the instrument frame through the interlocking cap and fitting on the left.

metal sheath in a manner similar to a sword sliding into a scabbard. This particular design is called a Bayonet Neil–Conselman (BNC) connector (see Figure 7.11a,b). The signal from the inner reference electrode for the combination electrode travels on the central copper wire while the outside reference electrode is in contact with the braided shield. Electrical noise is intercepted by the shield and directed to the ground.

There are many sources for noise that originate inside sensors, electronic circuits, and accessory equipment. Unfortunately, a robust discussion of the underlying phenomena and approaches for extracting signals from noise is beyond the scope of this book. Some good references for further study are given in the bibliography at the end of the chapter [1–4].

7.5. MAKING ELECTRODES AND REFERENCE BRIDGES

7.5.1. Voltammetric Working Electrodes

Working electrodes and reference electrodes for voltammetry are available commercially from several instrument makers for modest cost. They can also be made easily. Stationary working electrodes for cyclic voltammetry (CV) experiments or for flow-through detectors can be made in the lab from platinum or gold wire that is sealed in a glass or polymer insulating material (see Figure 7.12). Glassy carbon rod, typically 2 or 3 mm in diameter, can be cut into small disks with a glass saw and embedded in acrylic plastic with epoxy. Of course, the use of epoxy and polymer insulating material restricts the use of such electrodes to solvents that do not dissolve or swell the material or leach organic molecules from the device. Another option for gold or platinum working electrodes is to seal them in a glass tube by heating with a torch. However, the glass must have a thermal coefficient of expansion similar to that of the metal so that the glass does not crack or pull away from the metal as it cools.

A lathe can be used to turn the electrode in order to clean the working surface and make it flat. The final polishing of the surface is done by hand with progressively finer abrasives and polishing materials down to a grit size of 1 μm or less. Moving the electrode over the polishing surface in a circular motion will cause the electrode surface to take on a curvature. (That type of grinding is how lens were made originally.) Introducing curvature

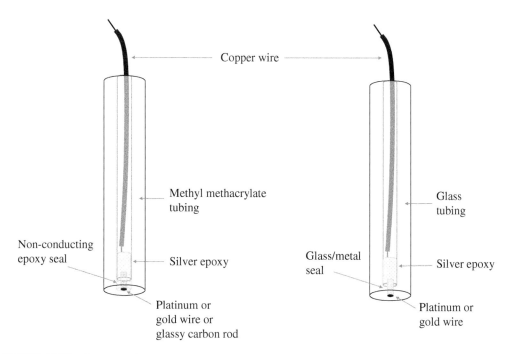

FIGURE 7.12 Some working electrode fabrication techniques. Soft glass can make a seal to platinum or gold without the need for epoxy.

should be avoided because the equations that have been derived to predict current levels assume a flat surface. In order to keep the electrode surface flat, the electrode should be moved in a "figure 8" pattern with respect to the polishing pad. The surface can be rinsed with high purity water and used directly, but for careful work it is wise to use a sonicating bath to remove polishing debris. Methods for removing trace organic molecules and electrochemically pretreating the surface of carbon electrodes are discussed in Chapter 5 (Section 5.5.2.3). The geometrical surface area is often calculated based on the diameter of the wire or rod used to make the disk-shaped electrode. However, surface roughness can make the actual electrochemically active surface area as much as twice as big as the geometrical area [5]. One can measure the active surface area by performing a chronoamperometry experiment with a well-studied electroactive species, such as ferricyanide. The diffusion coefficient of ferricyanide has been reported to have value of 0.762 ($\pm$0.01)$\times 10^{-5}$ cm^2/s (4.0 mM FeIII(CN)$_6{}^{-3}$ in 0.1 M KCl that is at 25 °C) [6]. (See the discussion on chronoamperometry in Section 5.3.5.1).

Electrical contact to the working electrode can be made by using commercially available silver-filled epoxy to attach a wire lead.

7.5.2.   Reference Electrodes

Silver/silver chloride reference electrodes can be made quite easily in the lab [7]. A 7–8 cm piece of 6 mm glass tubing serves as good housing for a reference electrode. About half the length of a 10-cm piece of small gage silver wire is coiled in a sufficiently tight coil that it can be placed inside the glass tube without rubbing the walls. Figure 7.13 shows a simple arrangement for coating a silver wire with silver chloride. The silver wire and a platinum wire are immersed in 0.1 M HCl. The two wires are connected through a multimeter in the ammeter mode, a battery, and a 2 kΩ variable resistor in series. The silver wire is attached to the positive side of the battery. Before starting the coating process, one should have a rough estimate of the surface area of the silver wire that will receive the coating. The area estimate is needed in order to control the current level for the optimum coating. When the silver wire is connected to the positive side of the battery, the resistance of the potentiometer is adjusted to give a current density of about 1–2 mA/cm^2 of silver wire. A film will appear at the silver surface that may range in color from white to a dark purple, while hydrogen gas

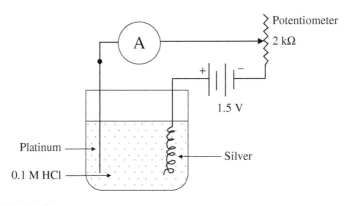

FIGURE 7.13   Arrangement for coating a silver wire with silver chloride [7].

bubbles will form on the platinum wire. Under these conditions a good coating forms in 10–15 minutes. The silver/silver chloride wire should be removed from the solution and rinsed gently with a stream of 18 MΩ deionized water. The coating should be kept wet, preferably, stored in 0.1 M KCl solution.

There are many ways to make a salt bridge, but a convenient method is to prepare a sintered glass plug at one end of a 6 mm glass tube. This can be done by first heating one end of the glass tube in a propane torch until the opening is too small to allow any glass beads ($\leq$0.5 mm) to pass. Then, enough beads are added to the inside of the tube to reach a length of about 1 cm. This end of the glass tube is placed back into the flame until the glass beads fuse together. After cooling the tube, a few millimeters are cut off the end to expose the middle of the segment with the sintered glass beads. This procedure creates a porous opening. Finally, the sintered glass opening is dipped into a hot solution of agar (0.4 g agar per 25 ml of 0.1 M KCl solution) to a depth of about 0.5 cm. After removing the glass tube and cooling the agar, the tube is filled with 0.1 M KCl. If the salt bridge is properly prepared, the solution will not drain from the tube. This salt bridge can be stored in a closed jar with 0.1 M KCl at a height similar to that inside the tube.

To prepare the electrode for an experiment, the coated silver/silver chloride wire is inserted into the tube for the salt bridge so that the coating is submerged (see Figure 7.14). A short section of the uncoated wire should extend beyond the open end of the glass. In order to inhibit evaporation, the opening should be sealed. The uncoated silver wire can be folded over the edge of the glass and held there with a short length of flexible tubing (such as latex rubber). The open end of the flexible tubing is closed with a glass bead or a short segment of 6 mm glass rod as a plug. (In order to avoid pressure that might push

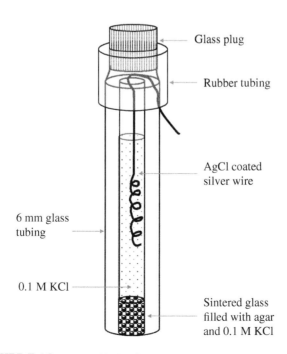

FIGURE 7.14 Assembled reference electrode made in the lab.

agar out the sintered end of the tube, one can slide a hypodermic needle into the opening before inserting the plug. The needle can easily be removed after the plug is in place.) An alligator clip can be used to connect the electrode to the potentiostat. The potential for this reference electrode is

$$E_{ref} = E^o_{\frac{Ag^+}{Ag}} - 0.059\,16\,\log(\gamma_{Cl^-}[Cl^-])$$

$$= 0.222 - 0.059\,16\,\log(0.755 \cdot 0.100) = 0.288\text{ V versus SHE} \qquad (7.18)$$

One can regenerate the silver/silver chloride coating if it dries out, discolors or is damaged in some way. The old coating is removed by soaking the wire in 4 M NH_3. Subsequently, the wire is rinsed with 18 MΩ deionized water, M HNO_3, and rinsed again with deionized water before re-chloridizing with the method depicted in Figure 7.13.

PROBLEMS

7.1 Draw a circuit and specify the resistances that will divide a voltage of 6 V to give 1 V at a point that you indicate.

7.2 What is the voltage at point A in the following circuit?

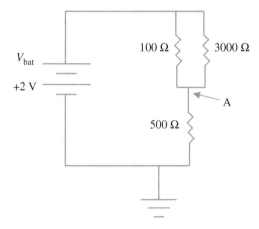

7.3 Calculate the current through the 100 Ω resistor in the diagram for Problem 7.2.

7.4 Derive the algebraic expression for the effective resistance in the circuit below.

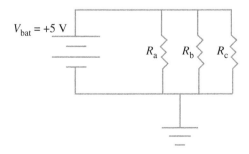

7.5 For the circuit in Problem 7.4, let $R_a = 500\,\Omega$, $R_b = 400\,\Omega$, and $R_c = 300\,\Omega$. Calculate the effective resistance in the circuit.

7.6 For the conditions given in Problems 7.4 and 7.5, calculate the current passing through each of the resistors R_a, R_b, and R_c.

7.7 For the circuit in the diagram below, calculate the output voltage for an input current, i_{in}, of 15 nA and a feedback resistance, R_f, of 200 kΩ.

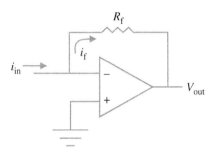

7.8 Draw an op amp circuit capable of producing an output of 0.1 V for a current signal of 20 nA using two stages.

7.9 In the circuit below,

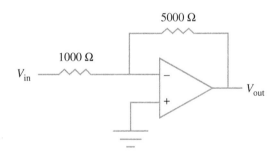

(a) What is the output voltage when $V_{in} = 0.100$ V for an ideal amplifier?

(b) What input voltage would give an output of -0.500 V?

7.10 Draw a modification of the circuit in Problem 7.9 so that it would be possible to amplify a signal from a chloride ion selective electrode (ISE) by a factor of 5 without draining significant current.

7.11 Describe the two most important considerations in choosing the concentration of the KCl solution for the internal filling solution and the salt bridge for a silver/silver chloride reference electrode such as in Figure 7.13.

7.12 In Figure 7.12 only two of the three leads to the potentiometer are connected. Explain what the potentiometer's purpose is and how it functions in that circuit.

7.13 Calculate the reference potential for a silver/silver chloride reference electrode for a solution that is 0.100 M chloride activity.

7.14 In a three-electrode potentiostat, with a silver/silver chloride reference electrode (with 3 M KCl), what voltage should be applied at V_1 in order to reduce ferricyanide at +0.336 V versus NHE(normal hydrogen electrode)?

7.15 In a three-electrode potentiostat, the counter or auxiliary electrode (AE) carries current so that the reference electrode (RE) does not need to. Under what circumstances does the auxiliary electrode carry current? What sort of electron-transfer process occurs at the auxiliary electrode?

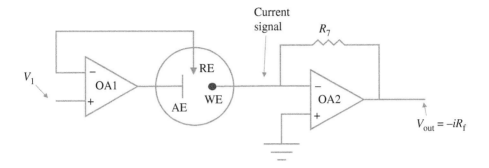

7.16 Derive the general mathematical equation that relates the output voltage, V_{out}, to the input signal, V_{in}.

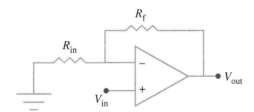

REFERENCES

1. Mancini, R. (ed.) (2002). *Op Amps for Everyone*. Dallas, Texas: Texas Instruments http://web.mit.edu/6.101/www/reference/op_amps_everyone.pdf.

2. Jung, W. (ed.) (2005). *Op Amp Applications Handbook*. Newnes/Elsevier https://www.analog.com/en/education/education-library/op-amp-applications-handbook.html.

3. Moore, J.H., Davis, C.C., and Coplan, M.A. (1989). *Building Scientific Apparatus*, 2e. Redwood City, CA: Addison-Wesley Publishing.

4. Carter, B. and Brown, T.R. (2016). *Handbook of Operational Amplifier Applications*. Texas Instruments http://www.ti.com/lit/an/sboa092b/sboa092b.pdf.

5. Hoogvliet, J.C., Dijksma, M., Kamp, B., and van Bennekom, W.P. (2000). *Anal. Chem.* 72: 2016–2021.

6. Sawyer, D.T., Sobkowiak, A., and Roberts, J.L. Jr. (1995). *Electrochemistry for Chemists*, 2e, 219. New York: Wiley.

7. Blaedel, W.J. and Meloche, V.W. (1963). *Elementary Quantitative Analysis*, 2e. New York: Harper and Row.

IONIC STRENGTH, ACTIVITY, AND ACTIVITY COEFFICIENTS

Many electrochemical phenomena are governed by ions interacting with charged surfaces. Because of the electrostatic fields associated with individual ions, ions influence each other in ways that are not observed with neutral molecules. These charge effects are more noticeable as the concentration of charge increases. The ionic strength of a solution μ, a measure of the concentration of charge for the solution as a whole, is defined by Eq. (A.1):

$$\mu = \frac{1}{2} \sum c_i z_i^2 \tag{A.1}$$

where z_i is the charge on an ion and c_i is the concentration on the same ion. The summation is done for all cations and anions. In order to appreciate how much influence the ionic strength can have, consider the dissolution of a sparingly soluble ionic crystal such as calcium sulfate:

$$CaSO_4 \rightleftarrows Ca^{2+} + SO_4^{2-} \tag{A.2}$$

In pure water at 25 °C, this process can be accurately modeled algebraically using the concentrations of the two ions:

$$K_{sp} = [Ca^{2+}][SO_4^{2-}] = 5 \times 10^{-5} \tag{A.3}$$

This equation predicts approximately 7.1 mmol of solid dissolves per liter of solution, assuming that the solvent is pure water.

If instead of pure water, one were to expose the solid to a solution of 0.10 M $NaNO_3$, then more than 2.5 times that quantity of $CaSO_4$ would dissolve per liter of solution than before. Why might this be the case? In each solution, one would expect all the ions to be surrounded by a cluster of solvent molecules. In order for a Ca^{2+} ion to recombine with the solid $CaSO_4$ lattice, it will need to strip away part of the water molecules surrounding it as well as water clinging to the crystal surface. Nevertheless, the recombination process is favorable due to the attractive forces of the oppositely charged ions acting to bring the calcium and sulfate ions back together in the lattice. However, in the $NaNO_3$ solution, the Ca^{2+} ion will also need to break through an adsorbed layer of either Na^+ ions near the negative charge site on the solid surface as well nitrate anions that are attracted to the positive charge on the calcium ion before it can recombine with a vacant lattice site. Furthermore, the sulfate anion will also experience more difficulty returning to the lattice because of the neighboring sodium cations getting in the way. This extra competition with spectator

Electroanalytical Chemistry: Principles, Best Practices, and Case Studies, First Edition. Gary A. Mabbott.
© 2020 John Wiley & Sons, Inc. Published 2020 by John Wiley & Sons, Inc.

TABLE A.1   Estimate of effective hydrated radius

Ion charge	Inorganic ions	Effective hydrated radius in Å
+1	H^+	9
	Li^+	6
	Na^+, $CdCl^+$, $[Co(NH_3)_4(NO_2)_2]^+$	4–4.5
	K^+	3
	Rb^+, Cs^+, NH_4^+, Tl^+, Ag^+	2.5
−1	ClO_2^-, IO_3^-, HCO_3^-, $H_2PO_4^-$, HSO_3^-, $H_2AsO_4^-$	4–4.5
	OH^-, F^-, NCS^-, NCO^-, HS^-, ClO_3^-, ClO_4^-, BrO_3^-, IO_4^-, MnO_4^-	3.5
	Cl^-, Br^-, I^-, CN^-, NO_2^-, NO_3^-	3
+2	Mg^{2+}, Be^{2+}	8
	Ca^{2+}, Cu^{2+}, Zn^{2+}, Sn^{2+}, Mn^{2+}, Fe^{2+}, Ni^{2+}, Co^{2+}	6
	Sr^{2+}, Ba^{2+}, Ra^{2+}, Cd^{2+}, Hg^{2+}	5
	Pb^{2+}, $[Co(NH_3)_5Cl]^{2+}$	4.5
	Hg_2^{2+}	4
−2	WO_4^{2-}	5
	CO_3^{2-}, SO_3^{2-}, MoO_4^{2-}, $[Fe(CN)_6NO]^{2-}$	4.5
	SO_4^{2-}, $S_2O_3^{2-}$, $S_2O_6^{2-}$, $S_2O_8^{2-}$, SeO_4^{2-}, CrO_4^{2-}, HPO_4^{2-}	4
+3	Al^{3+}, Fe^{3+}, Cr^{3+}, Sc^{3+}, Y^{3+}, La^{3+}, In^{3+}, Ce^{3+}, Pr^{3+}, Nd^{3+}, Sm^{3+}	9
	$[Co(ethylenediamine)_3]^{3+}$	6
	$[Co(NH_3)_5H_2O]^{3+}$, $[Co(NH_3)_6]^{3+}$, $[Cr(NH_3)_6]^{3+}$	4
−3	PO_4^{3-} $[Fe(CN)_6]^{3-}$	4
+4	Th^{4+}, Zr^{4+}, Ce^{4+}, Sn^{4+}	11
−4	$[Co(S_2O_3)(CN)_5]^{4-}$	6
	$[Fe(CN)_6]^{4-}$	5
−5	$[Co(SO_3)_2(CN)_4]^{5-}$	9

Ion charge	Organic ions	Effective hydrated radius in Å
+1	$(C_3H_7)_4N^+$	8
	$(C_3H_7)_3NH^+$	7
	$(C_2H_5)_4N^+$, $(C_3H_7)_2NH_2^+$	6
	$(C_2H_5)_3NH^+$, $(C_3H_7)NH_3^+$	5
	$(CH_3)_4N^+$, $(C_2H_5)_2NH_2^+$	4.5
	$(CH_3)_3NH^+$, $C_2H_5NH_3^+$, $(NH_3^+)CH_2COOH$	4
	$CH_3NH_3^+$, $(CH_3)_2NH_2^+$	3.5
−1	$(C_6H_5)_2CHCOO^-$	8
	$[OC_6H_2(NO_2)_3]^-$, $CH_3OC_6H_4COO^-$	7
	$C_6H_5COO^-$, $C_6H_4OHCOO^-$, $C_6H_4ClCOO^-$, $C_6H_5CH_2COO^-$	6
	$(CH_3)_2C{=}CHCOO^-$	6
	$CHCl_2COO^-$, CCl_3COO^-	5
−1	CH_3COO^-, CH_2ClCOO^-, $NH_2CH_2COO^-$	4.5
	$HCOO^-$, $H_2citrate^-$	3.5
−2	$[OOC(CH_2)_5COO]^{2-}$, $[OOC(CH_2)_6COO]^{2-}$, congo red anion^{2-}	7
	$C_6H_4(COO)_2^{2-}$, $H_2C(CH_2COO)_2^{2-}$, $(CH_2CH_2COO)_2^{2-}$	6
	$H_2C(COO)_2^{2-}$, $(CH_2COO)_2^{2-}$, $(CHOHCOO)_2^{2-}$	5
	$(COO)_2^{2-}$, $Hcitrate^{2-}$	4.5
−3	$Citrate^{3-}$	5

Source: From Kielland [2].

ions effectively costs energy for the ions returning to the lattice. Looking at it another way, the effective reaction concentration of the calcium and sulfate ions has decreased in the presence of the extra ions. Similar nonideal behavior is observed for ionic species reacting directly in a homogeneous solution (such as acetate combining with a proton to form a molecule of acetic acid). Normal concentrations do not seem to work in algebraic equations for equilibrium constants for reactions involving ions. Rather, the property that defines an ion's effective reactivity is its activity, a_i.

$$a_i = \gamma_i c_i \tag{A.4}$$

where γ_i is called the activity coefficient and c_i is the concentration of the ion.

Furthermore, the activity coefficient, γ_i, changes with ionic strength. Fortunately, there is a quantitative way of handling this nonideal behavior of ionic reactions in solution. In the early part of the 1920s, Peter Debye and Erich Hückel developed an electrostatic model to account for ion–ion interactions. Their theory led to a derivation of an equation for predicting the activity coefficient in terms of the ionic strength. The original expression for the activity coefficient was applicable at only low values of ionic strength. In a revised model, Debye and Hückel treated ions as hydrated particles with finite size rather than as point charges. Here is a form of the "Extended" Debye–Hückel equation that is used widely today:

$$\log(\gamma_i) = \frac{-0.511 z_i^2 \sqrt{\mu}}{1 + Ba\sqrt{\mu}} \tag{A.5}$$

In this equation, B and a are empirical parameters. B depends on the temperature and the dielectric constant of the solvent and has a value of 0.328 in aqueous solution at 25 °C. The term "a" is a function of the hydrated diameter of the ion, i. It might be best thought of as the "mean distance of closest approach" of the ion with an ion of the opposite charge in Å [1]. The values for the hydrated radii for many common ions have been estimated from experiment and appear in tables such as Table A.1.

Notice that the hydrogen ion has a large hydrated radius. At first that seems counterintuitive, since the ionic radius for hydrogen is quite small. However, because a solvent molecule can approach so closely, the relative strength of the field associated with the charge is relatively high making the bond between the ion and the water molecule very strong. Also, the number of solvent molecules per ion is greater for the hydrogen ion than a larger ion of the same charge, such as a sodium ion. The ion together with its complement of water molecules behave as though they were a single particle. Hence, a hydrogen ion with its shroud of water appears to be a bulkier ion in solution.

The extended Debye–Hückel equation is reasonably accurate for ions in solution up to ionic strength of 0.1 M. Other workers have tried to account for the formation of ion pairs (combinations of oppositely charged ions) at higher ionic strength (decreasing the effective concentration of charges). A model that is frequently used for ionic strength levels between 0.1 and 0.6 M is the modified Davies equation:

$$\log(\gamma_i) = \frac{-0.511 z_i^2 \sqrt{\mu}}{1 + 1.5\sqrt{\mu}} + 0.2 z_i^2 \mu \tag{A.6}$$

EXAMPLE *A.1*

Calculate the activity of Ca^{2+} ions in these two different solutions prepared by dissolving 0.007 mol $CaCl_2$ in a liter of

(a) 0.01 M $MgCl_2$ solution at 25 °C.
(b) 0.10 M $MgCl_2$ solution at 25 °C.

(a)
$$\mu = \frac{1}{2}\sum c_i z_i^2 = 0.5\{[Mg^{2+}](2^2) + [Cl^-](-1)^2 + [Ca^{2+}](2)^2\}$$

$$= (0.5)\{0.01(4) + [0.02 + 0.014](1) + (0.007)(4)\} = 0.051$$

Since the ionic strength is <0.1 M, the extended Debye–Hückel equation applies. From Table A.1, the hydrated radius for Ca^{2+} is 6 Å:

$$\log(\gamma_{Ca^{2+}}) = \frac{-0.511 z_i^2 \sqrt{\mu}}{1 + Ba\sqrt{\mu}} = \frac{-(0.511)(2^2)\sqrt{0.051}}{1 + (0.328)(6)\sqrt{0.051}} = -0.319$$

$$\gamma_{Ca^{2+}} = 10^{-0.319} = 0.479$$

$$aCa^{2+} = \gamma_{Ca^{2+}}[Ca^{2+}] = (0.479)[0.007] = 0.003\,35\text{ M}$$

(b)
$$\mu = \frac{1}{2}\sum c_i z_i^2 = 0.5\{[Mg^{2+}](2^2) + [Cl^-](-1)^2 + [Ca^{2+}](2)^2\}$$

$$= (0.5)\{0.1(4) + [0.2 + 0.014](1) + (0.007)(4)\} = 0.321$$

Since the ionic strength lies between 0.1 and 0.6 M, the modified Davies equation applies:

$$\log(\gamma_{Ca^{2+}}) = \frac{-0.511 z_i^2 \sqrt{\mu}}{1 + 1.5\sqrt{\mu}} + 0.2 z_i^2 \mu = \frac{(-0.511)(2^2)\sqrt{0.321}}{1 + 1.5\sqrt{0.321}} + 0.2(2^2)(0.321) = -0.369$$

$$\gamma_{Ca^{2+}} = 10^{-0.369} = 0.427$$

$$aCa^{2+} = \gamma_{Ca^{2+}}[Ca^{2+}] = (0.427)[0.007] = 0.002\,99\text{ M}$$

REFERENCES

1. Bockris, J.O'.M. and Reddy, A.K.N. (1970). *Modern Electrochemistry*, vol. 2, 225. New York: Plenum.
2. Kielland, J. (1937). Individual activity coefficients of ions in aqueous solutions. *J. Am. Chem. Soc.* 59 (9): 1675–1678.

THE NICOLSKY–EISENMAN EQUATION

It is enlightening to take a deeper look into the model of the surface of pH-sensitive glass. Some important insights can be gained by deriving the relationship between the interfacial potential and the free hydrogen ion activity. The process at each interface can be described as an equilibrium between the H^+ activities in the glass and in the adjacent solution. For the sample solution,

$$H^+_{(glass)} \rightleftharpoons H^+_{(soln)} \tag{B.1}$$

The electrochemical potential of the H^+ ion in solution, $\bar{\mu}_{H^+,soln}$, is the sum of the chemical potential or free energy of formation for a mole of H^+ ion in solution at standard state, $\mu^o_{H^+,soln}$, plus a term that adjusts for conditions that are not at standard state, namely, $RT \ln a_{H^+,spl}$ and the electrical work done in bringing a mole of ions with its charge, z, into solution, $zF\phi_{spl}$.

$$\bar{\mu}_{H^+,soln} = \mu^o_{H^+,soln} + RT \ln a_{H^+,spl} + (1)F\phi_{spl} \tag{B.2}$$

A similar equation can be written for the electrochemical potential of the H^+ ion on the glass surface, $\bar{\mu}_{H^+,glass}$.

$$\bar{\mu}_{H^+,glass} = \mu^o_{H^+,glass} + RT \ln a_{H^+,glass} + (1)F\phi_{glass} \tag{B.3}$$

At equilibrium, the electrochemical potential of the products in Eq. (B.1) is equal to the electrochemical potential of the reactants. Setting $\bar{\mu}_{H^+,glass} = \bar{\mu}_{H^+,soln}$ gives:

$$\mu^o_{H^+,glass} + RT \ln a_{H^+,glass} + F\phi_{glass} = \mu^o_{H^+,soln} + RT \ln a_{H^+,spl} + F\phi_{spl} \tag{B.4}$$

Rearranging to solve for the potential across this interface, $\phi_{glass} - \phi_{soln}$:

$$F\phi_{glass} - F\phi_{soln} = \mu^o_{H^+,soln} - \mu^o_{H^+,glass} + RT \ln a_{H^+,soln} - RT \ln a_{H^+,glass} \tag{B.5}$$

$$\phi_{glass} - \phi_{soln} = \frac{(\mu^o_{H^+,soln} - \mu^o_{H^+,glass})}{F} + \frac{RT}{F} \ln \frac{a_{H^+,soln}}{a_{H^+,glass}} \tag{B.6}$$

$$\phi_{glass} - \phi_{soln} = \phi^o + \frac{RT}{F} \ln \frac{a_{H^+,soln}}{a_{H^+,glass}} \tag{B.7}$$

Electroanalytical Chemistry: Principles, Best Practices, and Case Studies, First Edition. Gary A. Mabbott.
© 2020 John Wiley & Sons, Inc. Published 2020 by John Wiley & Sons, Inc.

This expression predicts the interfacial potential based on the hydrogen activities in the sample solution and in the glass. All the other terms are constants. The acid dissociation reaction of the silanol group is the principal interaction. Can we assume that $a_{H^+,glass}$ is a constant, too? No, the hydrogen ion does not diffuse uniformly throughout the glass. Nicolsky [1] pointed out that the hydrogen ion activity in the glass depends on the exchange process between hydrogen irons and sodium ions on the glass surface.

$$H^+_{(glass)} + Na^+_{(soln)} \rightleftarrows H^+_{(soln)} + Na^+_{(glass)} \tag{B.8}$$

(Nicolsky chose to describe the reaction as though the sodium is displacing hydrogen ions. That may be a better description once the hydrated layer has been formed.) The equilibrium constant for this reaction can be expressed as follows:

$$K_{ex} = \frac{(a_{H^+,soln})(a_{Na^+,glass})}{(a_{H^+,glass})(a_{Na^+,soln})} = \frac{(a_{H^+,soln})(\gamma_{Na^+,glass}C_{Na^+,glass})}{(\gamma_{H^+,glass}C_{H^+,glass})(a_{Na^+,soln})} \tag{B.9}$$

Then, assuming that the activity coefficients for the ions in the glass are both constant,

$$K'_{ex} = \frac{(\gamma_{H^+,glass})}{(\gamma_{Na^+,glass})}K_{ex} = \frac{(a_{H^+,soln})(C_{Na^+,glass})}{(C_{H^+,glass})(a_{Na^+,soln})} \tag{B.10}$$

where the activities of the ions in the glass are assumed to be proportional to their concentrations in the glass, $C_{H^+,glass}$ and $C_{Na^+,glass}$. In other words, $a_{H^+,glass} = kC_{H^+,glass}$ and $a_{H^+,glass} = kC_{H^+,glass}$. (Introducing concentrations allows one to make some simplifications based on mass balance arguments.) These ion concentrations in the glass are related. The concentration of all the exchangeable adsorption sites must sum to a constant value:

$$C_o = C_{H^+,glass} + C_{Na^+,glass} + C_{SiO^-,glass} \tag{B.11}$$

where $C_{SiO^-,glass}$ represents the concentration of anion sites without a compensating cation. The value of $C_{SiO^-,glass}$ is a measure of the excess charge density on the surface. This number must be small. Even 500 pmol of charge/cm^2 will produce a potential difference of about 0.5 V. We assume that $C_{SiO^-,glass} \ll C_{H^+,glass} + C_{Na^+,glass}$ and $C_o \approx C_{H^+,glass} + C_{Na^+glass}$ or

$$C_{Na^+,glass} = C_o - C_{H^+,glass} \tag{B.12}$$

Substituting this expression into the equilibrium equation for the ion exchange process gives us an opportunity to remove the ion concentrations in the glass from the equation:

$$K'_{ex} = \frac{(a_{H^+soln})(C_{Na^+glass})}{(C_{H^+glass})(a_{Na^+soln})} = \frac{(a_{H^+soln})(C_o - C_{H^+glass})}{(C_{H^+glass})(a_{Na^+soln})} = \frac{(a_{H^+soln})(C_o)}{(C_{H^+glass})(a_{Na^+soln})} - \frac{(a_{H^+soln})}{(a_{Na^+soln})} \tag{B.13}$$

Rearranging gives:

$$\frac{(a_{H^+,soln})}{(C_{H^+,glass})} = \frac{K'_{ex}(a_{Na^+,soln}) + a_{H^+,soln}}{C_o} \tag{B.14}$$

Substituting the result from Eq. (B.14) into Eq. (B.7) describing the interfacial potential, $\phi_{glass} - \phi_{soln}$:

$$\phi_{glass} - \phi_{soln} = \phi^o + \frac{RT}{F} \ln \frac{a_{H^+,soln}}{a_{H^+,glass}} = \phi^{o\prime} + \frac{RT}{F} \ln \frac{a_{H^+,soln}}{C_{H^+,glass}} \tag{B.15}$$

$$= \phi^{o\prime} + \frac{RT}{F} \ln \left\{ \frac{K'_{ex}(a_{Na^+,soln}) + a_{H^+,soln}}{C_o} \right\} \tag{B.16}$$

Because the total number of adsorption sites, C_0, in the glass is a constant, that term can be included with the other constants, $\phi^{o\prime}$.

$$\phi_{glass} - \phi_{soln} = \phi^{o\prime\prime} + \frac{RT}{F} \ln\{K'_{ex}(a_{Na^+,soln}) + a_{H^+,soln}\} \tag{B.17}$$

This result is telling us that the electrode potential will respond to sodium ion activity as well as hydrogen ion activity. Sodium is an interferent in measurements of pH based on a glass electrode. However, since the glass binds H^+ ions *much* more strongly than it does Na^+ ions, the equilibrium constant, K'_{ex}, for the displacement of hydrogen ions by sodium ions is a small number. We can expect that $a_{H^+,soln} \gg K'_{ex}(a_{Na^+,soln})$ except at very high sodium concentrations and very low hydrogen ion concentrations. This effect is sometimes referred to as the sodium error or alkaline error. Lithium and potassium can also compete with hydrogen ions for the negatively charged adsorption sites at high pH. Figure 3.12 shows the deviation in potential from the ideal response for a glass electrode as a function of pH. (The "ideal" in this case is $E = E_o - 0.059\,16pH$.)

Interestingly, the selectivity of the glass for H^+ ions over Na^+ ions is better (K'_{ex} is smaller) for glasses made with Li_2O instead of Na_2O. Lithium is still an interfering ion for pH sensors made from lithium glasses, but it provides an attractive option since high sodium, high pH conditions are much more common than samples containing high lithium ion concentrations. Commercial pH electrodes of this sort with less alkaline error are available.

Diffusion potential: The very tiny movement of charge through the glass membrane is carried mainly, but not completely, by sodium ions. The fact that more than one ion is moving through the membrane also leads to a voltage error. The mechanism depends on a difference in mobility of the ions and on a concentration gradient that drives a given ion through the membrane in one direction. However, mathematically modeling this diffusion potential results in an expression that is parallel in form to Eq. (B.17) accounting for ion exchange of interfering ions [2]. The fortunate thing about this outcome is that both of these effects can be summarized in a single equation by replacing the equilibrium constant for the exchange process with $K_{jH}^{POT} = \left(\frac{u_j}{u_{H^+}} \right) K'_{ex}$, where u_j and u_{H^+} are the relative mobilities of ion "j" and the hydrogen ion in the membrane. This relationship is often called the Nicolsky–Eisenman equation:

$$E_{memb} = E_o + \frac{RT}{F} \ln \left\{ a_{H^+_{spl}} + \sum_j K_{jH}^{POT} \times a_j^{z_H/z_j} \right\} \tag{B.18}$$

In this expression, E_o represents an "off-set" voltage or the voltage for a plot of the measured membrane potential versus pH extrapolated to pH = 0. The interfering ion has an activity of a_j and a charge of z_j. Divalent ions do not show much movement through glass, so z_j is usually 1. The term K_{jH}^{POT} is known as the selectivity coefficient for ion j compared to the analyte ion, H^+.

REFERENCES

1. Isard, J.O. (1967). The glass-electrode potential. In: *Glass Electrodes for Hydrogen and Other Cations* (ed. G. Eisenman), 63. New York: Marcel Dekker, Inc.
2. Eisenman, G. (ed.) (1967). *Glass Electrodes for Hydrogen and Other Cations*, 136. New York: Marcel Dekker, Inc.

THE HENDERSON EQUATION FOR LIQUID JUNCTION POTENTIALS[1]

APPENDIX

C

Liquid junction potentials are almost unavoidable in electrochemical measurement systems. However, understanding the conceptual model for their origin helps one make choices in experimental design that minimize the errors caused by junction potentials. A mathematical model helps one calculate the magnitude of the junction potential in order to decide whether it is a concern. If the error is small enough to be acceptable or it is constant for all sample and standard solutions so that it is accounted for by the calibration procedure, then no change in procedure is warranted.

The formation of a potential difference at the interface between two solutions stems from the difference in mobilities of the ions in the electrolyte. The boundary between a salt bridge and the sample solution is usually a porous frit made of glass, ceramic, or polymer material. Movement of an ion across the boundary is driven by a concentration gradient and/or an electric field and can be described by the flux of the ion, in units of $mol/cm^2/s$. It is sometimes helpful to remember that a flux has the same dimensions as the product of a concentration and a velocity. (Convection or bulk flow of liquid is another mechanism of material transport, but is assumed to have a negligible influence on the liquid junction potential and is not included in the model here.) The flux for an ion is described by the Nernst–Planck equation.

$$J_i = -D_i \frac{\partial C_i}{\partial x} - D_i C_i \frac{z_i F}{RT} \frac{\partial \phi}{\partial x} \tag{C.1}$$

In Eq. (C.1), D_i is the diffusion coefficient, C_i is the concentration, and z_i is the charge of the ion, i. The electrical potential energy is ϕ, F is Faraday's constant (96,485 C/mol), R is the molar gas constant (8.314 J/(K mol)), and the temperature, T, is given in Kelvin.

In this discussion, it is assumed that the movement (along the x-coordinate) is perpendicular to a flat plane that forms the boundary between the two solutions.

Although the Nernst–Planck equation is unfamiliar to most people, most chemists recognize the phenomena represented by the two terms on the right-hand side. The first term indicates that a concentration gradient will drive movement of a species by diffusion. The diffusion coefficient is merely the proportionality constant relating the flux of an uncharged species to the concentration gradient. (The equation without the second term

[1] This discussion is based on the derivation of Bakker [1].

Electroanalytical Chemistry: Principles, Best Practices, and Case Studies, First Edition. Gary A. Mabbott.
© 2020 John Wiley & Sons, Inc. Published 2020 by John Wiley & Sons, Inc.

is often called Fick's first law of diffusion.) The second term on the right accounts for an additional driving force for migration of charged species. The driving force is the potential gradient (or electric field), $\frac{\partial \phi}{\partial x}$. The movement of charge under the influence of an electrical potential energy difference is also an intuitively reasonable mechanism. It is the basis for separation of charged molecules by electrophoresis, for example.

The goal of this analysis is to find the difference in the potential between the two phases. This potential energy difference is the liquid junction potential. There are two keys to solving this problem. The first of these is the fact that the Nernst–Planck equation applies to each ion. The second key is an assumption that the boundary conditions reach a steady state, namely, a situation where the sum of all the fluxes equals zero. The last idea is a consequence of the restriction that the measurement must be made without drawing significant current. The zero-current condition was already a requirement imposed on all potentiometric measurements in order to avoid distorting the conditions at the sensor/sample solution interface. (Of course, a finite current does exist during real experiments, but it is usually on the picoampere level or lower and is usually negligible.)

An analysis of a very simple (but also very common) set of conditions reveals some important guidelines for experimenters to follow. Consider a salt bridge separating a reference electrode with a solution of 1 M KCl from a sample solution that was spiked with an ionic strength adjusting buffer so that the sample solution electrolyte is essentially 0.1 M KCl. (Assume that the other ionic components in the original sample solution are less than 1 mM so that their influence on the liquid junction potential can be neglected.) A further reasonable assumption is that the concentration gradients for K^+ and Cl^- are linear and that at any point the total cation concentration and total anion concentration are equal to each other. That is $C_{K^+} = C_{Cl^-} = C = C_{spl} + mx$, where C_{spl} is the ion concentration in the sample solution and the concentration at the maximum value of x is C_{ref}, the concentration of each ion in the reference solution.

In the first moment that the boundary is formed, there is no charge separation at the interface, so the electric field is zero. The concentration gradient drives the ion movement. The first term in Eq. (C.1) indicates that the ion with the bigger diffusion coefficient will have a bigger flux; its velocity across the boundary is greater than the other ion. In this case, chloride ions are faster than potassium ions. An excess of negative charge develops on the sample solution side and that creates an electric field which attracts potassium ions and slows down the flux of chloride. In a very short time, the forces associated with the electric field and the concentration gradient balance each other, and the anion flux equals the cation flux:

$$-D_{K^+}\frac{\partial C_{K^+}}{\partial x} - D_{K^+}C_{K^+}\frac{z_{K^+}F}{RT}\frac{\partial \phi}{\partial x} = -D_{Cl^-}\frac{\partial C_{Cl^-}}{\partial x} - D_{Cl^-}C_{Cl^-}\frac{z_{Cl^-}F}{RT}\frac{\partial \phi}{\partial x} \text{ or}$$

$$-D_{K^+}\frac{\partial C_{K^+}}{\partial x} - D_{K^+}C_{K^+}\frac{F}{RT}\frac{\partial \phi}{\partial x} = -D_{Cl^-}\frac{\partial C_{Cl^-}}{\partial x} + D_{Cl^-}C_{Cl^-}\frac{F}{RT}\frac{\partial \phi}{\partial x} \quad (C.2)$$

Since $C_{K^+} = C_{Cl^-} = C = C_{spl} + mx$, $\frac{\partial C_{K^+}}{\partial x} = \frac{\partial C_{Cl^-}}{\partial x} = m$, Eq. (C.2) simplifies to

$$(D_{Cl^-} - D_{K^+})m + (-D_{K^+} - D_{Cl^-})\left\{ C\frac{F}{RT}\frac{\partial \phi}{\partial x} \right\} = 0 \quad (C.3)$$

Rearranging gives

$$m\frac{(D_{Cl^-} - D_{K^+})}{(-D_{K^+} - D_{Cl^-})}\left\{\frac{RT}{CF}\right\} = \frac{\partial\phi}{\partial x} \tag{C.4}$$

Integrating Eq. (C.4) gives the potential difference between the two liquids.

$$\int_0^{x_{max}} m\frac{(D_{Cl^-} - D_{K^+})}{(-D_{K^+} - D_{Cl^-})}\left\{\frac{RT}{CF}\right\}\partial x = \int_{\phi_{spl}}^{\phi_{ref}}\partial\phi \tag{C.5}$$

$$= \int_0^{x_{max}} m\frac{(D_{Cl^-} - D_{K^+})}{(-D_{K^+} - D_{Cl^-})}\left\{\frac{RT}{(C_{spl} + mx)F}\right\}\partial x = \int_{\phi_{spl}}^{\phi_{ref}}\partial\phi \tag{C.6}$$

The right-hand side of Eq. (C.36) is the junction potential, $\phi_{ref} - \phi_{spl} = E_j$:

$$E_j = \phi_{ref} - \phi_{spl} = m\frac{(D_{Cl^-} - D_{K^+})}{(-D_{K^+} - D_{Cl^-})}\left\{\frac{RT}{mF}\right\}\ln(C_{spl} + mx)\big|_0^{x_{max}} \tag{C.7}$$

Recalling that $C_{spl} + m(x_{max}) = C_{ref}$,

$$E_j = \phi_{ref} - \phi_{spl} = -\frac{(D_{Cl^-} - D_{K^+})}{(D_{K^+} + D_{Cl^-})}\left\{\frac{RT}{F}\right\}\{\ln(C_{ref}) - \ln(C_{spl})\} \tag{C.8}$$

$$E_j = \phi_{ref} - \phi_{spl} = -\frac{(D_{Cl^-} - D_{K^+})}{(D_{K^+} + D_{Cl^-})}\left\{\frac{RT}{F}\right\}\left\{\ln\left(\frac{C_{ref}}{C_{spl}}\right)\right\} \tag{C.9}$$

In this case, $C_{spl} = 0.1\,M$, $C_{ref} = 1\,M$, $D_{Cl^-} = 2.03 \times 10^{-5}\,cm^2/s$, and $D_{K^+} = 1.96 \times 10^{-5}\,cm^2/s$. Evaluating the junction potential at 298 K gives

$$E_j = \phi_{ref} - \phi_{spl} = -\frac{(2.03 - 1.96)(10^{-6})}{(2.03 + 1.96)(10^{-6})}\left\{\frac{(8.314)(298)}{96\,485}\right\}\left\{\ln\left(\frac{0.1}{1}\right)\right\} = 0.001\,04\,V \tag{C.10}$$

The conclusion is that a salt bridge between a sample solution with a supporting electrolyte of 0.1 M KCl and a reference solution of 1.0 M KCl is only 1 mV. If a voltage error of that size were operating in an experiment with an ion selective electrode, it would lead to an error in the concentration of a singly charged analyte of about 0.3%. That seems quite reasonable. Of course, if the sample and standard solutions have the same electrolyte conditions, then the liquid junction potential is constant and becomes a part of the y-intercept in the calibration curve and the error is compensated for.

Other supporting electrolytes are less well matched in terms of the cation and anion diffusion coefficients. Note the difference in the diffusion coefficients for Na^+ and Cl^-, for example, in Table C.1. If one were to use NaCl in a similar arrangement, the junction potential would be $-0.012\,V$ or $-12\,m$ which corresponds to an error in the analyte concentration of 3% on an ISE experiment.

A simple 1 : 1 salt at different concentrations on the two sides of the boundary is the only combination that leads to a simple analytical solution of the Nernst–Planck equation for calculating the liquid junction potential. Even the situation where the junction between two phases share the same anion but have different cations is more difficult to solve. (The integral does not have an analytical solution.) Fortunately, Henderson introduced an equation (C.11) that yields an approximate solution for the junction potential for other combinations involving linear concentration gradients.

$$E_j = \frac{\sum_j z_j u_j (C_{j,s} - C_{j,r})}{\sum_j z_j^2 u_j (C_{j,s} - C_{j,r})} \left(\frac{RT}{F}\right) \ln \left\{ \frac{\sum_j z_j^2 u_j C_{j,s}}{\sum_j z_j^2 u_j C_{j,r}} \right\} \tag{C.11}$$

In Eq. (C.11), terms are summed for all ions, j, using their respective concentrations, $C_{j,r}$ and $C_{j,s}$, in the separate phases, r and s. The term u_j is the mobility of an ion. It is also called the electrophoretic mobility. It is the proportionality constant relating the velocity, v, that an ion reaches under the strength of an electric field, $\mathcal{E}$. Whenever an ion is accelerated the drag of the solution acts as a force in opposition to the electric force limiting the velocity of the ion. The ion mobility can be calculated by setting the forces equal to each other and solving for v/ε.

$$u_j = \frac{v}{\varepsilon} = \frac{\lfloor z_j \rfloor e}{6\pi \eta r} \tag{C.12}$$

In Eq. (C.12), e represents the charge on an electron, η is the viscosity of the solution, and r is the effective radius of the hydrated ion. Because data may be available for only the diffusion coefficient rather than the mobility of the ions of interest, it is helpful to be able to estimate one from the other. Here is Einstein's expression relating the electric mobility and the diffusion coefficient.

$$D_j = \left(\frac{RT}{F}\right)\left(\frac{u_j}{\lfloor z_j \rfloor}\right) \text{ or } u_j = \left(\frac{F}{RT}\right) \lfloor z_j \rfloor D_j \tag{C.13}$$

Thus, the Henderson equation can also be written using diffusion coefficients:

$$E_j = \frac{\sum_j z_j \lfloor z_j \rfloor D_j (C_{j,s} - C_{j,r})}{\sum_j z_j^2 \lfloor z_j \rfloor D_j (C_{j,s} - C_{j,r})} \left(\frac{RT}{F}\right) \ln \left\{ \frac{\sum_j z_j^2 \lfloor z_j \rfloor D_j C_{j,s}}{\sum_j z_j^2 \lfloor z_j \rfloor D_j C_{j,r}} \right\} \tag{C.14}$$

Here is an example calculation. Consider a reference bridge containing 1 M KCl and a sample solution containing 0.1 M HCl. From Table C.1, $D_{H^+} = 96.6 \times 10^{-6} \text{ cm}^2/\text{s}$, $D_{K^+} = 19.57 \times 10^{-6} \text{ cm}^2/\text{s}$, and $D_{Cl^-} = 20.32 \times 10^{-6} \text{ cm}^2/\text{s}$.

$$E_j = \frac{(96.6)(0.1 - 0) + 19.57)(0 - 1) + (-1)(20.32)(0.1 - 1)}{(96.6)(0.1 - 0) + 19.57)(0 - 1) + (1)(20.32)(0.1 - 1)} \left(\frac{8.314 \times 298}{96\,485}\right)$$

$$\times \ln \left\{ \frac{(96.6)(0.1) + (20.32)(0.1)}{19.57)(1) + (20.32)(1)} \right\}$$

$$= 0.008\,57 \text{ V or } 8.57 \text{ mV}$$

TABLE C.1   Diffusion coefficients for selected ions

Ion	Diffusion coefficient (cm²/s)
OH^-	52.73×10^{-6}
Na^+	13.34×10^{-6}
K^+	19.57×10^{-6}
SO_4^{2-}	10.65×10^{-6}
Ca^{2+}	7.92×10^{-6}
Cl^-	20.32×10^{-6}
Mg^{2+}	7.06×10^{-6}
H^{+a}	96.6×10^{-6}
NO_3^{-a}	19.46×10^{-6}

[a] Calculated from the electric mobility given by Bakker [1].
Source: From Bakker [1].

TABLE C.2   Junction potential as a function of [HCl] in the sample solution for a KCl salt bridge

Reference M KCl	Junction potentials (mV)			
	Sample = 0 M HCl	0.05 M HCl	0.1 M HCl	0.2 M HCl
1	−2.7	−5.7	−8.57	−12.5
2	−2.8	−3.8	−5.7	−8.57
3	−2.8	−3.1	−4.5	−6.8

Table C.2 shows the calculated junction potential for three different reference salt bridge solutions and different concentrations of HCl in the sample solution (Solutions varying in acid concentration represent situations where junction potentials are likely to vary). It is apparent that the junction potential for the more concentrated KCl solution is less influenced by the changes in the sample solution.

REFERENCE

1. Bakker, E. (2014). *Fundamentals of Electroanalysis: 1. Potentials and Transport*. Apple iBook Store (ebook) https://itunes.apple.com/us/book/fundamentals-electroanalysis-1-potentials-transport/id933624613?mt=11.

STANDARD ELECTRODE POTENTIALS FOR SOME SELECTED REDUCTION REACTIONS

APPENDIX

D

Half reaction	$E°$ versus standard hydrogen electrode (SHE) (V)
$Li^+_{(aq)} + e^- \rightleftarrows Li_{(s)}$	−3.045
$K^+_{(aq)} + e^- \rightleftarrows K_{(s)}$	−2.925
$Rb^+_{(aq)} + e^- \rightleftarrows Rb_{(s)}$	−2.925
$Cs^+_{(aq)} + e^- \rightleftarrows Cs_{(s)}$	−2.92
$Ba^{2+}_{(aq)} + 2e^- \rightleftarrows Ba_{(s)}$	−2.916
$Sr^{2+}_{(aq)} + 2e^- \rightleftarrows Sr_{(s)}$	−2.89
$Ca^{2+}_{(aq)} + 2e^- \rightleftarrows Ca_{(s)}$	−2.84
$Na^+_{(aq)} + e^- \rightleftarrows Na_{(s)}$	−2.714
$Mg^{2+}_{(aq)} + 2e^- \rightleftarrows Mg_{(s)}$	−2.356
$Ce^{3+}_{(aq)} + e^- \rightleftarrows Ce^{2+}_{(aq)}$	−2.34
$\frac{1}{2}H_{2(g)} + e^- \rightleftarrows H^-_{(aq)}$	−2.25
$Be^{2+}_{(aq)} + 2e^- \rightleftarrows Be_{(s)}$	−1.97
$Zr^{4+}_{(aq)} + 4e^- \rightleftarrows Zr_{(s)}$	−1.70
$Al^{3+}_{(aq)} + 3e^- \rightleftarrows Al_{(s)}$	−1.67
$Ti^{2+}_{(aq)} + 2e^- \rightleftarrows Ti_{(s)}$	−1.63
$Mn^{2+}_{(aq)}) + 2e^- \rightleftarrows Mn_{(s)}$	−1.18
$V^{2+}_{(aq)} + 2e^- \rightleftarrows V_{(s)}$	−1.13
$H_3BO_{3(aq)} + 3H^+_{(aq)} + 3e^- \rightleftarrows B_{(s)} + 3H_2O$	−0.890
$SiO_{2(vit)} + 4H^+_{(aq)} + 4e^- \rightleftarrows Si_{(s)} + 2H_2O$	−0.888
$TiO_{2(s,hydrated)} + 4H^+_{(aq)} + 4e^- \rightleftarrows Ti_{(s)} + 2H_2O$	−0.882
$Zn^{2+}_{(aq)} + 2e^- \rightleftarrows Zn_{(s)}$	−0.7626
$TlI_{(s)} + e^- \rightleftarrows Tl_{(s)} + I^-_{(aq)}$	−0.74
$TlCl_{(s)} + e^- \rightleftarrows Tl_{(s)} + Cl^-_{(aq)}$	−0.5568
$Sb_{(s)} + 3H^+_{(aq)} + 3e^- \rightleftarrows SbH_{3(g)}$	−0.510

(*Continued*)

Electroanalytical Chemistry: Principles, Best Practices, and Case Studies, First Edition. Gary A. Mabbott.
© 2020 John Wiley & Sons, Inc. Published 2020 by John Wiley & Sons, Inc.

Half reaction	$E°$ versus standard hydrogen electrode (SHE) (V)
$H_3PO_{2(aq)} + H^+_{(aq)} + e^- \rightleftarrows P_{(s,wht)} + 2H_2O$	−0.508
$H_3PO_{3(aq)} + 2H^+ + 2e^- \rightleftarrows H_3PO_{2(aq)} + H_2O$	−0.499
$Fe^{2+}_{(aq)} + 2e^- \rightleftarrows Fe_{(s)}$	−0.44
$Cr^{3+}_{(aq)} + e^- \rightleftarrows Cr^{2+}_{(aq)}$	−0.7626
$Cd^{2+}_{(aq)} + 2e^- \rightleftarrows Cd_{(s)}$	−0.4025
$Ti^{3+}_{(aq)} + e^- \rightleftarrows Ti^{2+}_{(aq)}$	−0.37
$PbI_{2(s)} + 2e^- \rightleftarrows Pb_{(s)} + 2I^-_{(aq)}$	−0.365
$PbSO_{4(s)} + 2e^- \rightleftarrows Pb_{(s)} + SO_4^{2-}_{(aq)}$	−0.3505
$Eu^{3+}_{(aq)} + e^- \rightleftarrows Eu^{2+}_{(aq)}$	−0.35
$In^{3+}_{(aq)} + 3e^- \rightleftarrows In_{(s)}$	−0.3382
$Tl^+_{(aq)} + e^- \rightleftarrows Tl_{(s)}$	−0.3363
$PbBr_{2(s)} + 2e^- \rightleftarrows Pb_{(s)} + 2Br^-_{(aq)}$	−0.280
$Co^{2+}_{(aq)} + 2e^- \rightleftarrows Co_{(s)}$	−0.277
$H_3PO_{4(aq)} + 2H^+_{(aq)} + 2e^- \rightleftarrows H_3PO_{3(aq)} + H_2O$	−0.276
$PbCl_{2(s)} + 2e^- \rightleftarrows Pb_{(s)} + 2Cl^-_{(aq)}$	−0.268
$Ni^{2+}_{(aq)} + e^- \rightleftarrows Ni_{(s)}$	−0.257
$V^{3+}_{(aq)} + e^- \rightleftarrows V^{2+}_{(aq)}$	−0.255
$SnF_6^{2-}_{(s)} + 4e^- \rightleftarrows Sn_{(s)} + 6F^-_{(aq)}$	−0.25
$N_{2(g)} + 5H^+_{(aq)} + 4e^- \rightleftarrows N_2H_5^+_{(aq)}$	−0.23
$As_{(s)} + 3H^+_{(aq)} + 3e^- \rightleftarrows AsH_{3(g)}$	−0.225
$Mo^{3+}_{(aq)} + 3e^- \rightleftarrows Mo_{(s)}$	−0.2
$CuI_{(s)} + e^- \rightleftarrows Cu_{(s)} + I^-_{(aq)}$	−0.182
$CuI_{(s)} + e^- \rightleftarrows Cu_{(s)} + I^-_{(aq)}$	−0.182
$CO_{2(g)} + 2H^+_{(aq)} + 2e^- \rightleftarrows HCOOH_{(aq)}$	−0.16
$AgI_{(s)} + e^- \rightleftarrows Ag_{(s)} + I^-_{(aq)}$	−0.152
$Si_{(s)} + 4H^+_{(aq)} + 4e^- \rightleftarrows SiH_{4(aq)}$	−0.143
$Sn^{2+}_{(aq)} + 2e^- \rightleftarrows Sn_{(s)}$	−0.136
$Pb^{2+}_{(aq)} + 2e^- \rightleftarrows Pb_{(s)}$	−0.125
$P_{(wht,s)} + 3H^+_{(aq)} + 3e^- \rightleftarrows PH_{3(g)}$	−0.063
$O_{2(g)} + H^+_{(aq)} + e^- \rightleftarrows HO_{2(aq)}$	−0.046
$Hg_2I_{2(s)} + 2e^- \rightleftarrows 2Hg_{(l)} + 2I^-_{(l)}$	−0.040
$Se_{(s)} + 2H^+_{(aq)} + 2e^- \rightleftarrows H_2Se_{(g)}$	−0.028
$2H^+_{(aq)} + 2e^- \rightleftarrows H_{2(g)}$	0.000
$CuBr_{(s)} + e^- \rightleftarrows Cu_{(s)} + Br^-_{(aq)}$	0.033
$HCOOH_{(aq)} + 2H^+_{(aq)} + 2e^- \rightleftarrows HCHO_{(aq)} + H_2O$	0.056
$AgBr_{(s)} + e^- \rightleftarrows Ag_{(s)} + Br^-_{(aq)}$	0.0711
$TiO^{2+}_{(aq)} + 2H^+_{(aq)} + e^- \rightleftarrows Ti^{3+}_{(aq)} + H_2O$	0.100
$CuCl_{(s)} + e^- \rightleftarrows Cu_{(s)} + Cl^-_{(aq)}$	0.121

(Continued)

Half reaction	$E°$ versus standard hydrogen electrode (SHE) (V)
$C_{(s)} + 4H^+_{(aq)} + 4e^- \rightleftarrows CH_{4(g)}$	0.132
$Hg_2Br_{2(s)} + 2e^- \rightleftarrows 2Hg_{(l)} + 2Br^-_{(aq)}$	0.1392
$S_{(s)} + 2H^+_{(aq)} + 2e^- \rightleftarrows H_2S_{(aq)}$	0.144
$Sn^{4+}_{(aq)} + 2e^- \rightleftarrows Sn^2_{(aq)}$	0.150
$SO_4^{2-}_{(aq)} + 4H^+_{(aq)} + 2e^- \rightleftarrows H_2SO_{3(aq)} + H_2O$	0.158
$Cu^{2+}_{(aq)} + e^- \rightleftarrows Cu^+_{(aq)}$	0.159
$BiOCl_{(s)} + 2H^+_{(aq)} + 3e^- \rightleftarrows Bi_{(s)} + H_2O + Cl^-_{(aq)}$	0.1697
$AgCl_{(s)} + e^- \rightleftarrows Ag_{(s)} + Cl^-_{(aq)}$	0.2223
$HCHO_{(aq)} + 2H^+_{(aq)} + 2e^- \rightleftarrows CH_3OH_{(aq)}$	0.232
$(CH_3)_2SO_{2(aq)} + 2H^+_{(aq)} + 2e^- \rightleftarrows (CH_3)_2SO_{(aq)} + H_2O$	0.238
$HAsO_{2(aq)} + 3H^+_{(aq)} + 3e^- \rightleftarrows As_{(s)} + 2H_2O_{(aq)}$	0.248
$SO_4^{2-}_{(aq)} + 8H^+_{(aq)} + 8e^- \rightleftarrows H_2S_{(aq)} + 3H_2O$	0.310[a]
$Cu^{2+}_{(aq)} + 2e^- \rightleftarrows Cu_{(s)}$	0.340
$AgIO_{3(s)} + e^- \rightleftarrows Ag_{(s)} + IO_3^-_{(aq)}$	0.354
$Fe(CN)_6^{3-}_{(aq)} + e^- \rightleftarrows Fe(CN)_6^{4-}_{(aq)}$	0.3610
$H_2SO_{3(aq)} + 4H^+_{(aq)} + 4e^- \rightleftarrows S_{(s)} + 3H_2O$	0.500
$2H_2SO_{3(aq)} + 4H^+_{(aq)} + 6e^- \rightleftarrows S_4O_6^{2-}_{(aq)} + 6H_2O$	0.507
$Cu^+_{(aq)} + e^- \rightleftarrows Cu_{(s)}$	0.520
$I_{2(aq)} + 2e^- \rightleftarrows 2I^-_{(aq)}$	0.5355
$I_3^-_{(aq)} + 2e^- \rightleftarrows 3I^-_{(aq)}$	0.536
$Cu^{2+}_{(aq)} + Cl^-_{(aq)} + e^- \rightleftarrows CuCl_{(s)}$	0.559
$H_3AsO_{4(aq)} + 2H^+_{(aq)} + 2e^- \rightleftarrows HAsO_{2(aq)} + 2H_2O_{(aq)}$	0.560
$MnO_4^-_{(aq)} + e^- \rightleftarrows MnO_4^{2-}_{(aq)}$	0.560
$S_2O_6^{2-}_{(aq)} + 4H^+_{(aq)} + 2e^- \rightleftarrows 2H_2SO_{3(aq)}$	0.569
$CH_3OH_{(aq)} + 2H^+_{(aq)} + 2e^- \rightleftarrows CH_{4(g)} + H_2O$	0.59
$Ag(CH_3CO_2)_{(aq)} + e^- \rightleftarrows Ag_{(s)} + CH_3CO_2^-_{(aq)}$	0.643
$Cu^{2+}_{(aq)} + Br^-_{(aq)} + e^- \rightleftarrows CuBr_{(s)}$	0.654
$Ag_2SO_{4(s)} + 2e^- \rightleftarrows 2Ag_{(s)} + SO_4^{2-}_{(aq)}$	0.654
$O_{2(g)} + 2H^+_{(aq)} + 2e^- \rightleftarrows H_2O_{2(aq)}$	0.695
$2NO_{(g)} + 2H^+_{(aq)} + 2e^- \rightleftarrows H_2N_2O_{2(aq)}$	0.71
$Fe^{3+}_{(aq)} + e^- \rightleftarrows Fe^{2+}_{(aq)}$	0.771
$Ag^+_{(aq)} + e^- \rightleftarrows Ag_{(s)}$	0.7991
$2NO_3^-_{(aq)} + 4H^+_{(aq)} + 2e^- \rightleftarrows N_2O_{4(g)} + 2H_2O$	0.803
$Cu^{2+}_{(aq)} + I^-_{(aq)} + e^- \rightleftarrows CuI_{(s)}$	0.861
$NO_3^-_{(aq)} + 3H^+_{(aq)} + e^- \rightleftarrows HNO_{2(aq)} + H_2O$	0.94
$NO_3^-_{(aq)} + 4H^+_{(aq)} + 3e^- \rightleftarrows NO_{(g)} + 2H_2O$	0.957
$PtO_{(s)} + 2H^+_{(aq)} + 2e^- \rightleftarrows Pt_{(s)} + H_2O$	0.98
$HNO_{2(aq)} + H^+_{(aq)} + e^- \rightleftarrows NO_{(g)} + H_2O$	0.996

(Continued)

Half reaction	$E°$ versus standard hydrogen electrode (SHE) (V)
$N_2O_{4(g)} + 4H^+_{(aq)} + 4e^- \rightleftarrows 2NO_{(g)} + 2H_2O$	1.039
$Br_{2(l)} + 2e^- \rightleftarrows 2Br^-_{(aq)}$	1.065
$ICl_2^-{}_{(aq)} + e^- \rightleftarrows 2Cl^-_{(aq)} + \frac{1}{2}I_{2(aq)}$	1.07
$Cu^{2+}_{(aq)} + 2CN^-_{(aq)} + e^- \rightleftarrows Cu(CN)_2^-{}_{(aq)}$	1.12
$H_2O_{2(aq)} + H^+_{(aq)} + e^- \rightleftarrows OH_{(aq)} + H_2O$	1.14
$ClO_3^-{}_{(aq)} + 3H^+_{(aq)} + 2e^- \rightleftarrows HClO_{2(aq)} + H_2O$	1.181
$ClO_{2(g)} + H^+_{(aq)} + e^- \rightleftarrows HClO_{2(aq)}$	1.188
$IO_3^-{}_{(aq)} + 6H^+_{(aq)} + 5e^- \rightleftarrows \frac{1}{2}I_{2(aq)} + 3H_2O$	1.195
$ClO_4^-{}_{(aq)} + 2H^+_{(aq)} + 2e^- \rightleftarrows ClO_3^-{}_{(aq)} + H_2O$	1.201
$O_{2(aq)} + 4H^+_{(aq)} + 4e^- \rightleftarrows 2H_2O_{(aq)}$	1.229
$MnO_{2(s)} + 4H^+_{(aq)} + 2e^- \rightleftarrows Mn^{2+}_{(aq)} + 2H_2O$	1.23
$N_2H_5^+{}_{(aq)} + 3H^+_{(aq)} + 2e^- \rightleftarrows 2NH_4^+{}_{(aq)}$	1.275
$2HNO_{2(aq)} + 4H^+_{(aq)} + 4e^- \rightleftarrows N_2O_{(g)} + 3H_2O$	1.297
$NH_3OH^+{}_{(aq)} + 2H^+_{(aq)} + 2e^- \rightleftarrows NH_4^+{}_{(aq)} + H_2O$	1.35
$Cl_{2(g)} + 2e^- \rightleftarrows 2Cl^-_{(aq)}$	1.3583
$Cr_2O_7^{2-}{}_{(aq)} + 14H^+_{(aq)} + 6e^- \rightleftarrows 2Cr^{3+}_{(aq)} + 2H_2O$	1.36
$2NH_3OH^+{}_{(aq)} + H^+_{(aq)} + 2e^- \rightleftarrows N_2H_5^+{}_{(aq)} + 2H_2O$	1.41
$HO_{2(aq)} + H^+_{(aq)} + e^- \rightleftarrows H_2O_{2(aq)}$	1.44
$PbO_{2(s)} + 4H^+_{(aq)} + 2e^- \rightleftarrows Pb^{2+}_{(aq)} + 2H_2O$	1.468
$BrO_3^-{}_{(aq)} + 6H^+_{(aq)} + 5e^- \rightleftarrows \frac{1}{2}Br_{2(aq)} + 3H_2O$	1.478
$Mn^{3+}_{(aq)} + e^- \rightleftarrows Mn^{2+}_{(aq)}$	1.5
$Au^{3+}_{(aq)} + 3e^- \rightleftarrows Au_{(s)}$	1.52
$H_5IO_{6(aq)} + H^+_{(aq)} + 2e^- \rightleftarrows IO_3^-{}_{(aq)} + 3H_2O$	1.603
$HClO_{(aq)} + H^+_{(aq)} + e^- \rightleftarrows \frac{1}{2}Cl_{2(aq)} + H_2O$	1.63
$MnO_4^-{}_{(aq)} + 4H^+_{(aq)} + 3e^- \rightleftarrows MnO_{2(s)} + 2H_2O$	1.70
$Ce^{4+}_{(aq)} + e^- \rightleftarrows Ce^{3+}_{(aq)}$	1.72
$H_2O_{2(aq)} + 2H^+_{(aq)} + 2e^- \rightleftarrows 2H_2O$	1.763
$Co^{3+}_{(aq)} + e^- \rightleftarrows Co^{2+}_{(aq)}$	1.92
$O_{3(g)} + 2H^+_{(aq)} + 2e^- \rightleftarrows O_{2(g)} + H_2O$	2.075
$OH_{(aq)} + H^+_{(aq)} + e^- \rightleftarrows H_2O$	2.38
$F_{2(g)} + 2H^+_{(aq)} + 2e^- \rightleftarrows 2HF_{(aq)}$	3.053

Selected half-reactions from Ref. [1].
[a]Calculated value from Ref. [2].

REFERENCES

1. Bard, A.J., Parsons, R., and Jordan, J. (1985). *Standard Potentials in Aqueous Solution, International Union of Pure and Applied Chemistry*. New York: Marcel Dekker.

2. Stumm, W. and Morgan, J.J. (1996). *Aquatic Chemistry: Chemical Equilibria and Rates in Natural Waters*, 3e, 465. New York: Wiley.

THE NERNST EQUATION FROM THE CONCEPT OF ELECTROCHEMICAL POTENTIAL[1]

APPENDIX

E

One can obtain a quantitative relationship between the activity of the ions in solution and the potential that develops at the electrode/solution interface. At equilibrium, the change in free energy with respect to changes in reactants and products is zero ($\Delta G_{rxn} = 0$). The free energy of formation for various species can often be found in the literature. These energies describe the system at standard state (a reference point for temperature, pressure, and activities of reactants and products). The standard conditions in most electrochemical work refer to 298.15 K, 1.00 atm pressure for gaseous species, and unit activities of each reactant and product. When the activities are expressed in molar quantities, the free energy of a reactant or product is also called the chemical potential, μ, for that species. For example, the chemical potential for the Fe^{3+} ion in an aqueous solution is given by the following:

$$\mu_{Fe^{3+}} = \mu^o_{Fe^{3+}} + RT \ln(a_{Fe^{3+}}) \tag{E.1}$$

In Eq. (2.2), R is the universal gas constant, 8.314 J/(K mol); T is the temperature in Kelvin. The activity, $a_{Fe^{3+}}$, is the product of the concentration, c, and the activity coefficient, $\gamma_{Fe^{+3}}$.

$$a_i = \gamma_i c_i \tag{E.2}$$

When the system is at the standard state, the activity of the Fe^{3+} is 1 M and the chemical potential, $\mu_{Fe^{3+}}$, is equal to the chemical potential for the standard state, $\mu^o_{Fe^{3+}}$. The term $RT\ln(a_{Fe^{+3}})$ adjusts for conditions that are not at standard state. The activity is the effective reaction concentration of the species. The presence of other ions in solution effects the behavior or the Fe^{3+}. The activity coefficient, γ, accounts for this effect. Other ions in the neighborhood tend to screen the electric field of a reactant species decreasing its ability to interact with other species that would, otherwise, exchange an electron or a proton or react in some other way had the interfering ions had not been there. Generally, the activity of an ion decreases as the concentration of the surrounding electrolyte increases. (A discussion of activity and activity coefficients appears in Appendix A at the end of this book.)

Consider the one-electron reduction of Fe^{+3} to Fe^{+2} at a Pt electrode. Because the Fe^{3+} ion carries a charge, a complete representation of the free energy associated with the ion

[1]Source: Bakker [1].

Electroanalytical Chemistry: Principles, Best Practices, and Case Studies, First Edition. Gary A. Mabbott.
© 2020 John Wiley & Sons, Inc. Published 2020 by John Wiley & Sons, Inc.

should also include the work done to bring this charge into the aqueous solution. The electrochemical potential, $\bar{\mu}$, of an ion is the sum of the chemical potential and the electrical work required to bring into the medium the charge accompanying a mole of ions.

$$\bar{\mu} = \text{electrochemical potential} = \text{chemical potential} + \text{electrical work done/mole of ions} \tag{E.3}$$

For Fe^{3+} ions, the electrochemical potential is given here:

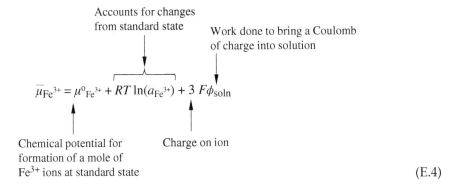

$$\bar{\mu}_{Fe^{3+}} = \mu^{o}_{Fe^{3+}} + RT\ln(a_{Fe^{3+}}) + 3\,F\phi_{soln} \tag{E.4}$$

The term ϕ_{soln} is used to represent the work required to bring a coulomb of charge from outer space into the solution (in volts, or joules per coulomb). Because the other energy terms are expressed in joules per mole, ϕ_{soln} must be multiplied by Faraday's constant, F (96 485 C/mol) to give units of J/mol. The same term is also multiplied by the charge on the ion, z, to give the work done for inserting a mole of ions with three charges/ion into the medium.

The strategy here is to write out the electrochemical potential for each reactant and product species. Then, because this is an equilibrium situation

$$(\Delta G_{rxn} = 0 = \bar{\mu}_{products} - \bar{\mu}_{reactants}) \tag{E.5}$$

the sum of the electrochemical potentials for the products must be equal to that of the reactants.

$$\bar{\mu}_{products} = \bar{\mu}_{reactants} \tag{E.6}$$

The electrochemical potential for the electron can be written as follows:

$$\bar{\mu}_{e(Pt)} = \mu^{o}_{e(Pt)} + RT\ln(a_{e(Pt)}) - F\phi_{Pt} \tag{E.7}$$

For Fe^{2+} ions, the electrochemical potential is

$$\bar{\mu}_{Fe^{2+}} = \mu^{o}_{Fe^{2+}} + RT\ln(a_{Fe^{2+}}) + 2F\phi_{soln} \tag{E.8}$$

Now, one can set $\overline{\mu}_{\text{products}} = \overline{\mu}_{\text{reactants}}$

$$\overline{\mu}_{Fe^{2+}} = \overline{\mu}_{Fe^{3+}} + \overline{\mu}_{e(Pt)} \tag{E.9}$$

$$\mu^o_{Fe^{2+}} + RT\ln(a_{Fe^{2+}}) + 2F\phi_{\text{soln}} = \mu^o_{e(Pt)} + RT\ln(a_{e(Pt)}) - F\phi_{Pt} + \mu^o_{Fe^{3+}} + RT\ln(a_{Fe^{3+}}) + 3F\phi_{\text{soln}} \tag{E.10}$$

The objective is to obtain an expression for the electric potential energy difference $\phi_{Pt} - \phi_{\text{soln}}$, so, it is helpful to group all of the ϕ terms on the left and everything else on the right side.

$$F\phi_{Pt} - F\phi_{\text{soln}} = \mu^o_{e(Pt)} + RT\ln(a_{e(Pt)}) + \mu^o_{Fe^{3+}} - \mu^o_{Fe^{2+}} + RT\ln\left\{\frac{(a_{Fe^{3+}})}{(a_{Fe^{2+}})}\right\} \tag{E.11}$$

$$\phi_{Pt} - \phi_{\text{soln}} = \underbrace{\frac{\mu^o_{e(Pt)} + RT\ln(a_{e(Pt)}) + \mu^o_{Fe^{3+}} - \mu^o_{Fe^{2+}}}{F}}_{\text{This is a constant}\, =\, E^o_{Fe^{3+}/Fe^{2+}}} - \frac{RT}{F}\ln\left[\frac{(a_{Fe^{2+}})}{(a_{Fe^{3+}})}\right] \tag{E.12}$$

It can be argued that the number of electrons that are transferred to or from the platinum will be a negligible fraction of the total electrons available in the conduction band of the metal. Consequently, the activity of the electron in the platinum is essentially constant, and it can be grouped together with other terms that are constant by definition to create a new constant, $E^o_{Fe^{3+}/Fe^{2+}}$. This new constant is known as the standard electrode potential (or the standard reduction potential) for the Fe^{3+}/Fe^{2+} electron transfer reaction. (Reaction equations in which electrons appear as a reactant or product are incomplete, since free electrons are not available as reagents, and the source of the electrons is not given. Nevertheless, the practice of conceptualizing the process in this form is very useful. Reaction equations in which electrons appear as a reactant or product are called "half-reactions.") The term on the left-hand side of Eq. (E.12) is the difference in the electric potential between the solution and the platinum wire. It represents the voltage drop created by the separation of charge that appears at the interface (the electrical double layer). It is also apparent that this voltage is a characteristic of the type of electroactive species in solution. The electrode voltage is known as the redox potential of the solution, $E_{\text{soln}} = \phi_{Pt} - \phi_{\text{soln}}$, and is dependent on the activities of the Fe^{3+} and Fe^{2+} ions in solution. The equation becomes

$$E_{\text{soln}} = E^o_{Fe^{2+}/Fe^{3+}} - \frac{RT}{F}\ln\left\{\frac{(a_{Fe^{2+}})}{(a_{Fe^{3+}})}\right\} \tag{E.13}$$

This format for the relationship between the redox potential and activities of redox components is known as the Nernst equation after Walther Nernst who introduced it in 1887. In writing Eq. (E.13) from the expression in (E.12), the ratio of ion activities was inverted, and the sign in front of the logarithm term was changed in order to compensate. Of course, the two versions are equivalent, but it may be a little easier to remember the latter form. Equation (E.13) uses the ratio of activities for the products over the reactants as is the convention for defining equilibrium constants. Because the convention is to write a half-reaction as a reduction process (electrons appear on the reactant side of the arrows), one can also write the ratio of activities for products (the reduced form) over reactants (the oxidized form) in the Nernst equation, if a negative sign is used in front of the logarithm term. In other words, each species on the reduced side of the equation appears in the numerator, and a negative sign appears before the coefficient. A general procedure for formulating the equation for the redox potential for any balanced electron transfer reaction can be stated as follows. Consider the reduction of a moles of an oxidized species, Ox, by n electrons to form b moles of the reduced product, Red

$$a \, \text{Ox} + n \, \text{e} \; \rightleftarrows b \, \text{Red} \tag{E.14}$$

The Nernst equation for the redox potential of a solution containing a mixture of the reactants and products can be written as:

$$E_{\text{soln}} = E^{\text{o}}_{\text{Ox/Red}} - \frac{RT}{nF} \ln \frac{(a_{\text{Red}})^b}{(a_{\text{Ox}})^a} \tag{E.15}$$

If one can write the Nernst equation easily using a direct procedure such as just demonstrated, then why spend time discussing electrochemical potentials? One reason is that working with electrochemical potentials is a more general approach. It covers all processes that develop a voltage at the interface between two media. The shortened rubric for writing the Nernst equation would appear to work only for electron transfer reactions. Oxidation–reduction reactions are merely a subset of charge transfer processes that are important to analytical chemistry and other fields. Ion transfer reactions are another. For example the surface of the special glass in a pH electrode develops a potential because hydrogen ions are adsorbed onto the glass from the sample solution; electrons are not a part of that reaction equation. However, starting with the electrochemical potentials for the participating ions leads to the Nernst equation, if certain assumptions are made. Furthermore, it is prudent to know what those assumptions are that lead to the simplified expressions that one might be using.

REFERENCE

1. Bakker, E. (2014) *Fundamentals of Electroanalysis: 1. Potentials and Transport*. Apple iBook Store. https://itunes.apple.com/us/book/fundamentals-electroanalysis-1-potentials-transport/ id933624613?mt=11 (accessed 9 June 2019).

SOLUTIONS TO PROBLEMS

1.1 Faraday's law:

$$Q = nFN = (2)(96\ 485\ \text{C/mol})(450\ \text{g})/(63.546\ \text{g/mol})$$

$$= 1.37 \times 10^6\ \text{C}$$

$$i_{\text{avg}} = Q/t = 1.37 \times 10^6\ \text{C}/(42\text{d} \times 24\ \text{h/d} \times 60\ \text{min/h} \times 60\ \text{s/min}) = 0.377\ \text{C/s}$$

$$= 0.377\ \text{A}$$

1.2 $Q = CV \rightarrow V = Q/C$

$$Q = (1.0 \times 10^{12}/6.02 \times 10^{23}\ \text{electrons/mol})(96\ 485\ \text{C/mol}) = 1.60 \times 10^{-7}\ \text{C}$$

$$C = (20 \times 10^{-6}\ \text{C/V/cm}^2)(\pi)(0.3\ \text{cm}^2) = 5.65 \times 10^{-6}\ \text{C/V}$$

$$V = 1.60 \times 10^{-7}\ \text{C}/5.65 \times 10^{-6}\ \text{C/V} = 0.0283\ \text{V}$$

1.3 Since there is a concentration gradient, there will be a junction potential. Nitrate ions will lead the sodium ions in diffusing from the reference solution toward the sample solution. Therefore, the solution side of the salt bridge will be more negative than the reference side.

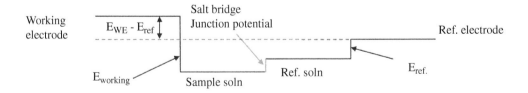

These conditions make the working electrode potential appear less positive than it would without the junction potential.

Electroanalytical Chemistry: Principles, Best Practices, and Case Studies, First Edition. Gary A. Mabbott.
© 2020 John Wiley & Sons, Inc. Published 2020 by John Wiley & Sons, Inc.

1.4   Ohmic loss $= iR_{soln}$. Since the current in a potentiometric experiment is ideally zero, the ohmic loss is zero.

1.5   Ohmic loss $= iR_{soln} = (250 \times 10^{-6}\,A)(152\,\Omega) = 0.038\,V$

1.6   (a) One effect of the addition of KCl to the solution is a change in the ionic strength and, therefore, the activity coefficient of the H^+ ion.
(b) A second effect of the additional KCl is a change in the concentration gradient for various ions across the salt bridge, and, therefore a change in the junction potential.

1.7
$$\Delta Q = C\Delta V; \Delta V = -0.500 - (-0.100) = -0.400\,V$$
$$C = (\pi)(0.1\,cm)^2(24 \times 10^{-6}\,F/cm^2) = 7.53_9 \times 10^{-7}\,C/V$$
$$\Delta Q = (-0.400\,V)(7.53_9 \times 10^{-7}\,C/V) = 3.02 \times 10^{-7}\,C$$
$$Q = nFN \text{ or } \Delta N = \Delta Q/(nF) = (3.02 \times 10^{-7}\,C)/(1 \times 96\,485\,C/mol)$$
$$= 3.12 \times 10^{-12}\,mol$$

1.8   Recall that $R = N_A k$, where $N_A = 6.022 \times 10^{23}$ molecules/mol.

$$R = 8.314\,J/(K\,mol)$$

then, $k = R/N_A = (8.314\,J/(K/mol))/6.022 \times 10^{23}$ molecules/mol $= 1.38 \times 10^{-23}\,J/$ (K molecule)
At 298 K, $(3/2)\,kT = (3/2)(1.38 \times 10^{-23}\,J/(K\,molecules))(298) = 6.17 \times 10^{-21}\,J/$ molecule

$$1\,eV = 1.60 \times 10^{-19}\,J \rightarrow 1\,eV/\{(3/2)kT\} = 1.60 \times 10^{-19}\,J/6.17 \times 10^{-21}\,J = 25.9$$

1 eV is about 26 times larger than the average thermal energy of a molecule at 25 °C.

1.9
$$1\,eV = 1.60 \times 10^{-19}\,J$$

$$E_{photon} = h\nu = hc/\lambda = (6.62 \times 10^{-34}\,J\,s)(3.00 \times 10^8\,m/s)/(400 \times 10^{-9}\,m)$$
$$= 4.96_5 \times 10^{-19}\,J$$

Therefore, a blue photon has about three times as much energy as 1 eV.

1.10
$$1\,eV = 1.60 \times 10^{-19}\,J$$

The bond-dissociation energy for the carbon–carbon bond is 90 kcal/mol

$$E_{C-C\,bond} = (90\,kcal/mol)(4.184\,kJ/kcal)(1000\,J/kJ)/6.022 \times 10^{23}\,molecules/mol$$
$$= 6.25 \times 10^{-19}\,J/molecules$$

So, a C—C bond in ethane is about 4 eV.

2.1

$$E = E^{\mathrm{o}}_{\frac{\mathrm{AgCl}}{\mathrm{Ag}}} - \frac{RT}{F} ln\{a_{\mathrm{Cl}^-}\} = 0.222 - 0.05\,916\ \log(0.10) = 0.281\ \mathrm{V}$$

2.2 $H_2O_2 + 2e^- + 2H^+ \rightleftharpoons 2H_2O$ Le Chatelier's principle indicates that the forward direction would be more favorable at higher $[H^+]$ or lower pH. How much more favorable can be calculated from the Nernst equation for this process.

$$E_{\mathrm{soln}} = E^{\mathrm{o}} - \frac{2.303RT}{nF} \log\left\{\frac{a_{\mathrm{Red}}}{a_{\mathrm{Ox}}}\right\} = E^{\mathrm{o}} - \frac{2.303RT}{2F} \log\left\{\frac{1}{a_{H_2O_2} a_{H^+}^2}\right\}$$

$$= E^{\mathrm{o}} - \frac{2.303RT}{F}\mathrm{pH} - \frac{2.303RT}{2F} \log\left\{\frac{1}{a_{H_2O_2}}\right\}$$

The potential, E_{soln} of the system becomes more positive (indicating more favorable for the reactants to oxidize another reactant by removing electrons) as the pH goes down at a rate of $2.303\,RT/F$ or $\sim 59\ \mathrm{mV/pH}$ unit.

2.3 The voltage spent as current passes through the meter is given by Ohm's law: $V = iR$, where R is the resistance in the meter. Therefore,

$$R_{\mathrm{max}} = \frac{V_{\mathrm{error}}}{i} = \frac{0.05 \times 10^{-3}\ \mathrm{V}}{1 \times 10^{-6}} = 50\ \Omega$$

2.4

$$\mathrm{No.of\ distinct\ settings} = (360^\circ/\mathrm{turn})(10\ \mathrm{turns}) = 3600\ \mathrm{values}$$

$$\mathrm{Precision} = 1.5000\ \mathrm{V}/3600\ \mathrm{values} = 0.000\,416\ \mathrm{V/step}$$

2.5 For 0.50 M activity:

$$E_1 = E^{\mathrm{o}}_{\mathrm{AgCl/Ag}} - \frac{RT}{F}\ ln\{a_{\mathrm{Cl}^-}\} = 0.222 - 0.05\,916 \log(0.50) = 0.239_8$$

Calculate the chloride activity for a 0.5 M concentration of KCl. $a_{\mathrm{cl}^-} = (\gamma_{\mathrm{cl}^-})(c_{\mathrm{cl}^-})$. To find the activity coefficient, γ_{cl^-}, apply the Davies equation

$$\log(\gamma_{\mathrm{cl}^-}) = \frac{-0.511z_i^2\sqrt{\mu}}{1 + 1.5\sqrt{\mu}} + 0.2z_i^2\mu$$

Since the electrolyte is a 1 : 1 salt with singly charged ions, $\mu = C_{\mathrm{KCl}} = 0.5$

$$\log(\gamma_{\mathrm{cl}^-}) = \frac{-0.511(1)\sqrt{0.5}}{1 + 1.5\sqrt{0.5}} + 0.2(1)(0.5) = -0.0753$$

$$\gamma_{cl^-} = 10^{-0.0753} = 0.841 \text{ and } a_{cl^-} = (\gamma_{cl^-})(c_{cl^-}) = (0.841)(0.5) = 0.421 \text{ M}$$

$$E_2 = E^o_{AgCl/Ag} - \frac{RT}{F} \ln\{a_{cl^-}\} = 0.222 - 0.05\,916 \log(0.421) = 0.244$$

2.6 (a) $E_{soln} = 1.70 - \frac{RT}{3F} \ln \left\{ \frac{1}{a_{MnO_4}(a_{H^+})^4} \right\}$

Note that the activity for solids and the solvent, water, are unity.

(a) $E^{o\prime} = 1.70 - \frac{RT}{3F} \ln \left\{ \frac{1}{(a_{H^+})^4} \right\} = 1.70 - \frac{RT}{3F}(2.303)(4)\text{pH}$

$= 1.70 - \frac{(8.314)(298)}{3(96\,485)}(2.303)(4)(7.00) = 1.14_8 \text{ V}$

2.7 Deep-well water does not have an opportunity to mix with oxygen in the air. When the bacteria in a water column consume all the oxygen, other redox species are sought by anaerobic bacteria for metabolizing food. Ferric ion is one of the stronger options. Hydrogen ions produced as a by-product of the oxidation of organic material also help dissolve the iron(III)hydroxide.

$$Fe(OH) + 3H^+ \rightleftharpoons Fe^{+3} + 3H_2O$$

Then, the ferric ion acts as an oxidizer:

$$nFe^{+3} + Red \rightleftharpoons nFe^{+2} + Ox$$

where Red represents an organic molecule that can be oxidized to Ox in the process of losing n electrons.

Fe(II)-rich water brought to the surface is oxidized by oxygen from the air.

$$4Fe^{+2} + O_2 + 2H_2O \rightleftharpoons 4Fe^{+3} + 4OH^-$$

Then, the ferric ion and hydroxide precipitate.

$$Fe^{+3} + 3OH^- \rightleftharpoons Fe(OH)_3$$

2.8 $Sn^{4+} + 2e^- \rightleftharpoons Sn^2, E^o = 0.150; Cr_2O_7^{2-} + 14H^+ + 6e^- \rightleftharpoons 2Cr^{3+} + 2H_2O, E^o = 1.36$

(a) $E_{endpt.} = \frac{n_s E^{o\prime}_s + n_t E^{o\prime}_t}{n_s + n_t} = \frac{2(0.150) + (6)(1.36)}{2 + 6} = 1.057 \text{ V}$. The dichromate reaction is very pH sensitive, but 0.5 M H_2SO_4 is 1 M in H^+, placing this at standard conditions for the half-reaction.

(b) $E_{Cr_2O_7} = 1.36 - (2.303)\frac{(8.314)(298)}{6(96\,485)} \log \left\{ \frac{(a_{Cr^{+3}})^2}{(a_{Cr_2O_7})(a_{H^+})^{14}} \right\}$ and at 0.05 M H_2SO_4, $[H^+]$

$= 0.1 \text{ M or pH} = 1$

$$E^{o\prime}_{Cr_2O_7} = 1.36 - (2.303)\frac{(8.314)(298)}{6(96\,485)} \log\left\{\frac{1}{(a_{H^+})^{14}}\right\} = 1.36 - \frac{0.05\,916}{6}(14)\text{pH}$$

$$= 1.36 - \frac{0.05\,916}{6}(14)(1) = 1.22_1\ \text{V}$$

$$E^\prime_{endpt.} = \frac{n_s E^{o\prime}_s + n_t E^{o\prime}_t}{n_s + n_t} = \frac{2(0.150) + (6)(1.22)}{2+6} = 0.954\ \text{V}$$

2.9

$$Hg_2Cl_2 + 2e^- \rightleftarrows 2Hg + 2Cl^-$$

$$E_{soln} = E^o - \frac{2.303RT}{nF} \log\{a_{Cl^-}\}^2 = E^o - \frac{2.303RT}{nF} \log\{(\gamma_{Cl^-})(c_{Cl^-})\}^2$$

(a) Temperature appears in the coefficient of the log-term.
(b) Temperature affects the activity coefficient
(c) Temperature affects the diffusion coefficients of the electrolyte ions and, indirectly, affects the junction potential.
(d) The standard potential, E^o also depends on temperature through the underlying equilibria.

$$Hg_2^{+2} + 2e^- \rightleftharpoons Hg_{(l)}$$

$K_{sp} \Big\uparrow\Big\downarrow Cl^-$

$$Hg_2Cl_2$$

Ksp is dependent on T.
The entropy increases in
dissolving the solid, so
lower T favors the solid,
shifting $E^{o\prime}$ in negative direction.

$$\Delta G^o = \Delta H - T\Delta S = -nFE^o = -RT\ \ln K$$

2.10

(a) $E_{NH_3OH^+} = 1.35 - (2.303)\frac{(8.314)(298)}{2(96\,485)} \log\left\{\frac{(a_{NH_4^+})}{(a_{NH_3OH^+})(a_{H^+})^2}\right\}$

(b) $E^{o\prime}_{NH_3OH^+} = 1.35 - (2.303)\frac{(8.314)(298)}{2(96\,485)} \log\left\{\frac{1}{(a_{H^+})^2}\right\} = 1.35 - \frac{0.05\,916}{2}(2)\text{pH}$

$$= 1.35 - (0.05\,916)(2) = 1.23_1\ \text{V}$$

2.11

$$E_1 = E^o_{AgCl/Ag} - \frac{RT}{F} \ln\{a_{Cl^-}\} = 0.222 - 0.05\,916 \log(0.10) = 0.281$$

Test soln.		$E^{o\prime} = E^o_{AgCl/Ag} - \frac{RT}{F} \ln\{a_{Cl^-}\}$
	$\updownarrow$ 0.150 $\downarrow E_2$	$= 0.222 - 0.05916 \log(0.10) = 0.281$
Ag/AgCl, 0.1 M KCl 0.281		
Saturated calomel, SCE 0.242		$E_2 = 0.150 - (0.281 - 0.242)$
		$= 0.111\ \text{V}$
SHE 0.000		

2.12
$$BrO_3^- + 6H^+ + 5e^- \rightleftarrows {}^{1\!/\!2}\,Br_2 + 3H_2O, E^\circ = 1.478\ V$$

$$E_{soln} = E^\circ_{\frac{BrO_3^-}{Br_2}} - \frac{0.05\,916}{5}\log\left\{\frac{(a_{Br_2})^{1/2}}{(a_{BrO_3^-})(a_{H^+})^6}\right\}$$

$$= 1.478 - \frac{0.05\,916}{5}\log\left\{\frac{(1.65\times10^{-3})^{1/2}}{(7.53\times10^{-4})(0.1)^6}\right\} = 1.38_6$$

2.13   The ionic strength, $\mu = 0.5\,\Sigma c_i(z_i)^2 = 0.5\{[H^+](1)^2 + [Cl^-](-1)^2\{[Na^+](1)^2 + [BrO_3^-](-1)^2\}$

$$\mu = 0.5\sum c_i(z_i)^2 = 0.5\{[H^+](1)^2 + [Cl^-](-1)^2 + [Na^+](1)^2 + [BrO_3^-](-1)^2\}$$

$$= 0.5\{[0.1](1)^2 + [0.1](-1)^2 + [0.00\,165](1)^2 + [0.00\,165](-1)^2\} = 0.102\ M$$

We need to calculate the activity coefficients for BrO_3^- and H^+.

$\log \gamma_i = \frac{-0.511 z_i^2\sqrt{\mu}}{1+0.328a\sqrt{\mu}}$, where a hydrated radius in angstroms for the ion. $a_{H^+} = 9$ and $a_{BrO_3^-} = 3.5$

$$\log \gamma_{H^+} = \frac{-0.511 z_i^2\sqrt{\mu}}{1+Ba\sqrt{\mu}} = \frac{-(0.511)\sqrt{0.101}}{1+(0.328)(9)\sqrt{0.101}} = 0.671;$$

$$\log \gamma_{BrO_3^-} = \frac{-0.511 z_i^2\sqrt{\mu}}{1+Ba\sqrt{\mu}} = \frac{-(0.511)\sqrt{0.101}}{1+(0.328)(6)\sqrt{0.101}} = 0.550$$

$$a_{H^+} = (\gamma_{H^+})(0.1) = (0.671)(0.1) = 0.0671\ M$$

$$a_{BrO_3^-} = (\gamma_{BrO_3^-})(BrO_3^-) = (0.550)(7.53\times10^{-4}) = 4.14\times10^{-4}\ M$$

$$E_{soln} = E^\circ_{BrO_3^-/Br_2} - \frac{0.05\,916}{5}\log\left\{\frac{(a_{Br_2})^{1/2}}{(a_{BrO_3^-})(a_{H^+})^6}\right\}$$

$$= 1.478 - \frac{0.05\,916}{5}\log\left\{\frac{(1.65\times10^{-3})^{1/2}}{(4.14\times10^{-4})(0.0671)^6}\right\} = 1.37_1$$

2.14   There is no concentration gradient for the chloride ion, so it does not contribute to the junction potential. For K^+ ions, the concentration is high in the reference solution and zero in the sample. That will lead to a net positive charge on the sample side of the salt bridge. The reverse is true of the Na^+ ion. The K^+ and Na^+ oppose each other. Because the diffusion coefficient for the K^+ is $19.57\times10^{-6}\ cm^2/s$ and the Na^+ is $13.34\times10^{-6}\ cm^2/s$, the potassium will give the salt bridge a net positive charge on the sample side.

2.15 From Eq. (C.14), $E_j = \dfrac{\sum_j z_j |z_j| D_j(C_{j,s}-C_{j,r})}{\sum_j z_j^2 |z_j| D_j(C_{j,s}-C_{j,r})}\left(\dfrac{RT}{F}\right)\ln\left\{\dfrac{\sum_j z_j^2 |z_j| D_j C_{j,s}}{\sum_j z_j^2 |z_j| D_j C_{j,r}}\right\} = \left(\dfrac{G}{H}\right)\left(\dfrac{RT}{F}\right)\ln\left\{\dfrac{M}{N}\right\}$,

where

$$G = \sum_j z_j |z_j| D_j(C_{j,s}-C_{j,r}); H = \sum_j z_j^2 |z_j| D_j(C_{j,s}-C_{j,r}); M = \sum_j z_j^2 |z_j| D_j C_{j,s}; N = \sum_j z_j^2 |z_j| D_j C_{j,r}$$

$C_{K,r} = C_{Cl,r} = 1; C_{K,s} = C_{Cl,s} = 0; C_{Na,s} = C_{OH,s} = 0.1;$

$D_K = 19.57 \times 10^{-6}\ D_{Cl} = 20.32 \times 10^{-6}\ D_{Na} = 13.34 \times 10^{-6}\ D_{OH^-} = 52.73 \times 10^{-6}\ \text{cm}^2/\text{s}$

$$G = \sum_j z_j |z_j| D_j(C_{j,s}-C_{j,r})$$

$\quad = (1)D_K(0-1) + (-1)D_{Cl}(0-1) + (1)D_{Na}(0.1-0) + (-1)D_{OH}(0.1-0)$

$\quad = (1)(19.57 \times 10^{-6})(-1) + (-1)(20.32 \times 10^{-6})(-1)$

$\qquad + (13.34 \times 10^{-6})(0.1) - 52.73 \times 10^{-6}(0.1)$

$\quad = -3.19 \times 10^{-6}$

$$H = \sum_j z_j^2 |z_j| D_j(C_{j,s}-C_{j,r})$$

$\quad = (1)D_K(0-1) + (1)D_{Cl}(0-1) + (1)D_{Na}(0.1-0) + (1)D_{OH}(0.1-0)$

$\quad = (19.57 \times 10^{-6})(-1) + (20.32 \times 10^{-6})(-1) + (13.34 \times 10^{-6})(0.1) + 52.73 \times 10^{-6}(0.1)$

$\quad = -3.33 \times 10^{-5}$

$$M = \sum_j z_j^2 |z_j| D_j C_{j,r} = (1)D_K(1) + (1)D_{Cl}(1) + (1)D_{Na}(0) + (1)D_{OH}(0)$$

$\quad = (19.57 \times 10^{-6})(1) + (20.32 \times 10^{-6})(1) = 3.99 \times 10^{-5}$

$$N = \sum_j z_j^2 |z_j| D_j C_{j,s} = (1)D_K(0) + (1)D_{Cl}(0) + (1)D_{Na}(0.1) + (1)D_{OH}(0.1)$$

$\quad = (13.34 \times 10^{-6})(0.1) + (52.73 \times 10^{-6})(0.1) = 6.61 \times 10^{-6}$

$$E_j = \left(\dfrac{G}{H}\right)\left(\dfrac{RT}{F}\right)ln\left\{\dfrac{M}{N}\right\} = \left(\dfrac{-3.19 \times 10^{-6}}{-3.33 \times 10^{-5}}\right)\left(\dfrac{8.314 \times 298}{96\,485}\right)ln\left\{\dfrac{3.99 \times 10^{-5}}{6.61 \times 10^{-6}}\right\}$$

$\quad = -0.00\,443\,\text{V} = -4.43\,\text{mV}$

3.1 In general $E_{cell} = E_{cnst} + \dfrac{0.05\,916}{2}\log[Ca^{2+}]$

$E_1 = 0.170\,\text{V}; E_2 = 0.193\,\text{V}; E_2 - E_1 = 0.193 - 0.170 = 0.023\,\text{V}$ also

$$E_2 - E_1 = \dfrac{0.05\,916}{2}\log C_2 - \dfrac{0.05\,916}{2}\log C_1 = \dfrac{0.05\,916}{2}\log\left(\dfrac{C_2}{C_1}\right) = 0.023$$

$\left(\dfrac{C_2}{C_1}\right) = 10^{(0.023)(2)/0.05\ 916} = 5.99_1$ or $C_2 = 5.99\ C_1$. Also, C_2 can be calculated from:

$C_2 = \dfrac{C_1 V_1 + C_s V_s}{V_1 + V_s}$ where C_S and V_S are the concentration and volume of the standard added. Then,

$$C_2 = 5.99 C_1 = \dfrac{C_1 V_1 + C_s V_s}{V_1 + V_s} \text{ or } 5.99 C_1 (V_1 + V_s) - C_1 V_1 = C_s V_s$$

$$C_1 \{ 5.99 (V_1 + V_s) - V_1 \} = C_s V_s$$

$$C_1 = \dfrac{C_s V_s}{\{ 5.99 (V_1 + V_s) - V_1 \}} = \dfrac{(0.125)(5.00)}{5.99(105) - 100} = 1.18 \times 10^{-3}$$

3.2

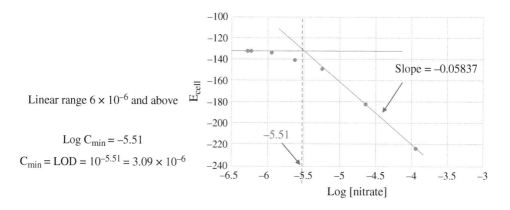

Linear range 6×10^{-6} and above

Log $C_{\min} = -5.51$

$C_{\min} = $ LOD $= 10^{-5.51} = 3.09 \times 10^{-6}$

Slope = −0.05837

3.3

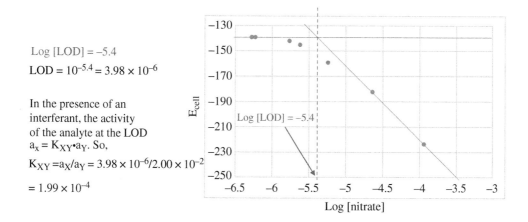

Log [LOD] = −5.4

LOD $= 10^{-5.4} = 3.98 \times 10^{-6}$

In the presence of an interferant, the activity of the analyte at the LOD $a_x = K_{XY} \cdot a_Y$. So,

$K_{XY} = a_X / a_Y = 3.98 \times 10^{-6} / 2.00 \times 10^{-2}$

$= 1.99 \times 10^{-4}$

Log [LOD] = −5.4

3.4 (a) The membrane may have been punctured, perhaps by something in the sediment. *This would leave the ISE unresponsive except to changes in the junction potential of the reference salt bridge.*

(b) The silver ions from the reference electrode may have reacted with something in solution to precipitate in the salt bridge. *This might lead to a changing junction potential that is consistent with the observation.*

(c) Your colleague may have forgotten to soak the electrode a standard solution of Pb^{2+} after putting a new membrane on the device. *A new membrane needs to be loaded with Pb^{2+}. If a different salt is used to introduce the ionic sites to the membrane, the potential will drift until the cations have been exchanged with Pb^2 ions and an internal activity stabilizes.*

(d) The temperature may have been changing during the measurement. *A changing temperature can cause a drift in the reference electrode potential and in the liquid junction potential.*

3.5
$$\Delta Q = C\Delta V,\ C = (0.090\ cm^2)(3 \times 10^{-6}\ F/cm^2) = 2.7 \times 10^{-7}\ F$$

$$\Delta Q = (2.7 \times 10^{-7}\ C/V)(0.050\ V) = 1.35 \times 10^{-8}\ C$$

$$\Delta N = \Delta Q/zF = 1.35 \times 10^{-8}\ C/(2)(96\,485\ C/mol)$$

$$= 6.99 \times 10^{-14}\ mol\ or\ (6.99 \times 10^{-14}\ mol)(6.022 \times 10^{23}) = 4.21 \times 10^{10}\ ions$$

3.6

[chloride]	log(chloride)	E/mV
1.00E–03	–3	–113
1.00E–04	–4	–54
1.00E–05	–5	5
5.00E–06	–5.30103	22.8
1.00E–06	–6	64
5.00E–07	–6.30103	81.8

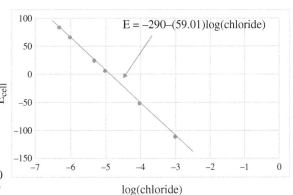

a) $E = -290-(59.01)\log(chloride)$
$15.5 = -290-(59.01)\log C$
$C = 10^{305/(-59.01)} = 6.65 \times 10^{-6}$

b) The concentration would have been more concentrated by a factor of 101/100
Or $C = (1.01)(6.65 \times 10^{-6}) = 6.71 \times 10^{-6}$
and the potential would have been:
$E = -290-(59.01)\log(6.71 \times 10^{-6}) = 15.27\ mV.$

However, there is also the error caused by a change in the ionic strength of the solutions.

$$E = E_1 - (slope)\log(a_{Cl-}) = (-290) - (59.01)\log(\gamma_{Cl-})(c_{Cl-})$$

$$= (-290) - (59.01\log(\gamma_{Cl-}) - 59.01\log(c_{Cl-}))$$

In the untreated sample solution, the ionic strength is very low, essentially 0, so the activity coefficient in that case is essentially unity. The difference in potential between a treated and an untreated sample will be caused by the change in the activity coefficient, namely,

$$\Delta E = 59.01 \log(\gamma_{Cl^-})$$

The γ_{Cl^-} value was lower in the treated solutions recall that $\log \gamma_i = \dfrac{-0.511 z_i^2 \sqrt{\mu}}{1 + Ba\sqrt{\mu}}$

In the treated solution, the ionic strength, $\mu = (4M)(1.00\,ml/101\,ml) = 0.0396\,M$

$$\log \gamma_{Cl^-} = \frac{-0.511(-1)^2 \sqrt{0.0396}}{1 + (0.328)(3)\sqrt{0.0396}} = -0.0997$$

Then, $\Delta E = 59.01(-0.0997) = -5.58\,mV$

That is, the untreated sample gave a signal that was 5.88 mV more negative than it would have if it had been treated with the same ionic strength adjusting buffer as the standards. In order to find the potential for the corresponding treated sample we need to add 5.88 mV to the observed value:

$$E_{treated} = E_{untreated} + 5.88 = 15.27 + 5.88 = 21.15 \approx 21.2\,mV$$

3.7
$$E = -0.273 = E_{cnst} + 0.05\,916\,\log(7.5 \times 10^{-6}) \rightarrow$$

$$E_{cnst} = -0.273 - 0.05\,916\,\log(7.5 \times 10^{-6}) = 0.0301\,V$$

$$E = 0.0301 + 0.05\,916\,\log\{[K^+] + K_{KH}[H^+]\}$$

$$= 0.0301 + (0.05\,916)\log\{7.5 \times 10^{-6} + (4.0 \times 10^{-4})(0.02)\}$$

$$= -0.254V$$

3.8 $E_{cell} = E_{cnst} + 0.05\,916\,\log\{a_K + K_{KNa}^{POT}\,a_{Na}\}$. From Table 3.1, $K_{KNa}^{POT} = 6 \times 10^{-5}$. The sodium will cause a 10% error when

$$K_{KNa}^{POT}\,a_{Na} = 0.1a_K = (0.1)(1 \times 10^{-4}) = 1 \times 10^{-5}$$

$$a_{Na} = \frac{1 \times 10^{-5}}{K_{KNa}^{POT}} = \frac{1 \times 10^{-5}}{6 \times 10^{-5}} = 0.166\,M$$

3.9 $E_1 = E_{cnst} + 0.05\,916\,\log[H^+]_1; \quad E_2 = E_{cnst} + 0.05\,916\,\log[H^+]_2 = E_1 + 0.005$

$$E_2 - E_1 = 0.005 = 0.05\,916\,\log[H^+]_2 - 0.05\,916\,\log[H^+]_1 = 0.05\,916\,\log\frac{[H^+]_2}{[H^+]_1}$$

$$\log\frac{[H^+]_2}{[H^+]_1} = \frac{0.005}{0.05\,916} = 0.0845 \quad \text{or} \quad \frac{[H^+]_2}{[H^+]_1} = 10^{0.0845} = 1.21 \quad \text{or}$$

$$121\% \rightarrow a\,21\%\text{change in } [H^+]$$

3.10 For a crystalline ISE, $LOD = (K_{sp})^{1/2} = (7.7 \times 10^{-13})^{1/2} = 8.7_7 \times 10^{-7} \, M$
With interfering Cl^- ion present, $E = E_{cnst} - 0.05916 \log\{a_{Br} + K_{BrCl}^{POT} a_{Cl}\}$, then

$$LOD = a_{Br} = K_{BrCl}^{POT} a_{Cl} = (3 \times 10^{-3})(0.01) = 3 \times 10^{-5}$$

3.11 From Eq. (C.14), $E_j = \frac{\sum_j z_j \lfloor z_j \rfloor D_j (C_{j,s} - C_{j,r})}{\sum_j z_j^2 \lfloor z_j \rfloor D_j (C_{j,s} - C_{j,r})} \left(\frac{RT}{F}\right) \ln\left\{ \frac{\sum_j z_j^2 \lfloor z_j \rfloor D_j C_{j,r}}{\sum_j z_j^2 \lfloor z_j \rfloor D_j C_{j,s}} \right\} = \left(\frac{G}{H}\right)\left(\frac{RT}{F}\right) \ln\left\{\frac{M}{N}\right\}$,
where

$$G = \sum_j z_j \lfloor z_j \rfloor D_j (C_{j,s} - C_{j,r}); H = \sum_j z_j^2 \lfloor z_j \rfloor D_j (C_{j,s} - C_{j,r}); M = \sum_j z_j^2 \lfloor z_j \rfloor D_j C_{j,r}; N$$

$$= \sum_j z_j^2 \lfloor z_j \rfloor D_j C_{j,s}$$

$C_{Na,r} = C_{Cl,r} = 3; C_{Na,s} = C_{Cl,s} = 0; C_{H,r} = C_{NO3,r} = 0; C_{H,s} = C_{NO3,s} = 0.2$
$D_{Na} = 13.34 \times 10^{-6} \, D_{Cl} = 20.32 \times 10^{-6} \, D_H = 96.6 \times 10^{-6} \, D_{NO3} = 19.46 \times 10^{-6}$

$$G = \sum_j z_j |z_j| D_j (C_{j,s} - C_{j,r}) = (1)D_{Na}(0-3) + (-1)D_{Cl}(0-3) + (1)D_H(0.2-0)$$

$$+ (-1)D_{NO_3}(0.2-0)$$

$$= (13.34 \times 10^{-6})(-3) + (-1)(20.32 \times 10^{-6})(-3) + (96.6 \times 10^{-6})(0.2)$$

$$+ (-1)(19.46 \times 10^{-6})(0.2) = 3.64 \times 10^{-5}$$

$$H = \sum_j z_j^2 |z_j| D_j (C_{j,s} - C_{j,r}) = (1)D_{Na}(0-3) + (1)D_{Cl}(0-3) + (1)D_H(0.2-0)$$

$$+ (1)D_{NO_3}(0.2-0)$$

$$= (13.34 \times 10^{-6})(-3) + (20.32 \times 10^{-6})(-3) + (96.6 \times 10^{-6})(0.2)$$

$$+ (19.46 \times 10^{-6})(0.2)$$

$$= -7.78 \times 10^{-5}$$

$$M = \sum_j z_j^2 |z_j| D_j C_{j,r} = (1)D_{Na}(3) + (1)D_{Cl}(3) + (1)D_H(0) + (1)D_{NO_3}(0)$$

$$= (13.34 \times 10^{-6})(3) + (20.32 \times 10^{-6})(3) = 1.01 \times 10^{-4}$$

$$N = \sum_j z_j^2 |z_j| D_j C_{j,s} = (1)D_{Na}(0) + (1)D_{Cl}(0) + (1)D_H(0.2) + (1)D_{NO_3}(0.2)$$

$$= (96.6 \times 10^{-6})(0.2) + (19.46 \times 10^{-6})(0.2) = 2.32 \times 10^{-5}$$

$$E_j = \left(\frac{G}{H}\right)\left(\frac{RT}{F}\right) \ln\left\{\frac{M}{N}\right\} = \left(\frac{3.64 \times 10^{-5}}{-7.78 \times 10^{-5}}\right)\left(\frac{8.314 \times 298}{96485}\right) \ln\left\{\frac{1.01 \times 10^{-4}}{2.32 \times 10^{-5}}\right\}$$

$$= 0.0177$$

4.1 Drinking water has a very low ionic strength on its own. Adding even a small amount of TISAB will dominate the ionic strength. A good target for the ionic strength is 0.1 M. That would be 10 ml of TISAB per 100 ml sample. That dilutes the original fluoride level by (100/110) or to ~91% of its original level. That is a good choice, since 1 mg/l = 0.001 g/17 g/(mol l) = 59 μM. This concentration is still well above the LOD of ~1 μM. Looking at Table 4.1, the junction potential for this reference electrolyte together with 0.1 M NaCl sample solution is only −0.13 mV. A slightly higher NaCl concentration might yield an even lower junction potential, but the activity coefficient and concentration of the fluoride, would both decrease using higher amounts of TISAB.

4.2 The ionic strength of the treated solution is ~0.5 M after treatment. See Figure 4.1, where a 3 M KCl salt bridge has been used. Log(0.5 M) = −0.30. The junction potential calculated from the Henderson Equation is ~ +1 mV for 0.5 M NaCl.

4.3 We would expect the cell potentials, E_{cell}, to follow a logarithmic relationship with the analyte activity, a_{Cl^-}, in the form:

$$E_{cell} = E_{constant} + \frac{RT}{zF}(2.303)\log(a_{Cl^-})$$

Both the constant term, $E_{constant}$, and the coefficient for the log term are functions of temperature. The constant term includes the reference potential for both the internal and external reference electrodes as well as the junction potential. All these are temperature dependent. He is using only one calibration solution to determine two variables. He needs a minimum of two. In principle, he could measure the temperature and calculate the coefficient, but he cannot be sure that the electrode will behave ideally. That is, the slope may not be 59.2 mV/decade even at 25°°C.

4.4 Solving the equation from the calibration curve, for the sample activity, a_{spl}, the spiked sample concentration, a_{spk}, and the duplicate spiked sample activity, $a_{dpl\ spk}$, are 2.32×10^{-4} M, 3.35×10^{-4} M, and 3.41×10^{-4} M nitrate.

$$E_{cell} = 0.265 - (0.0574)\log(a_{NO_3^-}) \text{ or } a_{NO_3^-} = 10^{(E_{cell}-0.265)/(-0.0574)}$$

$$\%\text{recovery} = \left(\frac{C_{spk\ spl} - C_{spl}}{C_{added}}\right)(100\%)$$

$$= \left(\frac{3.35 \times 10^{-4} - 2.32 \times 10^{-4}}{1.00 \times 10^{-4}}\right)(100\%) = 103\%$$

This is a very reasonable recovery, so we conclude that the calibration curve will be adequate in this case.

$$\%\text{discrepancy} = \left| \frac{C_{\text{spk spl}} - C_{\text{dspk spl}}}{C_{\text{spk spl}}} \right| (100\%)$$

$$= \left(\frac{3.35 \times 10^{-4} - 3.41 \times 10^{-4}}{3.35 \times 10^{-4}} \right) (100\%) = 1.8\%$$

This is also a very reasonably low discrepancy. The test appears reproducible.

4.5 Recall that $(V_0 + V_N) \times 10^{mE_{\text{cell}}} = \left(\frac{C_0}{C_s} V_0 + V_N \right) \times 10^{mE_{\text{cnst}}}$. Plot $(V_0 + V_N) 10^{mE_{\text{cell}}}$ versus V_N where $m = 1/(0.0577)$. The x-intercept is equal to $-V_e$ where $C_0 = C_s V_e / V_0$.

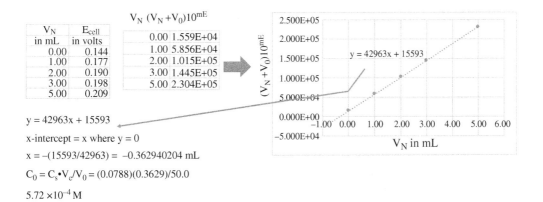

V_N in mL	E_{cell} in volts
0.00	0.144
1.00	0.177
2.00	0.190
3.00	0.198
5.00	0.209

V_N	$(V_N + V_0) 10^{mE}$
0.00	1.559E+04
1.00	5.856E+04
2.00	1.015E+05
3.00	1.445E+05
5.00	2.304E+05

$y = 42963x + 15593$

x-intercept $= x$ where $y = 0$

$x = -(15593/42963) = -0.362940204$ mL

$C_0 = C_s \cdot V_e / V_0 = (0.0788)(0.3629)/50.0$

5.72×10^{-4} M

4.6

$$E_{\text{memb}} = E^\circ + \frac{RT}{(+1)F} \ln \left(a_x + K_{\text{XNH}_4}^{\text{POT}} a_{\text{NH}_4^+} \right)$$

A 10% error will occur when

$$\left((0.1) a_x = K_{\text{XNH}_4}^{\text{POT}} a_{\text{NH}_4^+} \right) = (5.03 \times 10^{-3})(0.3) = 1.5 \times 10^{-4}$$

or $a_x = \frac{1.5 \times 10^{-4}}{0.1} = 1.5 \times 10^{-3}$ M creatinine.

4.7 Chlorate is at end of the Hofmeister series where the ions are the most easily extracted, while sulfate is at the most difficult to extract end of the list. Sulfate has two charges that increase the charge density on the ion. This creates a stronger ion-dipole interaction between sulfate and the molecules in its solvation sphere. Consequently, it costs more energy to remove these water molecules and bring the ion into the hydrophobic environment of the liquid membrane.

4.8   *Reagent blank*: Deionized water that is treated the same way as a field sample. This control checks for contamination of the reagents.

Initial calibration verification standard (ICV): A known concentration of the analyte made from a different source of reagents than the calibration standards. This solution is tested at the beginning of the day following the measurement of all the standard solutions. The measurement should accurately indicate the same concentration as what one would calculate from the weighed amount of reagent used to prepare the control solution. This is an initial test of the calibration curve and the measurement procedure.

Continuing calibration verification standard (CCV): This is a solution of known analyte concentration that has been prepared from the same stock solution as the calibration standards. It is separate from the calibration solutions in that it is not included as a point on the calibration curve. It is tested intermittently, perhaps after every 10 samples, in order to check that the system continues to give a response that indicates its concentration accurately.

Matrix spike: While in the field, a second sample is taken from the same location, and a known quantity of analyte is deliberately added to the solution. Based on the known addition of analyte and the response of the original sample, the calibration curve should accurately predict the response of the system to this solution. This test checks for differences in the matrix of the sample solutions compared to the standard solutions that may change the electrode response.

Matrix spike duplicate: This is a second spiked sample that is intended to be as close as possible to the composition as the matrix spike sample. This serves as a check on the reproducibility of the sampling process.

4.9

$$E_{memb} = E^o + \frac{RT}{(+1)F} \ln\left(a_x + \sum K_{XY}^{POT} a_Y\right)$$

The level where other ions interfere to cause a 10% error in the activity of the creatinine ion, X, corresponds to:

$$\left(\sum K_{XY}^{POT} a_Y = (0.1)a_x\right) \text{ or } a_x = \frac{\sum K_{XY}^{POT} a_Y}{0.1}$$

$$= (10)\{(2.00 \times 10^{-4})(0.16) + (3.16 \times 10^{-3})(0.06) + (5.01 \times 10^{-3})(0.035)$$

$$+ (1.58 \times 10^{-5})(0.005)\}$$

$$= 3.97 \times 10^{-3} \text{ M}$$

4.10   What is the expected range of concentrations in the blood?

How does the expected range compare with the LOD and LOQ of K^+ ion ISEs?

How to adjust the ionic strength to a reproducible value? What level of dilution will be workable?

What are the interfering ions for this ISE and will these be present in the sample?

What indirect factors such as pH, complexing agents or membrane fouling components might be present?

Can I use a standard calibration curve?

4.11 If the filling port were sealed, the pressure would drop inside the reference chamber as electrolyte drains out. That would slow down and eventually prevent electrolyte flow out of the reference bridge. The change in flow rate would lead to a change in the ion gradient at the interface of the reference and sample solutions causing a drift in the junction (diffusion) potential. We cannot eliminate the junction potential, so we try to keep it constant.

4.12

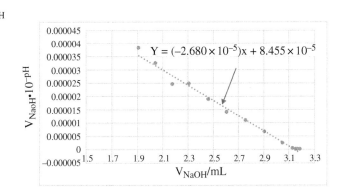

V_{NaOH}	pH	$V_{NaOH} \cdot 10^{-pH}$
1.912	4.7	3.8149E–05
2.046	4.8	3.2427E–05
2.18	4.95	2.446E–05
2.314	4.97	2.4795E–05
2.462	5.12	1.8676E–05
2.61	5.27	1.4017E–05
2.758	5.41	1.073E–05
2.906	5.64	6.6573E–06
3.055	6.14	2.2132E–06
3.132	7.16	2.1668E–07
3.154	9.86	4.3537E–10
3.162	10.42	1.2022E–10

X-intercept $\rightarrow 0 = (-2.680 \times 10^{-5})x + 8.455 \times 10^{-5} \rightarrow V_{endpoint} = -(8.455 \times 10^{-5})/(-2.680 \times 10^{-5}) = 3.154$ mL

5.1

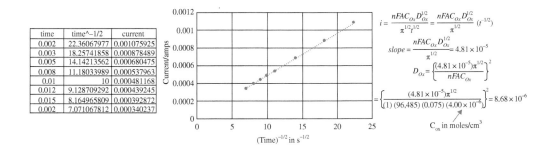

time	time^–1/2	current
0.002	22.36067977	0.001075925
0.003	18.25741858	0.000878489
0.005	14.14213562	0.000680475
0.008	11.18033989	0.000537963
0.01	10	0.000481168
0.012	9.128709292	0.000439245
0.015	8.164965809	0.000392872
0.002	7.071067812	0.000340237

$$i = \frac{nFAC_{Ox}D_{Ox}^{1/2}}{\pi^{1/2}t^{1/2}} = \frac{nFAC_{Ox}D_{Ox}^{1/2}}{\pi^{1/2}}(t^{-1/2})$$

$$slope = \frac{nFAC_{Ox}D_{Ox}^{1/2}}{\pi^{1/2}} = 4.81 \times 10^{-5}$$

$$D_{Ox} = \left\{\frac{(4.81 \times 10^{-5})\pi^{1/2}}{nFAC_{Ox}}\right\}^2$$

$$= \left\{\frac{(4.81 \times 10^{-5})\pi^{1/2}}{(1)(96,485)(0.075)(4.00 \times 10^{-6})}\right\}^2 = 8.68 \times 10^{-6}$$

C_{ox} in moles/cm^3

5.2 Net time required = 27.6 − 1.9 = 25.7 s.

$$Q = (i_{avg})(\text{time}) = (150.7 \times 10^{-3} \text{ A})(25.7 \text{ s}) = 3.87_3 \text{ C}$$

$$N = Q/(nF) = (3.87_3 \text{ C})/(2 \text{ mol electrons/mol } H_2O \times 96\,485 \text{ C/mol})$$

$$= 2.00_7 \times 10^{-5} \text{ mol } H_2O$$

$$\text{wt\%water} = (100\%)(2.00_7 \times 10^{-5} \text{ mol } H_2O)(18.01 \text{ g/mol})/0.0336 \text{ g spl}$$

$$= 0.107_5\%.$$

5.3

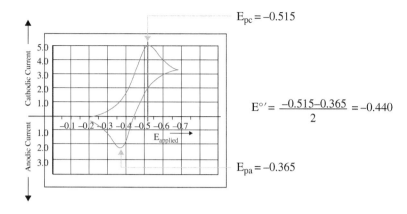

$$E_{pc} = -0.515$$

$$E^{\circ\prime} = \frac{-0.515 - 0.365}{2} = -0.440$$

$$E_{pa} = -0.365$$

5.4 For a reversible system, $\Delta E_p = (57\,mV)/n$. Here $n = 1$.

$$\Delta E_p = -0.365 - (-0.515) = 0.150\,V = 150\,mV \rightarrow Quasi-reversible.$$

5.5 If the two redox forms are stable on the time scale of the experiment, then the ratio of the peak currents will be approximately 1.0. Using the method outlined in Figure 5.13c, $\frac{i_b}{i_f} = \frac{i_{pa}}{i_{pc}} = \frac{(i_{pa})_0}{i_{pc}} + \frac{0.485(i_\lambda)_0}{i_{pc}} + 0.086 = \frac{2.15}{5.0} + \frac{(0.485)(3.25)}{5.0} + 0.086 = 0.831$. This looks a bit low and suggests that some of the reduction product is lost during the return scan. That is, the reduced form is unstable. A scan rate study would help support that idea.

5.6

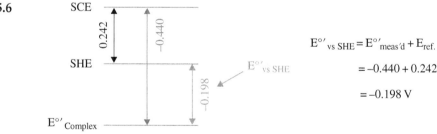

$$E^{\circ\prime}_{vs\,SHE} = E^{\circ\prime}_{meas'd} + E_{ref.}$$

$$= -0.440 + 0.242$$

$$= -0.198\,V$$

5.7

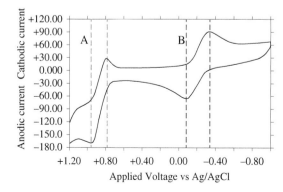

$E^{\circ\prime} = 0.5(E_{pa} + E_{pc})$

$E^{\circ\prime}$ (vs. SHE) = $E^{\circ\prime}$(vs. Ag/AgCl) + 0.199 V

System A	System B	
$E_{pa} = 0.970$	$E_{pa} = -0.316$	
$E_{pc} = 0.795$	$E_{pc} = -0.070$	
$E^{\circ\prime} = 0.882$	$E^{\circ\prime} = -0.193$	vs. Ag/AgCl
$E^{\circ\prime} = 1.081$	$E^{\circ\prime} = +0.006$	vs. SHE

System A is the stronger oxidizer because it has the more positive $E^{\circ\prime}$.

5.8 Both methods avoid most of the double layer charging current by sampling the current a few microseconds after the voltage is stepped. On the negative-going scan, the square wave voltammogram recycles some of the Fe(II) product on each step by stepping part of the way back. Therefore, more Fe(III) reactant is available to reduce on the following cycle than there is at the same place in the staircase voltammogram. Furthermore, the signal that is used in a square wave voltammogram is not just the cathodic current, but the difference in the cathodic and the anodic current – a much bigger signal.

5.9

$$E_{\text{soln}} = 0.220 - \frac{0.05\,916}{2} \log \left\{ \frac{a_{H_2Q}}{a_Q (a_{H^+})^2} \right\}$$

$$= 0.220 - \frac{0.05\,916}{2} \log \left\{ \frac{1}{(a_{H^+})^2} \right\} - \frac{0.05\,916}{2} \log \left\{ \frac{a_{H_2Q}}{a_Q (a_{H^+})^2} \right\}$$

$$E^{\circ\prime} = 0.220 - \frac{0.05\,916}{2} \log \left\{ \frac{1}{(a_{H^+})^2} \right\}$$

$$= 0.220 - \frac{0.05\,916}{2} (2) \log \left\{ \frac{1}{(a_{H^+})} \right\} = 0.220 - 0.05\,916 \text{ pH}$$

Therefore, a decrease in pH would lead to a more positive $E^{\circ\prime}$.

5.10 The peak separation decreased and the peak height increased. This is an indication of faster electron transfer kinetics. One possibility for the improvement is that the electrochemical pretreatment cleaned off some material that was adsorbed on the electrode surface and inhibited electron transfer. Another possibility is that the oxides produced in the carbon surface by the pretreatment play a role in the electrode reaction. Perhaps, they hold the reactant by dipole interaction or hydrogen bonding. Assisting with hydrogen transfer in a coupled chemical step is another possible role for the surface oxides. One test of the cleaning hypothesis is to repeat the experiment with an electroactive reactant, such as $Fe(CN)_6^{3-}$, that is sensitive to inhibition by adsorbed molecules but does not need surface oxides. Soaking the electrode briefly in propanol that has been treated with activated charcoal is a way of cleaning that will lead to (temporary) improvement in the electron transfer kinetics, if adsorption is the problem.

5.11 (a) Replace the HMDE with a polished glassy carbon electrode and deposit a thin mercury film on the electrode with the sample (after spiking the sample to a level of 1×10^{-5} M Hg^{2+}). The tiny droplets in the film have higher surface area and smaller volume; both enhance the signal.

(b) Use a longer deposition time in order to preconcentrate more Pb.

(c) Use a rotated disk or a flowing system to enhance the deposition current and gather more Pb.

(d) Use square wave voltammetry for the stripping step for better discrimination of the charging current and signal enhancement due to recycling of the Pb in the stepping process.

5.12 Au and Pt have partially filled d-orbitals that can form bonds with organic free radicals that are sometimes intermediate species in the oxidation of organic molecules. The carbon surface is not able to utilize d-orbitals and would need to break bonds to form new ones with an organic radical. On the Pt or Au surface, these organometallic compounds can block further electrolysis. Stepping the voltage to an extreme positive value breaks the bonds holding the organic species in favor of forming metal oxides. The metal oxides can be reduced to a clean metal surface by briefly stepping back to a negative voltage.

5.13 A freely diffusing reactant produces a peak current that follows the Randles–Sevcik equation which indicates that the peak current is proportional to the square root of the scan rate.

$$i_p = (269)n^{2/3}AD^{1/2}v^{1/2}C$$

A reactant that is confined to the surface does not depend on transport to reach the surface. It produces a current peak in a voltammogram because the amount of material on the surface is finite, and there is nothing to replace it. Its peak height is directly proportional to scan rate.

5.14 The cathodic peak at 0.2 V has been assigned to the reduction of quinone. If it is quinone, it should show a reverse peak when scanning a second time. One could also spike the solution with authentic quinone. If the assignment is correct, the peak should grow in height. Bulk electrolysis of the solution using a voltage of +0.7 should give two electrons per aminophenol molecule. ($Q = \int i \, dt = nFN$, where N is the number of moles of reactant.) A UV-visible spectrum of the electrolysis product should show the formation of the quinone group. A GC/MS of an extract of the electrolyzed solution may show mostly quinone. Depending on the rate of the hydrolysis of the quinone imine intermediate, some of the intermediate may be observable.

5.15 The lower curve has a return peak indicating that the one-electron product can be readily reoxidized to the starting material. In the upper curve, the return peak is almost gone indicating that less aryl radical is present. Furthermore, the reduction peak (on the forward scan) is much bigger than without CO_2 present suggesting that the CO_2 is reacting with the aryl radical to regenerate the starting material. Indeed, this is the type of behavior for an electrocatalytic process. CO_2 is being reduced by the aryl radical.

5.16 For a rotated disk electrode, the plateau current on the mass-transport-limited wave is proportional to the square root of the rotation rate. So, doubling the rotation rate increases the sensitivity by a factor of the square root of 2.

$$i_l = 0.62z_iFAD\omega^{1/2}v^{-1/6}C_o = SC_o$$

where S is the sensitivity of the method. For 500 rpm,

$$i_l = SC_o = 2.73 \, \text{nA} = S(0.1 \, \mu\text{M}) \text{ or } S = 2.73\frac{\text{nA}}{0.1}\mu\text{M} = 27.3 \, \text{nA}/\mu\text{M}$$

At 1000 rpm, $S' = S(2)^{1/2} \approx 1.41S = 1.41(27.3 \, \text{nA}/\mu\text{M}) = 38.4_9 \, \text{nA}/\mu\text{M}$

The limit of detection, LOD, is defined as the concentration corresponding to a signal that is three times the size of the noise.

$$\text{LOD} = \frac{3(\text{noise})}{\text{sensitivity}} = \frac{3(0.5\,\text{nA})}{38.49\,\text{nA}/\mu\text{M}} = 0.0389_7\,\mu\text{M} = 39.0\,\text{nM}$$

7.1

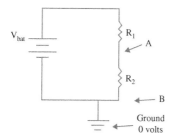

Let $V_{\text{bat}} = 6$ V.

$$V_A = (V_{\text{bat}})\,\frac{R_2}{R_1 + R_2} = 1\text{V}$$

There are lots of possible choices that will give a ratio of 1/6. A reasonable choice is $R_2 = 1\,\text{k}\Omega$. Then, $R_1 = 5\,\text{k}\Omega$.

7.2

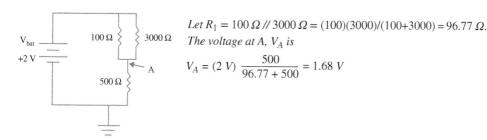

Let $R_1 = 100\,\Omega\,/\!/\,3000\,\Omega = (100)(3000)/(100+3000) = 96.77\,\Omega$. The voltage at A, V_A is

$$V_A = (2\ V)\,\frac{500}{96.77 + 500} = 1.68\ V$$

7.3 The voltage drop through the parallel resistors $= 2 - 1.68 = 0.32$ V. Ohm's law gives the current, i, through the $100\,\Omega$ resistor. $V = iR$ or $I = V/R = 0.32\,V/100\,\Omega = 3.2\times 10^{-3}$ A.

7.4 The net resistance, R_{Net} is related to the individual resistances by:

$$\frac{1}{R_{\text{Net}}} = \frac{1}{R_a} + \frac{1}{R_b} + \frac{1}{R_c} = \left(\frac{R_a + R_b}{R_a R_b}\right) + \frac{1}{R_c} = \frac{(R_a + R_b)R_c + R_a R_b}{R_a R_b R_c}$$

$$= \frac{R_a R_c + R_b R_c + R_a R_b}{R_a R_b R_c}\quad \text{or}\quad R_{\text{Net}} = \frac{R_a R_b R_c}{R_a R_c + R_b R_c + R_a R_b}$$

7.5

$$R_{\text{Net}} = \frac{R_a R_b R_c}{R_a R_c + R_b R_c + R_a R_b} = \frac{(500)(400)(300)}{(500)(300) + (400)(300) + (500)(400)} = 127\,\Omega$$

7.6

$$V = iR, i = V/R.\, i_a = 5\,V/500\,\Omega = 0.01\ \text{A};\, i_b = 5\,V/400\,\Omega = 0.0125\ \text{A};\, i_c$$

$$= 5\,V/300\,\Omega = 0.0167\ \text{A}$$

7.7
$$V_{out} = -i_{in}R_f = -(15 \times 10^{-9}\,\text{A})(200 \times 10^3\,\Omega) = 3 \times 10^{-3}\,\text{V}$$

7.8 The total gain needed $= 0.1\text{V}/20\,\text{nA} = (0.1\,\text{V})/(20 \times 10^{-9}\,\text{A}) = 5 \times 10^6\,\text{V/A}$. Consider a current-to-voltage op amp with a gain of $\times 10^4\,\text{V/A}$ followed by a second op amp in a simple gain mode to give an additional gain of 500.

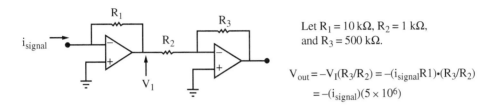

Let $R_1 = 10\,\text{k}\Omega$, $R_2 = 1\,\text{k}\Omega$, and $R_3 = 500\,\text{k}\Omega$.

$$V_{out} = -V_1(R_3/R_2) = -(i_{signal}R_1)\cdot(R_3/R_2)$$
$$= -(i_{signal})(5 \times 10^6)$$

7.9 (a) $V_{out} = -V_{in}(R_f/R_{in}) = -(0.1\,\text{V})(5000/1000) = -0.5\,\text{V}$

(b) $V_{out} = -V_{in}\left(\dfrac{R_f}{R_{in}}\right) = -V_{in}\left(\dfrac{5000}{1000}\right) = -V_{in}(5) = -0.5$ Then $-V_{in} = -\dfrac{-0.5}{5}$

$$= +0.1\,\text{V}$$

7.10

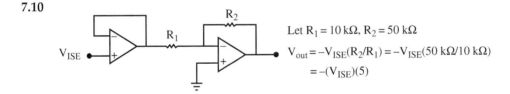

Let $R_1 = 10\,\text{k}\Omega$, $R_2 = 50\,\text{k}\Omega$

$V_{out} = -V_{ISE}(R_2/R_1) = -V_{ISE}(50\,\text{k}\Omega/10\,\text{k}\Omega)$
$= -(V_{ISE})(5)$

The initial voltage follower op amp has a very high input impedance and draws very little current.

7.11 The KCl concentration sets the reference electrode potential for the AgCl reduction reaction through the Nernst equation:

$$E_{ref} = E^o_{Ag^+/Ag} - 0.05\,916\log(\gamma_{Cl^-}[Cl^-])$$

The KCl concentration also will influence the salt bridge junction potential.

7.12 The resistance between the sliding contact to the potentiometer and either end changes as the position of the contact is moved. This creates a simple way of varying the value of the resistance in the circuit with the electrochemical cell which, in turn, limits the amount of current that flows through the circuit and the rate at which the silver surface oxidizes and AgCl solid is deposited.

7.13
$$E_{ref} = E^o_{Ag^+/Ag} - 0.05\,916\log(a_{Cl^-}) = 0.222 - 0.05\,916\log(0.100)$$

$$= 0.222 - (-0.05\,916) = 0.281_{16}\,\text{V}$$

7.14 The driving force for an electrode reaction is the difference between the potential of the working electrode, $\varnothing_{WE}$, and the potential of the solution, $\varnothing_{soln}$, as shown in diagram here. Because the working electrode potential is fixed at the ground potential defined by the outside circuit, $\varnothing_{WE}$ is fixed.

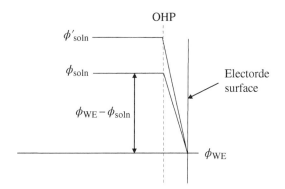

In order to change the potential across the interface the potential of the solution must be moved. That is done, by applying a voltage to the counter (or auxiliary) electrode. As illustrated in the figure here, the potential of the solution must be driven in the positive direction (up on the diagram) in order for the working electrode to appear at a more negative potential. Two different levels are shown for $\varnothing_{soln}$. The higher one creates a larger potential difference with respect to $\varnothing_{WE}$. From the perspective of the working electrode, $\varnothing_{WE}$ is more negative in that situation. When $\varnothing_{soln} = +1\,V$, $\varnothing_{WE}$ will be at a potential 1 V less than the solution at the point where the reference electrode is located. In other words, the potential at the working electrode will be 1 V less than the reference electrode potential. In a voltammogram, we plot the working electrode potential, $E_{WE} = \varnothing_{WE} - \varnothing_{soln}$, on the horizontal axis.

The op amp circuit is set up so that a voltage of V_1 applied to the non-inverting input will cause a voltage of V_1 to appear at the inverting input of OA1 in Figure 7.9. Consequently, the voltage, $\varnothing_{soln}$, in the solution near the reference electrode will be $V_1 + E_{ref}$. In general,

$$E_{WE} = \varnothing_{WE} - \varnothing_{soln} = \varnothing_{WE} - (V_1 + E_{Ref}) = -(V_1 + E_{Ref}).$$

7.15 Whenever current crosses the solution/working electrode (WE) interface, then the auxiliary electrode (AE) must carry current to match it. That includes charging current as well as Faradaic current. If an oxidation process occurs at the WE, then a reduction process must occur at the AE in order to keep electrons moving in the same direction through the cell. It will drive to whatever voltage that is needed to produce current at the same rate. The usual source for reduction is often the reduction of the solvent or the supporting electrolyte. (When the current is moving in the other direction, an oxidation process occurs at the AE. Again, the oxidation of the solvent or the supporting electrolyte is usually the source of the current.)

7.16

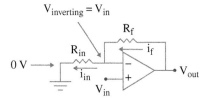

For the sake of discussion assume that $V_{in} > 0$.
The voltage at the inverting input is also equal to V_{in}.
In this case a current will go toward ground thru R_{in}.
$i_{in} = V_{in}/R_{in} = i_f$.
$V_{out} = i_f R_f + V_{in} = V_{in}{\cdot}R_f/R_{in} + V_{in} = V_{in}(R_f/R_{in}+1)$
$\qquad = V_{in}(R_f + R_{in})/R_{in}$

INDEX

activated carbon, 198

activation barrier, 165–166

active metal, 41–43

activity, 3, 13, 42–44
  coefficient, 3, 49, 52, 54, 96, 99, 111, 119–120, 122, 128, 137

adenosine monophosphate, 80

alkaline error, 88

allantoin, 248, 256, 259

all solid-state ISEs, 103–106

ampere, 3, 8

amperometric oxygen sensor, 214, 215

amperometric techniques, 159

amperometry, 10

analog signal, 269, 273–274

aniline, 183–185

anionic site, 76, 78

anode, 11

anodic stripping voltammetry (ASV), 209–212, 223

aquo-complexes, 39

Arrhenius equation, 164

asbestos wick, 89

ascorbic acid, 198, 206, 216, 218

asymmetry potential, 146

auxiliary electrode, 46, 279

background current, 175, 182–183, 186–187, 199, 201, 205, 209, 218, 221

Barker, Geoffrey, 187

basal surfaces, 193–194

Bayonet Neil–Conselman (BNC) connector, 282

Beckman, Arnold, 83

biamperometric endpoint detection, 220

bipotentiometric mode of detection, 220–221

bis(methyl styryl)benzene), 196–197, 199

boron-doped diamond (BDD), 192, 199–200

boundary potential, 72, 74, 77, 95

Bradley, A. Freeman, 90

bubbles, 142

bulk electrolysis, 237, 247, 256

Butler–Volmer equation, 167

calibration, 73, 91–92, 96–102, 122–125, 128–129, 132, 138–141, 145–148
  verification standards, 139

calomel, 47–48, 52

capacitance, 4, 21
  of double layer, 79, 105

carbon dioxide reduction, 260–261

carbon nanotubes, 106, 199, 200

carbon paste electrode, 194

catalytic mechanism, 237, 254, 260–263, 265–267

cathode, 11

charge, 4–10

charging current, 160, 177, 186–189, 201, 205

chloro-complexes, 121, 142

chronoamperometry, 176–177, 284

chronocoulometry, 177, 214

Clark, Leland, 213

coaxial cable, 281–282

combination pH Electrode, 88–90

concentration gradient, 171, 180, 202

conductance, 22–24

conductivity detectors, 24

convection, 24, 170, 179, 214

convective systems, 168

Cottrell equation, 176–177, 214, 252

coulomb, 7, 9

coulometric titration, 219, 222

coulometry, 163

counter electrode, 279

coupled chemical reaction, 179, 195, 196, 218

cracked glass bead, 89

creatinine, 130–135

creatininium cation, 131–132

crystalline membrane electrodes, 93–96

Electroanalytical Chemistry: Principles, Best Practices, and Case Studies, First Edition. Gary A. Mabbott.
© 2020 John Wiley & Sons, Inc. Published 2020 by John Wiley & Sons, Inc.

CHEMICAL ANALYSIS

A SERIES OF MONOGRAPHS ON ANALYTICAL CHEMISTRY AND ITS APPLICATIONS

Series Editor

MARK F. VITHA

Editorial Board

Stephen C. Jacobson, Stephen G.Weber